Photovoltaic Systems Engineering

The primary purpose of this textbook is to provide a comprehensive set of PV knowledge and understanding tools for the design, installation, commissioning, inspection and operation of PV systems. In recent years, more PV systems have been installed worldwide than any other electricity source. New, more efficient, more reliable and more cost-effective components and processes are rapidly appearing, along with continuously changing codes and standards. To keep up with the rapid changes, understanding the underlying principles is essential.

In addition to practical system design and installation information, this edition includes explanations of the basic principles upon which the design and operation of PV systems are based, along with a consideration of the economic and environmental impact of the technology. Numerous design examples are presented to assist the reader in incorporating the basic principles, components, codes and standards.

The book begins with basic sunlight parameters, system electronic components, wiring methods, structural considerations and energy storage methods Emphasis is on grid-connected systems, but a chapter on stand-alone systems is also included. Homework problems in each chapter focus on basic principles of the chapter but also include open-ended design problems to challenge the reader's creativity and understanding.

Photovoltaic Systems Engineering
Fifth Edition

Roger Messenger and Homayoon "Amir" Abtahi

CRC Press is an imprint of the
Taylor & Francis Group, an **informa** business

Fifth edition published 2025
by CRC Press
2385 NW Executive Center Drive, Suite 320, Boca Raton FL 33431

and by CRC Press
4 Park Square, Milton Park, Abingdon, Oxon, OX14 4RN

CRC Press is an imprint of Taylor & Francis Group, LLC

© 2025 [Roger Messenger and Homayoon "Amir" Abtahi]

First edition published by CRC 2000
Fourth edition published by CRC 2017

Reasonable efforts have been made to publish reliable data and information, but the author and publisher cannot assume responsibility for the validity of all materials or the consequences of their use. The authors and publishers have attempted to trace the copyright holders of all material reproduced in this publication and apologize to copyright holders if permission to publish in this form has not been obtained. If any copyright material has not been acknowledged please write and let us know so we may rectify in any future reprint.

Except as permitted under U.S. Copyright Law, no part of this book may be reprinted, reproduced, transmitted, or utilized in any form by any electronic, mechanical, or other means, now known or hereafter invented, including photocopying, microfilming, and recording, or in any information storage or retrieval system, without written permission from the publishers.

For permission to photocopy or use material electronically from this work, access www.copyright.com or contact the Copyright Clearance Center, Inc. (CCC), 222 Rosewood Drive, Danvers, MA 01923, 978–750–8400. For works that are not available on CCC please contact mpkbookspermissions@tandf.co.uk

Trademark notice: Product or corporate names may be trademarks or registered trademarks and are used only for identification and explanation without intent to infringe.

ISBN: **978-1-032-72621-2** (hbk)
ISBN: **978-1-032-74787-3** (pbk)
ISBN: **978-1-003-47089-2** (ebk)

DOI: 10.1201/9781003470892

Typeset in Times
by Apex CoVantage, LLC

Access the Instructor Resources: www.routledge.com/9781032726212

Contents

Preface xiii
Disclaimers xvii
Acknowledgments xix
Authors xxi

Chapter 1 Background 1

        1.1   Introduction 1
        1.2   Urgent Attention to World Population Forecasts 3
        1.3   Energy Demand and Carbon Dioxide Emissions 4
               1.3.1   Current and Projected Scenarios 4
               1.3.2   Solar PV Contribution and Demand 5
               1.3.3   Emissions 6
        1.4   A Brief Overview of Thermodynamics 8
        1.5   A Brief History of Photovoltaics 10
        1.6   Energy Units 11
        Homework Problems 12
        References 14
        Suggested Reading 16

Chapter 2 The Sun 17

        2.1   Introduction 17
        2.2   The Solar Spectrum 17
        2.3   The Effect of Atmosphere on Sunlight 19
        2.4   Sunlight Specifics 20
               2.4.1   Introduction 20
               2.4.2   Definitions 21
               2.4.3   The Orbit and Rotation of the Earth 22
               2.4.4   Tracking the Sun 24
               2.4.5   Measuring Sunlight 28
        2.5   Capturing Sunlight 29
               2.5.1   Maximizing Irradiation on the Collector 29
               2.5.2   Shading 32
               2.5.3   Special Orientation Considerations 35
        Homework Problems 38
        References 40
        Suggested Reading 40

Chapter 3 Introduction to PV Systems 41

        3.1   Introduction 41

v

	3.2	The PV Cell	43
	3.3	The PV Module—Essentials and Improvements	46
	3.4	The PV Array	48
	3.5	Energy Storage	49
	3.6	PV System Loads and MPPT	49
	3.7	PV System Availability—Traditional Concerns and New Concerns	50
		3.7.1 Traditional Concerns as Applied to Stand-Alone Systems	50
		3.7.2 New Concerns About PV System Availability	54
	3.8	Inverters—Conversion of DC to AC	55
		3.8.1 Introduction	55
		3.8.2 Square Wave Inverters	56
		3.8.3 Multilevel H-Bridge Inverters	57
		3.8.4 PWM Inverters	60
		3.8.5 Transformerless Inverters	64
		3.8.6 Other Desirable Inverter Features	64
	3.9	BOS Components	66
		3.9.1 Introduction	66
		3.9.2 Charge Controllers	67
		3.9.3 MPPTs and Linear Current Boosters	70
		3.9.4 Optimizers, Gateways, Communications and Rapid Shutdown	74
		3.9.5 Fossil Fuel Generators	77
		3.9.6 Miscellaneous Smaller But Important Components	78
	Homework Problems		81
	References		84
Chapter 4	Grid-Connected Utility Interactive PV Systems		85
	4.1	Introduction	85
	4.2	Applicable Codes and Standards	86
		4.2.1 The *NEC*	86
		4.2.2 IEEE Standard 1547–2018	90
		4.2.3 Other Issues	97
	4.3	Design Considerations for Straight Grid-Connected PV Systems	99
		4.3.1 Determining System Energy Output	99
		4.3.2 Array Selection and Installation	101
		4.3.3 Inverter Selection	102
		4.3.4 Other Installation Considerations	103
	4.4	Utility Interconnection Options	104
		4.4.1 Supply-Side (Line Side) Connections	104
		4.4.2 Load Side Connections	105
	4.5	Design of a System Based on Desired Annual System Performance Using Microinverters	107

		4.5.1	Introduction	107
		4.5.2	Array Sizing	108
		4.5.3	Module Selection	108
		4.5.4	Inverter Selection and Connection	109
		4.5.5	Utility Interconnection	110
	4.6	Design of an Optimizer System Based Upon Available Roof Space		112
		4.6.1	Introduction	112
		4.6.2	Array Selection	113
		4.6.3	Inverter Selection and Matching Strings to the Inverter	114
		4.6.4	Inverter Output to Utility Interconnection	117
	4.7	Design of a String Inverter-Based System		119
		4.7.1	Introduction	119
		4.7.2	System Design	119
		4.7.3	Other Options	122
	4.8	Design of a Nominal 100-kW Commercial Rooftop System That Feeds a Three-Phase Distribution Panel		123
		4.8.1	Introduction	123
		4.8.2	Inverter	123
		4.8.3	Modules and DC System Wiring	123
		4.8.4	AC System Wiring	126
		4.8.5	Final System Design and Operation	127
	4.9	Design of a Nominal 5-MW Agrivoltaic System		128
		4.9.1	Introduction	128
		4.9.2	Grid Stability Analysis	129
		4.9.3	Inverters	129
		4.9.4	Modules, Inverters and Array	130
		4.9.5	AC Wire Sizing, Voltage Drop, Disconnects and Overcurrent Protection	135
		4.9.6	Arc-Fault Calculations	140
		4.9.7	System Commissioning	141
	Homework Problems			143
	References			147
	Suggested Reading			148
Chapter 5	Structural Considerations			149
	5.1	Introduction		149
	5.2	Important Properties of Materials		149
		5.2.1	Column Buckling	149
		5.2.2	Thermal Expansion and Contraction	150
		5.2.3	Chemical Corrosion and Ultraviolet Degradation	150
		5.2.4	Properties of Steel	152
		5.2.5	Properties of Aluminum	153
	5.3	Design and Installation Guidelines		155

		5.3.1	Standards and Codes	155
		5.3.2	Building Code Requirements	156
		5.3.3	Aesthetics	157
	5.4	Forces Acting on PV Arrays		157
		5.4.1	Structural Loading Considerations	157
		5.4.2	Dead Loads	158
		5.4.3	Live Loads	158
		5.4.4	Wind Loads	159
		5.4.5	Snow Loads	167
		5.4.6	Other Loads	168
	5.5	Rooftop Mounting System Design		168
		5.5.1	Minimizing Installation Costs	169
		5.5.2	Enhancing Rooftop Array Performance	169
		5.5.3	Standoff Mounting	169
		5.5.4	Rack Mounting on Rooftops	176
		5.5.5	Integrated Mounting	178
	5.6	Large-Scale Ground Mount Arrays		178
		5.6.1	Introduction	178
		5.6.2	Large Fixed Tilt Arrays	180
		5.6.3	Single Axis, East–West Trackers	184
		5.6.4	Dynamic Effects	187
	Homework Problems			187
	References			189
	Suggested Reading			190
Chapter 6	Energy Storage			191
	6.1	Introduction		191
	6.2	Lithium Batteries		192
		6.2.1	LFP Batteries	192
		6.2.2	NCA Batteries	195
	6.3	Nickel-Based Battery Systems		195
		6.3.1	Nickel–Cadmium (Ni–Cd)	195
		6.3.2	Nickel–zinc Batteries	196
		6.3.3	NIMH Batteries	196
	6.4	Flow Batteries		197
		6.4.1	Introduction	197
		6.4.2	Sodium Sulfur	198
		6.4.3	Zinc Bromine	198
		6.4.4	Vanadium	199
		6.4.5	All Iron Redox	200
	6.5	Emerging Battery Technologies		200
	6.6	Hydrogen Storage		201
	6.7	The Fuel Cell		202
	6.8	Mechanical, Thermal and Other Storage Options		202
		6.8.1	Introduction	202

Contents ix

		6.8.2	Pumped Hydro 203
		6.8.3	Compressed Air...204
		6.8.4	Molten Salts..205
		6.8.5	Flywheels 206
		6.8.6	Weight Lifting—Lifted Weight Storage.................. 206
		6.8.7	Supercapacitors and Superconducting Magnets....... 207
	6.9	AC and DC Batteries 208
Homework Problems 210
References 211
Suggested Reading 212

Chapter 7 Grid-Connected PV Systems With Energy Storage (ESS) 213

 7.1 Introduction 213
 7.2 BESS Design Basics 214
  7.2.1 AC-Coupled Systems 214
  7.2.2 DC-Coupled Systems 216
  7.2.3 Load Determination 218
  7.2.4 Inverter Selection 219
  7.2.5 Storage Selection 219
  7.2.6 PV Array Selection... 219
  7.2.7 Fossil Fuel Generators.. 219
 7.3 A Microinverter-Based 120/240 Volt AC-Coupled Partial
  Home Battery Backup System..220
  7.3.1 Introduction 220
  7.3.2 ESS Selection 221
  7.3.3 Module and Inverter Selection 222
  7.3.4 System Controller Selection 222
  7.3.5 BOS Selection and Completion of the Design 223
  7.3.6 Array Mounting Equipment 224
  7.3.7 Wire and Circuit Breaker Sizing...............................226
  7.3.8 Wiring of Standby Loads 228
  7.3.9 Equipment Grounding Conductor and Grounding
   Electrode Conductor Sizing 230
 7.4 A Whole House AC-Coupled Backup System
  Using a String Inverter 230
  7.4.1 Introduction 230
  7.4.2 Module Selection and Source Circuit Design 232
  7.4.3 Inverter Selection and String Sizing 233
  7.4.4 Determination of Standby Loads and
   Storage Selection 234
  7.4.5 Selection of Remaining Major System
   Components..235
  7.4.6 Wire, Circuit Breaker and Conduit Sizing 235
 7.5 A 10-kW DC-Coupled PV/ESS Partial Backup System 236
  7.5.1 Introduction 236

		7.5.2	Array and Optimizers 237
		7.5.3	Inverter 238
		7.5.4	BESS 238
		7.5.5	Wiring, Conduit and Overcurrent Protection 238
	7.6	A 45-kW Three-Phase PV with BESS Using Inverters in Tandem 239	
		7.6.1	Inverter Selection 239
		7.6.2	Module Selection 240
		7.6.3	Array Performance 241
		7.6.4	ESS Selection 241
		7.6.5	Wire Sizing 242
	7.7	Large BESS Design Considerations 243	
		7.7.1	Introduction 243
		7.7.2	Challenges of Various Storage Options 245
		7.7.3	System Control and Communication Options 246
		7.7.4	Utility Interconnection 246
	Homework Problems 247		
	References 249		
	Suggested Reading 249		

Chapter 8 Stand-Alone PV Systems 250

	8.1	Introduction 250	
	8.2	The Simplest Configuration: Module and Fan 251	
	8.3	A PV-Powered Water Pumping System 253	
		8.3.1	Introduction 253
		8.3.2	Selection of System Components 254
		8.3.3	Design Approach for Simple Pumping System 255
	8.4	A PV-Powered Parking Lot Lighting System 258	
		8.4.1	Determination of the Lighting Load 258
		8.4.2	Lighting Design 261
	8.5	A PV-Powered Mountain Cabin 265	
		8.5.1	Introduction 265
		8.5.2	Load Determination 267
		8.5.3	Array Sizing, Array Tilt and Microinverter Selection 271
		8.5.4	Energy Storage Selection 274
		8.5.5	Excess Electrical Production 276
		8.5.6	Balance of System Component Selection 277
	8.6	Summary of Design Procedures 278	
	Homework Problems 281		
	References 283		
	Suggested Reading 283		

Contents xi

Chapter 9 Economic and Environmental Considerations 284

    9.1   Introduction 284
    9.2   Life Cycle Costing 285
        9.2.1   The Time Value of Money 285
        9.2.2   Present Worth Factors and Present Worth 287
        9.2.3   Life Cycle Cost 289
        9.2.4   Annualized LCC 291
        9.2.5   Unit Electrical Cost 292
        9.2.6   LCOE Analysis 292
        9.2.7   LCOS Analysis 294
    9.3   Borrowing Money 298
        9.3.1   Introduction 298
        9.3.2   Determination of Annual Payments on Borrowed Money 298
        9.3.3   The Effect of Borrowing on LCC 300
    9.4   Payback Analysis 301
    9.5   Externalities 303
        9.5.1   Introduction 303
        9.5.2   Subsidies 304
        9.5.3   Renewables and the Environment 304
        9.5.4   Health and Safety as Externalities 307
        9.5.5   Externalities Associated With PV Systems 308
    Homework Problems 312
    References 315
    Suggested Reading 316

Chapter 10 The Physics of Photovoltaic Cells 317

    10.1   Introduction 317
    10.2   Optical Absorption 317
        10.2.1   Introduction 317
        10.2.2   Semiconductor Materials 317
        10.2.3   Generation of EHP by Photon Absorption 319
        10.2.4   Photoconductors 321
    10.3   Extrinsic Semiconductors and the pn Junction 323
        10.3.1   Extrinsic Semiconductors 323
        10.3.2   The pn Junction 325
    10.4   Maximizing PV Cell Performance 333
        10.4.1   Introduction 333
        10.4.2   Minimizing the Reverse Saturation Current 334
        10.4.3   Optimizing Photocurrent 334
        10.4.4   Minimizing Cell Resistance Losses 342
    10.5   Exotic Junctions 344
        10.5.1   Introduction 344

		10.5.2	Graded Junctions	344
		10.5.3	Heterojunctions	345
		10.5.4	Schottky Junctions	346
		10.5.5	Multijunctions	348
		10.5.6	Tunnel Junctions	349
	Homework Problems			350
	References			351

Chapter 11 Evolution of Photovoltaic Cells and Systems ... 353

	11.1	Introduction		353
	11.2	Silicon PV Cells		355
		11.2.1	Production of Pure Silicon	355
		11.2.2	Single Crystal Silicon Cells	356
	11.3	Gallium Arsenide Cells		362
		11.3.1	Gallium Arsenide	362
		11.3.2	Fabrication of the Gallium Arsenide Cell	362
		11.3.3	Cell Performance	363
	11.4	CIGS Cells		364
		11.4.1	Introduction	364
		11.4.2	Fabrication of the CIS Cell	364
		11.4.3	Cell Performance	365
	11.5	Cadmium Telluride Cells		367
		11.5.1	Introduction	367
		11.5.2	Production of the CdTe Cell	367
		11.5.3	Cell Performance	368
	11.6	Emerging Technologies		369
		11.6.1	Intermediate Band Solar Cells	369
		11.6.2	Hot Carrier Cells	369
		11.6.3	Optical Up- and Down-Conversion	369
		11.6.4	Organic PV Cells	370
		11.6.5	Concentrating PV	370
		11.6.6	Perovskites	372
	11.7	Micro Grids		374
	11.8	Summary		374
	Homework Problems			375
	References			376
	Suggested Reading			378

Index ... 379

Preface

The goal of this textbook has always been to present a comprehensive engineering basis for photovoltaic (PV) system design, so the engineer can understand the *what*, the *why* and the *how* associated with electrical, mechanical, economic and aesthetic aspects of PV system design. The book is intended to *educate* the engineer in the design of PV systems so that when engineering judgment is needed, the engineer will be able to make intelligent decisions based on a clear understanding of the parameters involved. This goal differentiates this textbook from the many design and installation manuals that are currently available that *train* the reader *how* to do it but not *why*.

Widespread acceptance of the first four editions, coupled with accelerating growth and new ideas in the PV industry, along with additional years of experience with PV system design for the authors, and, for that matter, a bit of nudging from the publisher, has led to the publication of this fifth edition. In recent years, annual installed renewable capacity has exceeded annual installed nonrenewable capacity in the United States and most of the rest of the world. Who knows, there might even be jobs waiting out there for many who learn the material in this book.

The *what* question is addressed in the first three chapters, which present an updated background of energy production and consumption, some mathematical background for understanding energy supply and demand, a summary of the solar spectrum, how to locate the sun and how to optimize the capture of its energy, as well as the various components that are used in PV systems. Chapter 3 has been shortened a bit so actual design work can begin by about the fifth week of a three-credit course, but it also introduces lithium and a few other emerging battery technologies.

The *why* and *how* questions are dealt with in the remaining chapters in which every effort is made to explain why certain PV designs are done in certain ways, as well as how the design process is implemented. Included in the *why* part of the PV design criteria are economic and environmental issues that are discussed in Chapter 9. Chapter 5 has been embellished with additional practical considerations added to the theoretical background associated with mechanical and structural design.

Since the last edition, energy storage has transitioned from useful to essential. Chapter 6 now deals exclusively with energy storage not only as a backup power/ energy source but as an essential means of transitioning the world's energy supply from nonrenewable to renewable.

Chapters 4 and 7 have once again been updated, including revised homework problems, to incorporate the most recently available technology and design and installation practice. In particular, Chapter 4 incorporates more emphasis on higher voltage systems, new developments in IEEE 1547, new fire code requirements and *National Electrical Code* changes in the 2020 Edition, such as rapid shutdown, microinverters, optimizers, utility interconnects and advances in inverter, module and structural technology.

By the end of Chapter 4, instructors can assign relevant design problems earlier in the course to avoid having designs due at final exam time. We have found when significant design projects are due just before final exams, the students tend to spend time on design projects that might better be spent studying for the final exam, especially if the final exam counts more toward the overall course grade. Paradoxically, the better designers have sometimes ended up with lower course grades.

Since the publication of the previous edition in 2016, a nearly overwhelming wealth of new research on old and new technologies has been underway. We try to cover the highlights of some of this activity in Chapter 11 but have eliminated some of the historical content to make room for the new stuff.

A modified top-down approach is used in the presentation of the material. The material is organized to present a relatively quick exposure to all of the building blocks of the PV system, followed by design, design, and design. Even the physics of PV cells in Chapter 10 and the material on emerging technologies in Chapter 11 are presented with a design flavor. The focus is on adjusting the parameters of PV cells to optimize their performance, as well as on presenting the physical basis of PV cell operation.

Homework problems are incorporated that require both analysis and design, since the ability to perform analysis is the precursor to being able to understand how to implement good design. Many of the problems have multiple answers, such as "Calculate the number of daylight hours on the day you were born in the city of your birth." We have eliminated a few homework problems based on old technology and added a number of new problems based on contemporary technology. Hopefully, there are a sufficient number to enable students to test their understanding of the material.

We recommend that the course be presented so that by the end of Chapter 4, students will be able to engage in a comprehensive, relevant design project, and, by the end of Chapter 7, they will be able to design relatively complex systems. We like to assign two design projects—a straight grid-connected system based on Chapter 4 material and a battery backup grid-connected system based on Chapter 7 material. At the discretion of the instructor, an additional design of a stand-alone system might be considered at the end of Chapter 8, or the stand-alone design might be assigned rather than the battery backup grid-connected system.

While it is possible to cover all the material in this textbook in a three-credit semester course, it may be necessary to skim over some of the topics. This is where the discretion of the instructor enters the picture. For example, each of the design examples of Chapter 4 introduces something new, but a few examples might be left as exercises for the reader with a preface by the instructor as to what is new in the example. Alternatively, by summarizing the old material in each example and then focusing on the new material, the *why* of the new concepts can be emphasized.

The order of presentation of the material actually seems to foster a genuine reader interest in the relevance and importance of the material. Subject matter covers a wide range of topics, from chemistry to circuit analysis to electronics, solid-state device theory, structural analysis and economics. *The material is presented at a level that can best be understood by those who have reached upper division at the engineering undergraduate level and have also completed coursework in circuits and electronics.*

Preface

We recognize that the movement to reduce credit toward the bachelor's degree has left many programs with less flexibility in the selection of undergraduate elective courses and note that the material in this textbook can also be used for a beginning graduate-level course.

While the primary purpose of this material is for classroom use, with an emphasis on the electrical components of PV systems, we have endeavored to present the material in a manner sufficiently comprehensive that it will also serve the practicing engineer as a useful reference book.

The course can be successfully taught as an internet course, with a preference for live participation open to all. Those remote students who were actually sufficiently motivated to keep up with the course generally reported that they found the text to be very readable and a reasonable replacement for lectures. We highly recommend that if the internet is tried, quizzes be given frequently to coerce the student into feeling that this course is just as important as her/his linear systems analysis course. Informal discussion sessions can also be useful in this regard.

The PV field is evolving rapidly. While every effort has been made to present contemporary material in this work, the fact that the preparation of this edition has evolved over a period of a year almost guarantees that, by the time it is adopted, some of the material will be outdated. For the engineer who wishes to remain current in the field, many of the references and web sites listed will keep him/her up-to-date. Proceedings of the many PV conferences, symposia, and workshops, along with manufacturers' data, are especially helpful.

This textbook should provide the engineer with the intellectual tools needed for understanding new technologies and new ideas in this rapidly emerging field. The authors hope that at least 1 in every 4.6837 students will make his/her own contribution to the PV knowledge pool.

We apologize at the outset for the occasional presentation of information that may be considered to be practical or, perhaps, even interesting or useful. We fully recognize that engineering students expect the material in engineering courses to be of a highly theoretical nature with little apparent practical application. We have made every effort to incorporate heavy theory to satisfy this appetite whenever possible.

We also wish to note that we have endeavored to base the contents of this work on human intelligence as opposed to artificial intelligence (AI). We sincerely apologize if any part may appear as AI, despite our efforts to avoid it.

Furthermore, since the publication of the previous edition of this book, we recognize that the printed word has begun to disappear and be replaced by material on line. This creates a major challenge for publishers to adapt to this evolution. In particular, referencing the work of others is coming under more intense scrutiny, and obtaining permission to incorporate the results of the work of others in our attempt to provide a coherent presentation of the PV system design process has become more difficult. We have made every effort to credit the work of others, as we have done in the previous four editions. We sincerely hope that authors of cited works will appreciate that we are doing our best to inform students about the information they have so diligently prepared. Our sincere thanks to the authors who have generated the materials we have referenced.

Disclaimers

NREL

It's been said that no two snowflakes are the same. And, for that matter, it is unlikely that the weather on any two days at any location will be exactly the same. The minute-by-minute temperature and cloud cover will differ between any two days, whether they are one after the other or the same date in two successive or non-successive years. As a result, estimating the performance of a photovoltaic (PV) system is at best, an inexact science, since performance is critically weather dependent.

However, the longer the period that is included in a prediction or simulation of PV system performance, the more reliable the prediction becomes, since weather conditions tend to average out over the long term. Still, predicting the future is not nearly as reliable as recording the past. As a result, *when system performance is simulated in this text* using the System Advisory Model (SAM) developed by the U.S. National Renewable Energy Laboratory, because of the careful modeling tools used by NREL, SAM is one of the most reasonable tools to use to estimate annual energy production for a PV system. But no claim has been made that SAM is exact or perfect, and thus when SAM is used to estimate system performance, it must be done with the understanding of the following disclaimer:

> USER AGREES TO INDEMNIFY DOE/NREL/ALLIANCE AND ITS SUBSIDIARIES, AFFILIATES, OFFICERS, AGENTS, AND EMPLOYEES AGAINST ANY CLAIM OR DEMAND, INCLUDING REASONABLE ATTORNEYS' FEES, RELATED TO USER'S USE OF THE DATA. THE DATA ARE PROVIDED BY DOE/NREL/ALLIANCE "AS IS," AND ANY EXPRESS OR IMPLIED WARRANTIES, INCLUDING BUT NOT LIMITED TO THE IMPLIED WARRANTIES OF MERCHANTABILITY AND FITNESS FOR A PARTICULAR PURPOSE ARE DISCLAIMED. IN NO EVENT SHALL DOE/NREL/ALLIANCE BE LIABLE FOR ANY SPECIAL, INDIRECT OR CONSEQUENTIAL DAMAGES OR ANY DAMAGES WHATSOEVER, INCLUDING BUT NOT LIMITED TO CLAIMS ASSOCIATED WITH THE LOSS OF DATA OR PROFITS, THAT MAY RESULT FROM AN ACTION IN CONTRACT, NEGLIGENCE OR OTHER TORTIOUS CLAIM THAT ARISES OUT OF OR IN CONNECTION WITH THE ACCESS, USE OR PERFORMANCE OF THE DATA.

In other words, using SAM for educational purposes is just fine, but blaming SAM if a system is built and does not work as predicted by SAM is not OK.

NFPA/NEC, ASCE, AND PVSE

Codes and standards are essential to the safe implementation of all technologies, including photovoltaics. In addition to local, statewide and regional codes, two national codes play especially important roles in the safe design of photovoltaic systems—NFPA 70, National Electrical Code and ASCE Standard 7, Minimum Design Loads and Associated Criteria for Buildings and Other Structures. Since technology continues to advance, these codes need to keep up with these advances. As a result, just as the design professionals get to know them pretty well, they come out with a new edition.

The reader should understand that the *enforcement* of these codes and standards is the job of local building departments. As a result, there is generally a delay between the *publication* of a standard and the *adoption* of a standard. For example, in Florida, enforcement of the 2020 National Electrical Code began in December 2023, even though a 2023 edition is now available. Similarly, the 2016 Edition of ASCE 7 was used through the end of 2023, even though a 2022 Edition was available in 2022.

This edition of *Photovoltaic Systems Engineering* uses NEC 2020, ASCE 7–16 and ASCE 7–22 as references for PV system design, recognizing that by the time it is published and after it is in use for a while, these publications will change again, and again, and . . . . This is why it is so important to *understand* how to *use* these important documents, so when they are updated, it will be a routine exercise to dig into the new edition and recognize the changes that will affect the design process. And this is why we make every effort to explain how to use these documents as we move through the design of numerous systems.

Thus, as in the case of NREL SAM, the authors and publishers emphasize that the final authority on any design intended for construction lies with the local code enforcement jurisdiction. The reader must understand that reference to the materials in the design examples is intended to introduce a process of incorporating currently accepted (2024) design practices into a design. Neither code, standard nor this publication can be held responsible for accurately portraying any particular code or ordinance.

Acknowledgments

Thanks to a long list of photovoltaic (PV) pioneers for making this publication possible. Only a few decades ago, photovoltaics was considered to be an emerging technology. Some thought it would never make it on a large scale. But others with more persistence have persevered over the years such that PV is no longer a hobby but is a mature, widely accepted, large-scale, worldwide technology that is growing faster than most advocates ever imagined.

One of these PV pioneers is Jerry Ventre, who coauthored the first three editions of this text. Jerry has been retired for a while now and has spent much of his time doing volunteer work with organizations that focus on PV job growth. It has been an honor and a privilege to work with Jerry over the years, and we wish him the best in his current and future activity.

We are convinced that it is virtually impossible to undertake and complete a project such as this without the encouragement, guidance, and assistance from a host of friends, family, and colleagues. For this edition, we again asked many questions of many people as we rounded up information for the wide range of topics contained herein. A wealth of information flowed our way from the National Renewable Energy Laboratory (NREL) and the Florida Solar Energy Center as well as from many manufacturers and distributors of a diverse range of PV system components. We feel confident that, had we asked, we would have been able to obtain additional cooperation and materials from many more.

Special thanks to Kimandy Lawrence of Lord and Lawrence Consulting Engineers for offering Roger the opportunity to review more than 6000 PV system designs since December 2020. Since his introduction to PV about 20 years ago, Kimandy has been determined to "get it right." His philosophy has been to teach his competition how to get it right to be sure PV remained a respected and trusted energy source. "Look around, there are plenty of rooftops for everyone."

Last, but not least, we thank Mimi Abtahi and Jane Caputi for encouraging us to work on this manuscript rather than play golf, tennis, watch TV or pay attention to them.

Roger Messenger and Homayoon "Amir" Abtahi

Authors

Roger Messenger is Professor Emeritus of Electrical Engineering at Florida Atlantic University in Boca Raton, FL. He received his Ph.D. in Electrical Engineering from the University of Minnesota and is a Registered Professional Engineer, a former Certified Electrical Contractor and a former NABCEP Certified PV Installer, who has enjoyed working on a field installation as much as he enjoys teaching a webinar or working on the design of a system or contemplating the theory of operation of a system. His research work has ranged from electrical noise in gas discharge tubes to deep impurities in silicon to energy conservation to PV system design and performance. He worked on the development and promulgation of the original Code for Energy Efficiency in Building Construction in Florida and has conducted extensive field studies of energy consumption and conservation in buildings and swimming pools.

Since his retirement from Florida Atlantic University in 2005, he has worked as Vice President for Engineering at VB Engineering, Inc., in Boca Raton, FL, and as Senior Associate at FAE Consulting in Boca Raton. While at VB Engineering, he directed the design of several hundred PV designs, including the 1 MW system on the roof of the Orange County Convention Center in Orlando, FL. While at FAE Consulting, he led the design of an additional 6 MW of systems that were installed. Since 2020 he has reviewed over 60 MW of residential systems, including more than 8 MWh of residential Battery Energy Storage. He has also been active in the Florida Solar Energy Industries Association and the Florida Alliance for Renewable Energy, has served as a peer reviewer for the U.S. Department of Energy and has served on the Florida Solar Energy Center Advisory Board. He has conducted numerous seminars and webinars on designing, installing and inspecting PV systems. In May 2024 he received a Florida Solar Energy Industries Association Hall of Fame award for extraordinary contributions to the Florida Solar Industry.

Homayoon "Amir" Abtahi is currently Associate Professor of Mechanical Engineering at Florida Atlantic University. He received his Ph.D. in Mechanical Engineering from M.I.T. in 1981 and joined Florida Atlantic University in 1983. In addition to his academic activity, he has a wealth of practical experience, much of which has been obtained as a volunteer. He is a Registered Professional Engineer in Florida and is a member of ASCE and SAE. He has held LEED Certification since 2007, is ESTIDAMA Certified in the United Arab Emirates and is a Certified General Contractor and a Certified Solar Contractor in the State of Florida. His interests range widely from PV to PEM Fuel Cells, integrated capacitor/battery power modules and Atmospheric Water Generation.

In 1985, he installed the first-known solar power system in Venezuela and was responsible for the first-known application of solar power for post-hurricane emergency power and lighting and Ham Radio communication operations in the aftermath of Hurricane Hugo in St. Croix in 1989 and Hurricane Marilyn in St. Thomas in 1995. In 1989, he published the first comprehensive catalog of 12-V appliances for use with PV systems.

Recently, he has been involved with PV installations in the Caribbean, South America, Bangladesh and India. In the 2008–2010 time period, he was responsible for the design and installation of over 100 residential and 20 commercial/industrial PV systems. Over the past 15 years, he has had responsibility for the design and installation of 1 million BTUD of solar hot water and solar process heat. Along with the PV and thermal applications, he has had experience with heat exchangers, MEP plan review, LEED projects, tracking PV, micro-turbines, parabolic trough solar and other hybrid applications.

We did not inherit this world from our parents....
We are borrowing it from our children.
(Author unknown)

We believe the future of PV is as bright as the sun.
It is our fervent hope that the engineers who read this book
will dedicate themselves to the creation of a world
where children and grandchildren will be left
with air they can breathe and water they can drink,
where humans and the rest of nature
will nurture one another.

1 Background

1.1 INTRODUCTION

It took just 20 years from 1954 when Bell Labs in New Jersey announced the first successful testing of a photovoltaic (PV) cell [1] to 1974 when the worldwide shipment of PV modules reached 500 kW [2]. This was particularly huge for the time because, in 1974, a 5-Watt solar panel was treated as if it were a piece of jewelry. On the application side of things, NASA deployed the first PV application with a six-cell power module for the Vanguard satellite in 1958 [3]. Solar-powered wristwatches designed by Roger Riehl were offered to the public as "Synchronar 2100" in 1972 [4]. The energy crisis of 1973, alongside the geopolitical sense of urgency to be independent of fossil fuels [5], pushed investments in solar PV research and manufacturing to the front burner for the United States. Weary of conflicts at or near the largest reserves of oil and gas, the United States and especially Japan, which depended on oil imports for nearly all of its energy needs, spearheaded PV manufacturing that resulted in a significant drop in the cost of PV from $300/Watt in 1956 [6] to less than $5/Watt by the mid-1980s.

Despite the advances in efficiency and the prospect of lower manufacturing costs, solar panels were considered a novelty and a power source for mostly niche applications in the mid-1980s [7]. Remote microwave repeater stations, cell towers, solar-powered streetlights, solar-powered irrigation pumps, remote cabins and specialty soft panels for sailboats made up most of solar PV sales to the late 1990s. The entrenched energy suppliers correctly argued that PV was still too expensive to compete with conventional power generation, but then they incorrectly predicted that it would be a long time before solar (or wind) energies could even be conceived as "real" utility-grade power sources. Solar engineers and advocates were considered rebels in the serious "electrical grid world," and the thought of solar or wind energies ever replacing coal, nuclear and oil was considered heresy by utilities and fossil fuel producers. After all, solar was good enough only to power the remote cabins of the "hippies" in California. It was only when MW-sized wind generators started to be deployed in the United States and Europe, and the annual worldwide energy output of the wind sector surpassed 10 terawatt-hours (TWh = 10^9 kW-hours) in 1995, that utilities started paying attention [8].

It would take solar energy another decade to prove the naysayers wrong. First Solar Corporation announced a $1/Watt goal for the price of its amorphous CdTe solar panels in 2008 [9]. One can argue that solar energy was not taken seriously until "total installed" solar PV dipped just under $5/Watt ($5000/kW) in 2005. Supply chain and raw material shortages resulted in price stagnation until 2009 when PV manufacturing capacity managed to keep up with demand [10]. With a stable solar module supply and installation costs on par with conventional generation, solar suppliers started to talk the language that utilities and power engineers were

DOI: 10.1201/9781003470892-1

comfortable with. For comparison, consider that, in 2007, Florida Power and Light reported a forecast for the "Total Cost" of a nuclear power plant construction to be between $5492/kW to $8081/kW, which is close to the Moody Investor Services of estimated $5000 to $6000] [11]. However, the optimism for the solar advocates was short-lived. Solar energy skeptics soon found another excuse to oppose the widespread deployment of solar, and this one would not be easy to overcome.

It was and still is an undeniable characteristic of solar and wind generators: "intermittent." With a lack of efficient and cost-effective storage technologies and batteries to resolve the variability of solar energy during the day, and the simple fact that there is no solar energy to collect at night, this was a very substantial barrier for solar energy. Those who had thought reaching parity was enough to give solar a foot in the door of the utilities realized, just as many had before them, that being equal in performance and price a third of the time was just not good enough. Solar had to be more reliable and compete with low-cost hydropower from Canada or the Pacific Northwest to be allowed into the club. This was not asked of nuclear or any other competitors in the energy supply chain; in fact, the environmental costs of coal and nuclear power were to be paid by taxpayers, no questions asked.

The existing grid in the United States that helped provide power through projects like Tennessee Valley Authority, the Hoover Dam and many others was paid for by governmental subsidies and taxpayers [12]. When solar energy was over-generated in California [13] and Germany [14] and utilities offered electricity at negative prices to shed load, politicians criticized the investments in solar as "a waste of tax-payer dollars," when, in fact, the real culprit was the power grid. It was not designed for a bi-directional transmission of power or distributed generation so that excess renewable energy can be dispatched to where it's needed, when it's needed, with integrated storage to accommodate intermittent supplies.

The lack of solar energy at night, as well as daily and seasonal peaks in energy use, makes it obvious that energy storage, energy management and other clean energy sources will need to be a part of the energy mix of the future. In addition to the development of advanced batteries and other storage technologies, to be discussed in detail in Chapter 6, including the deployment of utility-scale MWh and GWh storage facilities, improved dispatch and transmission of renewable energy, along with innovative load management and control, is helping to integrate solar and other renewable energy sources into the grid.

Despite optimism that solar is finding its footing in the overall power supply space, recent events at COP-28 [15] and the resistance of fossil fuel-producing countries to cut the cord of dependence on fossil energy sources have shown once again that parity alone will not be sufficient to dislodge entrenched beliefs and loyalties. Solar needs to be twice as reliable and stable as any other source of energy and yet, even with storage, be cheaper than any other energy source to dislodge the status quo and make serious inroads—not just with utilities that have now accepted solar because it makes economic sense but also to convince society and policy makers at large. Does this sound familiar and remind us of how difficult it can be to be accepted by the comfortably entrenched old guards?

In the remaining sections of this chapter, we will discuss and dive into the justification and reasoning for the argument that solar and other "near-zero emission"

Background

sources are the "urgent" solutions. Global warming was re-labeled "climate change" to avoid offending certain groups and then renamed again as climate crisis for a reason. With more than 20 nations [16] dealing with severe droughts or flooding, numerous island nations begging for help, and many of the poorest and most vulnerable populations simultaneously fighting wars and facing climate change, it is time to pull the climate alarm lever and stop the narrative of entrenched interests intent on pushing our collective heads into the sand of denial: denial that Antarctica and Greenland's ice sheet losses will contribute from 1 to 2 meters to sea level rise [17], denial that the South pole and Greenland's ecosystem will substantially change before 2100 [18] and denial that sea level rise will lead to the displacement of more than a billion climate refugees by 2050 [19]. Do we really have to wait until the warm river of the Gulf Stream stops flowing from Florida Straights to England and Norway, freezing Northern Europe and overheating the U.S. southeastern seaboard before we act?

1.2 URGENT ATTENTION TO WORLD POPULATION FORECASTS

Population is clearly the most direct driver of energy requirements and thus potential emissions from energy generation. Population forecast models agreed upon by the UN [20], INED [21], U.S. Census [22], Worldometers [23] and several other international organizations vary in details, but the general consensus, as shown in Figure 1.1, adopted from PopulationPyramid.net [24] data, suggests that the world population will peak sometime between 2070 and 2100. The peak numbers vary

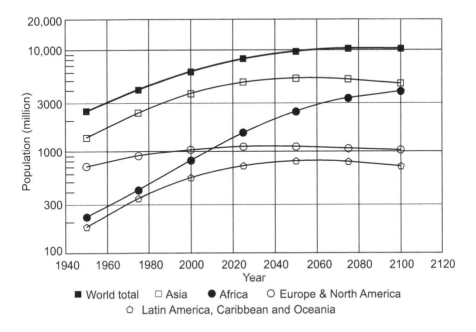

FIGURE 1.1 World and regional population data.

[Adapted from Population Pyramid data: www.populationpyramid.net/world/]

from source to source, but they are all in the range of 9.5 to 10.5 billion. Assuming 2075 is a good compromise when the world population peaks, it will be used as the reference, providing a 50-year window to assess and project energy demand and the resulting emissions.

The population trends of the major contributors to the world population are also shown. The regional data shows that Europe and North America's combined population is peaking. Remarkably, Asia's population, including China and India's populations, is also approaching a plateau and is expected to begin decreasing in the next few decades. Africa's population however does not show any indications of approaching a diminishing rate, let alone a peak. This uncertainty, combined with the fact that many African nations will see the biggest growth in energy demand in the next 50 years and still do not have the resources to switch over to renewables, may be the biggest challenge for moderating worldwide emissions.

To summarize, the following assumptions are made throughout this text:

- The year 2075 is when the world's population will plateau, giving us approximately a 50-year time frame to address energy demand and emissions.
- It will be optimistically assumed that developed regions of the world, including North America, Europe, China and India, will meet the energy emissions targets that have been laid out by each.
- The focus of this chapter is to address electricity power supply, not other forms of energy demand such as industrial heating, or transportation, except electricity demand for electricity-based cars or trains.
- It is also recognized that clean electricity is only one of numerous challenges facing this and future generations. Hopefully, clean electricity will also be part of solutions for other challenges such as adequate clean water supply.

To satisfy the energy needs of a peaking world population in 2075, no single form of energy, including renewable, nuclear or hydro, will be sufficient. The existing forms of clean energy must all be harnessed thoughtfully and as sustainably as possible to provide energy for the world of 2075. There should also be a concerted and honest effort to consider equity and the needs of underserved populations and countries, most of whom are innocent victims of climate change. Specifically, effort must be made to ensure that regions and peoples affected most severely by climate change receive urgent investment assistance to transition to clean renewable resources and eliminate the use of fuels and sources such as coal and kerosene and the associated unhealthy emissions, especially in proximity of vulnerable sectors of the population. Renewable engineering must be implemented with care to not disrupt ongoing human activity, while always cognizant of impact on nature.

1.3 ENERGY DEMAND AND CARBON DIOXIDE EMISSIONS

1.3.1 Current and Projected Scenarios

At the time of this writing, there is a good sense of the energy demand projections of the Organization for Economic Cooperation and Development (OECD) countries including Europe, North America, Japan, China, India and similarly developed regions of the world. At least on paper, Europe and North America have some very clear plans on how they're going to control emissions, and their emissions have either reached the maximum levels or are going down. Similarly, China and India have plans in place, but they are simultaneously under pressure to provide more power to their population—so there is uncertainty about whether those strides will be met. The key to sustainability is to achieve a balance between harmful emissions produced by human activity and neutralizing them. And even though emissions are decreasing in a number of countries with advanced economies, they are still nearly an order of magnitude greater than the removal rate.

1.3.2 Solar PV Contribution and Demand

On December 13, 2023, the United Nations (UN) Climate Change Convention (COP28) in Dubai came to an end. This date was a few days later than anticipated, mostly because of negotiations over language around transitioning away from fossil fuels; there were definite signs that fossil-producing nations were not yet willing to accept that the world has to be more aggressive with phasing fossil fuels out. In fact, the words "phase out" were considered too harsh, so the terminology "phase down" was used instead. In response to the back-and-forth, UN Secretary-General António Guterres addressed the nations that opposed more "direct" language around transitioning to cleaner energy by saying that whether countries wanted to believe it or not, phasing out of fossil fuels is inevitable—the question is only when.

As far as solar engineering and planning is considered, it is obvious that plans resulting from the 2015 Paris Agreement and 2021 Glasgow Climate Pact need to be accelerated, if nothing else; projections around how much renewables will need to be incorporated into energy planning will need more solar than anticipated. To see what energy engineers and planners will be facing, a review of the roadmaps that were developed following these agreements is helpful.

The International Energy Agency (IEA) has a very helpful publication entitled "World Energy Outlook" [25] that outlines three scenarios for the global reduction of fossil fuel emissions. These road maps will help energy engineers and planners to understand what will be anticipated up to 2050, including, specifically NZE-2050, and other less specific long-term goals:

1. The first scenario is the **Stated Policies Scenario (STEPS)**, which is reflective of current climate change policies.
2. The second is the **Announced Pledges Scenario (APS)**, which assumes that each country's climate targets are met in full, but without guarantee that this will happen.
3. The third scenario is **Net Zero Emissions by 2050 (NZE)**, in which half of the total energy (and two-thirds of useful energy) will have transitioned to zero-emission renewables. This scenario is based on more ambitious agreements of NZE and is a signed commitment from about 38 OECD countries in the world that have begun to put reliable plans in place.

A look at the latest projections for energy transition away from fossil fuels suggests that those will now be minimums based on what was just reached COP 28 in Dubai, meaning transitions will have to be accelerated.

In Table 1.1, the impact of the three scenarios adopted at COP 25 on PV generation for 2050 is shown.

To update the information or look at these projections and see what changes might occur beyond requirements or any government programs, it is worth reviewing recent coverage by NationalNews.com [26] and Bloomberg NEF [27] regarding increased PV installations above and beyond what was projected even at this early stage.

There was a projected 26% increase in PV installations from 2022 to 2023. This has gone through a great transformation, with China installing more than 250 GW in 2023, resulting in a 27% increase globally. The projected increase should therefore be revised to 64% for 2024. In addition, the world PV module production capacity, which was reported as 839 GW in 2023, is now expected to reach 1 TW in 2024 [28]. Assuming all PV production is installed, the goals of NZE could be reached as much as ten years earlier than anticipated.

1.3.3 Emissions

Regional Emissions associated with power generation are directly proportional to population, per capita energy demand and the average emissions generated by kWh generated.

$$\text{Regional Emissions} = \text{Regional Population} \times \text{Energy per Capita} \times \text{Emissions per kW-hour} \quad (1.1)$$

The relationship between CO_2 and global temperature rise is well understood. Globally, CO_2 concentration is measured at the Mauna Loa Observatory in Hawaii, which has set the global standard by which CO_2 levels are calculated [29].

As outlined in Figure 1.2, CO_2 has increased from a concentration of just under 320 parts per million (ppm) in 1960 to almost 420 ppm in 2020. Currently, 33 GT of CO_2 is annually emitted into the atmosphere globally. As can be observed from the exponentially appearing curve, the seasonal absorption of CO_2 during high seasonal

TABLE 1.1
GEC Model Scenarios: Anticipated 2050 Outcomes

Scenario	Contribution to World Energy Supply (%)	Solar/PV Component of the Renewables (% of Total Electricity Demand	Solar PV Energy Generation (TWh)
STEPS	65	24	12,118
APS	80	30	18,761
NZE	88	37	27,006

(Adapted from table 6.1, "World Energy Outlook" published by IEA, 2022.)

Background 7

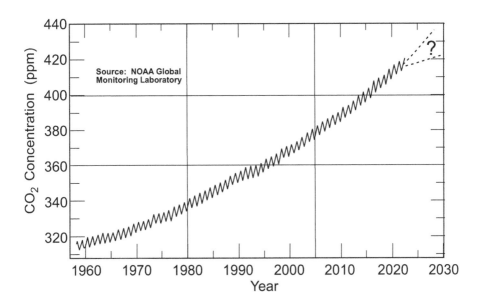

FIGURE 1.2 Historic atmospheric carbon dioxide concentration as measured at Mauna Loa. [Adapted from data from the NOAA Global Monitoring Laboratory as presented by Scripps Institution of Oceanography, UC San Diego.]

bio-absorption of CO_2 is no match for the CO_2 load and is just a small secondary wave on the primary increasing trend in atmospheric CO_2 percentage. In fact, the percent of CO_2 in the air we breathe increases by approximately 0.00035% or 3.5 ppm annually. This increase is particularly troubling due to the rate of increase over the past 60 years. The impact of the average global temperature is shown in Figure 1.3, adapted from NOAA-Goddard Institute [30]. Note that each data point in this figure represents the average CO_2 concentration and average global temperature for a particular year. The average global temperature is compared with baseline data between 1850 and 1900. Although specific years are not associated with each data point, the data is grouped into periods as shown on the graph. Beginning in 1940, data is grouped by 20-year increments, clearly showing the increasing average temperature as CO_2 levels continue to rise at an approximate rate of 0.1°C per 10 ppm increase. At this rate, the global average temperature will increase by 2°C above pre-industrial levels in just 25 years. After the 2015 Paris Agreement, the treaty's 196 adoptive countries aligned on pursuing efforts to limit the temperature increase to 1.5°C and prevent temperatures from rising to 2°C [31].

The impact of carbon emissions resulting from human activity on the global temperature rise has been rigorously established, giving us no choice but to trust the data and ensure we can curb rising temperatures—or the consequences will be disastrous. There is, however, another factor to also consider, and this might finally get some attention. At the current rates, the global atmospheric CO_2 levels will reach 0.08% by 2100. This will be very close to 0.1% CO_2 in the air (approximately 1100 ppm of CO_2),

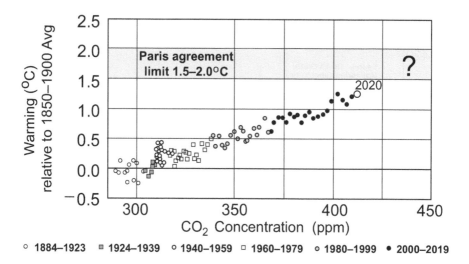

FIGURE 1.3 Correlation between atmospheric carbon dioxide levels and global temperature rise.

[Based upon data presented by Fakta o klimatu, Version 2022-05-12: factsonclimate.org/concentration-warming-relationship. Data source: NOAA, NASA Goddard Institute for Space Studies.]

considered unhealthy per current Standard-62 of ASHRAE's (American Society of Heating Refrigeration and Air Conditioning Engineers) air quality code [32].

In view of the discussion of the world population trends, the energy mix requirements and the consequences of emissions, there is justification and rationale for increased workforce development and education of energy engineers, especially in solar/PV. Hopefully, this textbook will help lead us toward that goal.

1.4 A BRIEF OVERVIEW OF THERMODYNAMICS

Most engineers are familiar with the first law of thermodynamics, which is the concept of energy conservation. Energy cannot be generated or destroyed. In essence, this means that energy can only be converted or transferred—so, in practice, first law efficiency is a function of the quality of the equipment being used as the ultimate objective is to convert all of that energy. An example of this is that a fluid-to-fluid heat exchanger can theoretically transfer close to 100% of the energy content from one fluid to another. However, converting heat energy to work energy is limited by the second law of thermodynamics.

The second law of thermodynamics governs how much heat energy can be converted to work or electric energy. The efficiency of this conversion from heat to mechanical work/electric energy is a function of the temperature difference between the high-temperature source and a low-temperature sink where the heat can be rejected. Probably the most common everyday application of the second law is the operation of hundreds of millions of internal combustion engines in

Background

vehicles worldwide. Most people have some understanding that without the heat (and pressure) of burning gasoline in the cylinders and the water circulating in the radiator and coolant system, the engine will not function properly. A reasonable fraction of the population has experienced the boiling over of a faulty radiator and the possible accompanying failure of the engine itself. In the winter, some of this "waste" heat is recirculated into the interior of the vehicle after being "captured" by the air circulating between the vehicle interior and the heating coils.

Another example is that a hydroelectric plant is not possible unless the discharge of the turbines is at a lower level than the height of the water behind the dam, because the difference in height is what creates the potential energy of the water at the higher level. Inversely, the water will not flow upward by itself. A pump with energy input is required to achieve that. Similarly, a heat pump requires energy to pump heat (measured in kilo-joules or BTUs) from a lower-temperature reservoir such as a freezer to a higher-temperature space such as the outside of the refrigerator.

To quantify the second law of thermodynamics, when a heat engine converts heat into work (where work can be shaft work or electricity or raising of a weight), conservation of energy still applies. Thus, the starting energy at the high temperature, or, more precisely, the energy supplied by the high temperature source must equal the sum of the heat converted to work plus the heat rejected, or, "waste heat." "Waste heat" is actually a misnomer because the loss of heat is a requirement for heat engines. The French physicist Sade Carnot quantified the limits of heat-to-work conversion with:

$$\text{Ideal Thermal Efficiency} = 1 - T_{Low}/T_{High}, \qquad (1.2a)$$

where temperatures are in Kelvin.

Since Kelvin is the temperature measured from absolute zero, to convert T in °C to Kelvin simply:

$$T (\text{Kelvin}) = T (°C) + 273.15. \qquad (1.2b)$$

For example, to convert 900°F to Kelvin:

1. Convert °F to °C: $(900 - 32)(5/9) = 482.2°C$.
2. Convert T°C to Kelvin: $482.2 + 273.15 = 755.37$ K.

Assuming that the average cooling temperature available around a power plant is 75°F equivalent to 297 K, it can be concluded that a powerplant with a heat source of 900°F has a maximum or ideal, Carnot efficiency of

$$1 - (297/755.37) \text{ or } 57.5\%.$$

Without getting too deep into proofs, a corollary is that:

$$\text{Ideal Thermal Efficiency} = (Q_H - Q_L)/Q_H, \qquad (1.2c)$$

where Q_H is heat supplied by the high-temperature source and Q_L is heat rejected to the environment.

$(Q_H - Q_L)$ is also equal to the net work out of the engine. This equation can also be written as a rate equation so that:

$$\text{Heat Input in kW} - \text{Heat Rejected in kW} = \text{Work Out in kW} \tag{1.3}$$

The requirement that all heat-to-work conversions have a low-temperature sink results in the need to reject heat to the environment surrounding a thermal power plant. This is true for any power plant including geothermal, nuclear, solar thermal, coal or any other fossil-fueled power plant. There is no exception to this rule, the result of which is that heat *must* be released into the environment for a thermal power plant to function. The lack of understanding of the second law often leads people to believe that there is *no* impact on the environment from what we call "cleaner" forms of energy. In all fairness, if the thermal losses of all the power plants in the world are added up, they probably don't reach even one or 2% of the global heating caused by all greenhouse gases (GHGs), including carbon dioxide in the atmosphere—but there is thermal pollution from these cleaner sources, even geothermal power plants because they also need cooling. The second law of thermodynamics limitation has not been exceeded since Sadi Carnot quantified it in 1824 [33], and there is no evidence that it ever has.

1.5 A BRIEF HISTORY OF PHOTOVOLTAICS

Edmond Becquerel [34] first discovered that sunlight could be converted directly into electricity in 1839, when he observed the photogalvanic effect. Then, in 1876, Adams and Day [35] found that selenium has PV properties. When Planck postulated the quantum nature of light in 1900 [36], the door was opened for other scientists to build on this theory. It was in 1930 that Wilson [37] proposed the quantum theory of solids, providing a theoretical linkage between the photon and the properties of solids. Ten years later, Mott and Schottky [38] developed the theory of the solid-state diode, and, in 1949, Bardeen, Brattain and Shockley invented the bipolar transistor [39]. This invention, of course, revolutionized the world of solid-state devices. The first solar cell, developed by Chapin, Fuller and Pearson, followed in 1954 as mentioned in the introduction to this chapter.

One might wonder why it took so long to develop the PV cell. The answer lies in the difficulty of producing sufficiently pure materials to obtain a reasonable level of cell efficiency. Prior to the development of the bipolar transistor and the advent of the space program, there was little incentive for preparing highly pure semiconductor materials. Coal and oil were meeting the world's need for electricity, and vacuum tubes were meeting the needs of the electronics industry. But since vacuum tubes and conventional power sources were impractical for space use, solid state gained its foothold.

Background

As will be discussed in detail in later chapters, PV cells are made of semiconductor materials and are assembled into modules of 36 or more cells. This observation is significant, since this means the same industry that has, in the past 60 years, progressed from the development of the bipolar transistor to integrated circuits containing millions of transistors is also involved in the development of PV cells.

1.6 ENERGY UNITS

Energy is measured in a number of ways, including the calorie, the BTU, the quad, the foot-pound (ft-lb) and the kilowatt hour (kWh). When pounds (lb) and kilograms (kg) are used in this text, they will be treated as weight rather than mass. Thus, a 220 lb person can also be described as a 100 kg person. For the benefit of those who may not have memorized the appendices of their freshman physics books, we repeat the definitions of a number of these energy-related quantities for an earth-based system at or about a temperature of 27°C.

1 calorie is the heat needed to raise the temperature of 1 gram of water 1°C.
1 BTU (British Thermal Unit) is the heat needed to raise the temperature of 1 lb of water by 1°F.
1 quad is 1 quadrillion (10^{15}) BTUs.
1 ft-lb is the energy expended in raising 1 lb through a height of 1 ft.
1 kWh is the energy expended by 1 kilowatt operating for 1 hour.
1 MTOE (million tons of oil equivalent) is the energy available from 1 million tons of crude oil.
1 Mbd (million barrels per day) is the energy available from 1 million barrels of crude oil.
1 Bcm (billion cubic meters) is the energy available from 1 billion cubic meters of natural gas.

With these definitions, and with the numerical values of MTOE, MBOE and Bcm, the following equivalencies can be determined:

1 BTU = 252 calories
1 kWh = 3413 BTUs = 2,655,000 ft-lb
1 ft-lb = 0.001285 BTU
1 quad = 10^{15} BTUs = 2.930×10^{11} kWh
1 MTOE = 0.0397 quad
1 MBOE = 0.00541 quad
1 Bcm = 0.0357 quad

It should be noted that MTOE, MBOE and Bcm represent the *total* amount of energy stored in the specified quantities of coal, oil and natural gas before they are converted into other forms of work or energy, such as electricity. Net energy is another term often used to describe the resulting useful energy obtained from a primary source.

Homework Problems

1.1 A common rating of a typical-size PV panel (also called module) is 400 Watts. An array or a cluster of ten solar panels therefore generates 4000 Watts of power during peak solar conditions. If peak solar conditions are 4.2 hours per day for a specific water pumping application in northern Oklahoma, how much energy can be expected to be collected in one day in kWh units? Convert this answer to kJ.

1.2 Solar PV panels are labeled at the manufacturing facility with a sticker that indicates the power rating of each module. These ratings, however, are based on a test of the module in a "solar test chamber" where the module is subjected to an industry standard of 1000 Watts/m^2-simulated solar light intensity. As will be discussed in later chapters, a more realistic solar energy intensity on the surface of the earth is closer to 800 Watts/m^2 maximum except in higher elevations with low humidity conditions. Based on this information, recalculate the energy collected in Problem 1.1.

1.3 If 20 kWh of energy is stored in batteries from solar PV generation, and only 80% of this stored energy can be used to run the electric loads of a remote microwave repeater station, what is the maximum power that can be used for a 24-hour constant load?

1.4 The CO_2 emissions plot (Figure 1.2) shows an increasing slope from 1960 to 2020. Extrapolating the graph, what should the slope be after 2020 to keep atmospheric CO_2 below 430 ppm by 2050?

1.5 The Paris Climate Accord of 2015 among 196 countries sets a limit of 1.5°C on the average global temperature rise. Figure 1.3 shows the relationship between the global temperature rise and the global carbon dioxide levels measured in ppm. How close is the global temperature rise to this 1.5°C limit? Is there still time to avoid reaching this level?

1.6 Based on the brief discussion of the second law of thermodynamics, calculate the heat loss of an ideal engine to the environment for a 50 MW (Net output) coal plant with a 50% ideal efficiency. If an actual coal plant runs at 30% efficiency, how much more heat will be rejected to the environment?

1.7 An engineer is suggesting an increase in the temperature of the boiler in a power plant from 500°C to 600°C and claims that this change could result in reducing the heating of the local atmosphere by at least 20%. Assume the ambient temperature is 300 K. Is she correct?

1.8 An engineering intern is asked to compare the theoretical thermal environmental impact of a pressurized water reactor (PWR) nuclear power plant with a combined cycle natural gas-fired power plant. Both plants are to have a net electrical power output of 1 GW and operate for an average of 80% of the time on an annual basis. PWR plants are designed to run at a maximum of 647 K for safety reasons, and the combined cycle gas turbine

runs at 1500°F at its highest temperature point in the gas turbine section. Using these temperatures as T_H for each cycle and assuming 300 K as T_L for both cycles, determine:
  a. The Carnot efficiency of each plant.
  b. The total energy input for each plant on an annual basis, assuming a year of power generation to be 80% of 8760 hours.
  c. The thermal impact of each plant on the environment for a year and for the expected life cycle of 30 years.

1.9 The Hellisheidi geothermal power plant near Hengill, Iceland, operates with hot water extracted from 44 wells, providing the heat source at 180°C. The average ambient temperature (T_L) in that region of Iceland can be assumed to be 0°C in winter and 12°C in the summer.
  a. Determine the ideal Carnot efficiency of the plant in the winter and the summer.
  b. If the plant generates a net electric power output of 90 MW, calculate the total amount of energy that needs to be extracted from the hot water wells for the ideal Carnot cycle.
  c. An advantage of operating a power plant in cold regions is that the "waste" can be actually piped to homes and businesses to be used as heat for homes and businesses. This is called district heating or co-generation. Based on the ideal Carnot cycle for winter, calculate how much heat will be available for district heating.

1.10 In one of the IEA (International Energy Agency-Ref. 25-table A3) scenarios, solar PV electricity generation will increase from 821 TW-hours in 2020 to 23,469 TWh by 2050. Assuming that solar PV panels generate an average of 80% of nameplate rating, determine the PV-installed capacities for 2020 and 2050 that would be required to provide the stated goals. Assume that the energy generated per day is equivalent to 4.2 hours per day of rated output for 365 days a year.

1.11 When dealing with projections of future numbers, it is often very useful to develop a general spreadsheet for which certain variables can be changed to see the long-term effect of the change. This problem deals with the development of a spreadsheet to use in exploring the effect of changing various variables on the yearly and cumulative values of available PV generation capacity as well as the projections of total annual energy generation by the projected capacity.

Specifically, the proposed spreadsheet will include existing worldwide PV generation capacity in GW at the beginning of 2020, annual installations and annual growth rate of installations, the cumulative worldwide capacity in GW on an annual basis between 2020 and 2050 and the estimated energy generated in TWh on an annual basis based upon an assumed ratio of TWh produced to GW installed. The following shows

Year	GW Start	GW Ann Add	GW End	Rate	End TWh	TWh/GW 1.5
2020	585	125	710	0.2	1065	
2021	710	150	860	0.25	1290	
2022	860	188	1048	0.25	1571	
2023	1048	234	1282	0.3	1923	
2024	1282	305	1587	0.2	2380	

the first five years of spreadsheet entries resulting from the assumptions made by the user in the shaded cells.

a. Using the numbers in the shaded cells, determine how the numbers in the unshaded cells are calculated.
b. Extend the spreadsheet out to the year 2050, using the formulas you have figured out for part a.
c. Use an annual increase rate of 20% for all years (0.2 is 20%) up to 2050 and note the year where TWh generated exceeds 23,469.
d. Vary the user-entered numbers to see the effect on the cumulative worldwide GW of generation capacity.
e. Vary the TWh/GW figure between 1.0 and 2.2 and observe the effect on total TWh. Note that TWh/GW is the variable that has significant dependence upon array location and module type.

1.12 Currently, 33 GT (33 billion metric tons) of the estimated total 50 GTs annual emission of total GHGs is made up of CO_2. It is estimated that coal plants generate 820 grams of CO_2 per kWh of electricity generated versus solar PV, accounting for emissions associated with the manufacturing of panels, the structure and earthworks, resulting in 40 grams of CO_2 emissions per kWh. Calculate the reduction in CO_2 emissions achieved if all solar PV generation was targeted to reducing coal power generation. Use the 23,469 TWh (per year) for 2050 estimate by IEA [25].

REFERENCES

[1] Bell Labs 1954: www.aps.org/publications/apsnews/200904/physicshistory.cfm
[2] 1977-Worldwide PV Manufacturing Exceeds 500 kW: www1.eere.energy.gov/solar/pdfs/solar_timeline.pdf
[3] First Use of Solar in Space-Vanguard 1958: www.nasa.gov/image-article/vanguard-satellite-1958/
[4] Wrist Watch: www.gearpatrol.com/watches/a575261/watchmaking-firsts-solar-powered-watch/
[5] Energy Crisis of 1973: www.energypolicy.columbia.edu/publications/the-1973-oil-crisis-three-crises-in-one-and-the-lessons-for-today/
[6] PV $300/Watt: www.energysage.com/about-clean-energy/solar/the-history-and-invention-of-solar-panel-technology/
[7] Niche Applications: www1.eere.energy.gov/solar/pdfs/solar_timeline.pdf

Background

[8] https://ourworldindata.org/grapher/wind-energy-consumption-by-region
[9] $1/Watt PV Cost-First Solar Sets a Goal of $1/Watt August 2008 Edition: https://spectrum.ieee.org/first-solar-quest-for-the-1-watt
[10] www.nrel.gov/docs/fy13osti/56776.pdf
[11] www.synapse-energy.com/sites/default/files/SynapsePaper.2008-07.0.Nuclear-Plant-Construction-Costs.A0022_0.pdf
[12] www.tva.com/about-tva/our-history/the-tva-act
[13] https://cleanenergygrid.org/california-has-too-much-solar-power-it-needs-another-grid-to-share-with/
[14] www.greentechmedia.com/articles/read/germanys-stressed-grid-is-causing-trouble-across-europe
[15] https://unfccc.int/news/cop28-agreement-signals-beginning-of-the-end-of-the-fossil-fuel-era
[16] www.uneca.org/stories/17-out-of-the-20-countries-most-threatened-by-climate-change-are-in-africa%2C-but-there-are
[17] https://sealevel.nasa.gov/faq/11/how-much-rise-should-we-expect-from-greenland-and-antarctica/
[18] www.nationalgeographic.com/environment/article/antarcticas-ice-could-cross-this-scary-threshold-within-40-years
[19] 2 Billion Climate Refugees: www.zurich.com/en/media/magazine/2022/there-could-be-1-2-billion-refugees-by-2050-here-s-what-you-need-to-know
[20] https://population.un.org/wpp/
[21] www.ined.fr/en/everything_about_population/
[22] www.census.gov/data-tools/demo/idb/#/dashboard?COUNTRY_YEAR=2060&COUNTRY_YR_ANIM=2060
[23] www.worldometers.info/world-population/world-population-projections/
[24] www.populationpyramid.net/world/
[25] www.iea.org/reports/world-energy-outlook-2023
[26] www.thenationalnews.com/business/energy/2023/07/12/solar-power-on-track-to-meet-net-zero-targets-by-2050-iea-says/
[27] https://about.bnef.com/blog/global-pv-market-outlook-4q-2023/
[28] www.pv-magazine.com/2023/06/08/global-pv-manufacturing-capacity-to-reach-1-tw-by-2024/
[29] https://gml.noaa.gov/ccgg/trends_ch4/
[30] https://factsonclimate.org/infographics/concentration-warming-relationship
[31] https://unfccc.int/process-and-meetings/the-paris-agreement.
[32] www.ashrae.org/technical-resources/standards-and-guidelines/standards-interpretations/interpretations-for-standard-62-2001
[33] Sadi Carnot: https://geosci.uchicago.edu/~moyer/GEOS24705/Readings/Carnot_article_1998.pdf
[34] www.solarenergyworld.com/solar-history-alexandre-edmond-becquerellar/
[35] Adams, W. G. and Day, R. E. , "The Action of Light on Selenium, " *Proc Royal Society, London*, vol. A25, p. 113, 1877.
[36] Quantum Nature of Light: www.history.com/this-day-in-history/the-birth-of-quantum-theory
[37] Quantum Theory of Solids: https://en.wikipedia.org/wiki/Alan_Herries_Wilson
[38] Streetman, Ben G., *Solid State Electronic Devices*, 4th ed., Prentice-Hall, Englewood Cliffs, NJ, 1995.
[39] Invention of Bipolar Transistor: https://en.wikipedia.org

SUGGESTED READING

.1% or .2% CO_2 in Air Drowsiness or Poor Air Quality Complaints: www.co2meter.com/blogs/news/carbon-dioxide-indoor-levels-chart

An Alternative CO_2 Reference: Levels If Current Trends Continue: https://clintonwhitehouse5.archives.gov/Initiatives/Climate/next100.html

Indoor Air Below 1100 ppm: www.ashrae.org/technical-resources/standards-and-guidelines/standards-interpretations/interpretations-for-standard-62-2001

2 The Sun

2.1 INTRODUCTION

Optimization of the performance of photovoltaic (PV) and other systems that convert sunlight into other useful forms of energy depends upon the knowledge of the properties of sunlight. This chapter provides a synopsis of important solar phenomena, including the solar spectrum, atmospheric effects, solar radiation components, determination of the sun position, measurement of solar parameters and positioning of the solar collector. In this chapter, an attempt is made to quantify the obvious and, perhaps, some not-so-obvious observations: Why is it light during the day and dark at night? Why are there more daylight hours in summer than in winter? Why is the sun higher in the sky in summer than in winter? Why are there more summer sunlight hours in northern latitudes than in latitudes closer to the equator? Why does less energy reach the surface of the earth on cloudy days? Why is it better for a solar collector to face the sun? What happens if a solar collector does not face the sun directly? How much energy is available from the sun? Why is the sky blue?

2.2 THE SOLAR SPECTRUM

The sun provides the energy needed to sustain life in our solar system. In one hour, the earth receives enough energy from the sun to meet its energy needs for nearly a year. In other words, this is about 7500 times the input to the earth's energy budget from all other sources. To maximize the utilization of this important energy resource, it is useful to understand some of the properties of this "ball of fire" in the sky.

The sun is composed of a mixture of gases with a predominance of hydrogen. As the sun converts hydrogen to helium in a massive thermonuclear fusion reaction, mass is converted to energy according to Einstein's famous formula, $E = mc^2$. As a result of this reaction the surface of the sun is maintained at a temperature of approximately 5800 K. This energy is radiated away from the sun uniformly in all directions, in close agreement with Planck's blackbody radiation formula. The energy density per unit area, w_λ, as a function of wavelength, λ, is given by

$$w_\lambda = \frac{2\pi h c^2 \lambda^{-5}}{e^{\frac{hc}{\lambda kT}} - 1} \text{ (W/m}^2\text{/unit wavelength in meters),} \qquad (2.1)$$

where $h = 6.63 \times 10^{-34}$ watt sec^2 (Planck's constant),
$c = 3.00 \times 10^8$ m/sec (speed of light in a vacuum),
$k = 1.38 \times 10^{-23}$ joules/K (Boltzmann's constant) and
T = absolute temperature of a blackbody in K (degrees Kelvin, where 0 K
= –273.16°C).

Equation (2.1) yields the energy density at the *surface* of the sun in W/m²/unit wavelength in meters. By the time this energy has traveled 93 million miles (150 million km) to the earth, the total extraterrestrial energy density decreases to 1367 W/m² and is often referred to as the solar constant (see Problem 2.1) [1].

Figure 2.1 shows plots of Planck's blackbody radiation formula for the sun @ 5800K and the earth @ 300K. Note that, at lower temperatures, all of the spectrum lies outside the visible range in the infrared range, whereas at 5800 K, the characteristic white color of the sun is obtained due to the mix of wavelengths in the visible spectral range. At even higher temperatures, the color shifts toward blue, and, at lower temperatures, the color shifts toward red. Light-emitting diode (LED) lights, for example, are typically operated at equivalent color temperatures of 2700 K in various steps up to 5000K and, hence, emit approximate blackbody spectra for these temperatures, depending on the choice of the user. Depending on the color temperature of the light source, older photographic equipment had to be compensated to obtain true colors, unless appropriate filters were used. The extraterrestrial solar spectrum indicates that the sun can be reasonably approximated as a blackbody radiator.

Most sources of light are not perfect blackbody sources. Reasonable approximations of blackbody spectra are obtained from sources that emit light as a result of heating a filament to a high temperature. But to be a perfect radiator, the object must also be a perfect absorber of light, which is not the case for common light sources. Non-filament light sources, such as gas discharge lamps, for example, emit either

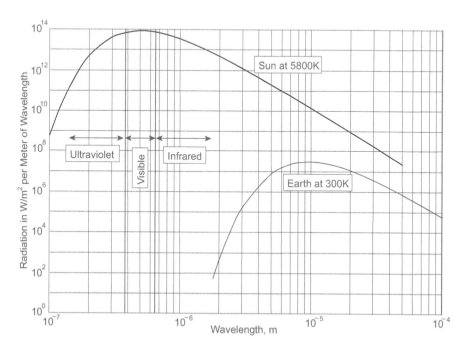

FIGURE 2.1 Radiation spectra of the sun and the earth.

the discrete spectral characteristic of the specific gas or, alternatively, emit light containing many discrete spectral lines from emission from phosphorescent materials. Thus fluorescent, high-pressure sodium and high-intensity discharge lamps have spectral characteristics of gas discharge systems. LEDs, depending on design, may also have discrete spectra or a combination of spectral lines. These are important facts to recognize, since, in the testing of PV systems, it is important to be able to produce standardized spectral testing conditions.

Knowledge of the spectral composition of the sun is important for understanding the effects of the atmosphere on the radiation from the sun and for understanding which materials should offer the best performance in the conversion of sunlight to electricity.

2.3 THE EFFECT OF ATMOSPHERE ON SUNLIGHT

As sunlight enters the earth's atmosphere some is absorbed, some is scattered and some passes through unaffected by the molecules in the atmosphere and is either absorbed or reflected by objects at ground level. Different molecules do different things. Water vapor, carbon dioxide and ozone, for example, absorb significant amounts of sunlight at certain wavelengths. Ozone plays an important role by absorbing a significant amount of radiation in the ultraviolet region of the spectrum, while water vapor and carbon dioxide absorb primarily in the visible and infrared parts of the spectrum.

Absorbed sunlight increases the energy of the absorbing molecules, thus raising their temperature. Scattered sunlight is responsible for light entering north-facing windows when the sun is in the south. Scattered sunlight, in fact, is what makes the sky blue. Without the atmosphere and its ability to scatter sunlight, the sky would appear black, such as it does on the moon. Direct sunlight consists of parallel rays, which are necessary if the light is to be focused. The reader may have experimented with a magnifying glass and found that it is not possible to burn holes in paper when a cloud covers the sun. This is because the diffuse light present under these conditions is coming from all directions and cannot be focused.

All of these components of sunlight have been given names of their own. Sunlight that reaches the earth's surface without scattering is called *direct* or *beam* radiation. Scattered sunlight is called *diffuse* radiation. Sunlight that is reflected from the ground is called *albedo* radiation, and the sum of all three components of sunlight is called *global* radiation.

The amount of sunlight either absorbed or scattered depends on the length of the path through the atmosphere. This path length is generally compared with a vertical path directly to sea level, which is designated as *air mass* = 1 (AM1). Hence, the air mass at a higher altitude will be less than unity for the sun directly overhead and the air mass will be generally more than unity when the sun is not directly overhead. In general, the air mass through which sunlight passes is proportional to the secant of the zenith angle, θ_z, which is the angle measured between the direct beam and the vertical. At AM1, after absorption has been accounted for, the intensity of the global radiation is generally reduced from 1367 W/m^2 at the top of the atmosphere to just over 1000 W/m^2 at sea level. Hence, for an AM1 path length, the intensity of sunlight is reduced to 70% of its original AM0 value. In equation form, this observation can be expressed as, assuming that the absorption constant depends directly upon air mass,

$$I = 1367(0.7)^{AM}. \qquad (2.2)$$

This equation is, of course, obvious for AM1. But does it hold for air masses different from unity? According to Meinel and Meinel [2], a better fit to observed data is given by

$$I = 1367(0.7)^{(AM)^{0.678}}. \qquad (2.3)$$

On average, over the surface of the earth, an amount of heat is reradiated into space at night that is just equal to the amount absorbed from the sun during the day. As long as this steady-state condition persists, the average temperature of the earth will remain constant. However, if for any reason, the amount of heat absorbed is not equal to the amount reradiated, the planet will either cool down or heat up. This delicate balance can be upset by events such as volcanoes that fill the atmosphere with a fine ash that reflects the sunlight away from the earth, thus reducing the amount of incident sunlight. The balance can also be upset by gases such as carbon dioxide and methane, which are mostly transparent to short wavelength (visible) radiation but more absorbing to long wavelength (infrared) radiation. Since incident sunlight is dominated by short wavelengths characteristic of the 5800 K sun surface temperature, and since the reradiated energy is dominated by long wavelengths characteristic of the earth's surface temperature of approximately 300 K, these *greenhouse gases* tend to prevent the earth from reradiating heat at night. To reach a new balance, it is necessary for the earth to increase its temperature since radiation is proportional to T^4, where T is the temperature in K.

The natural compensation mechanism is green plants on land and under water. Through the process of photosynthesis, they use sunlight and carbon dioxide to produce plant fiber and oxygen, which is released into the atmosphere. Hence, the replacement of green plants with concrete and asphalt and adding carbon dioxide to the atmosphere by burning fossil fuels have a combined negative effect on the stability of the concentration of carbon dioxide in the atmosphere. Global warming (global climate change) is being discussed extensively in other literature, some of which are listed in the chapter references.

2.4 SUNLIGHT SPECIFICS

2.4.1 INTRODUCTION

Nearly everyone in the Northern Hemisphere has noticed that the sun shines longer in the summer than in the winter. Nearly everyone also knows that the Sahara Desert receives more sunshine than London. Another obvious observation is that cloudy places receive less sunlight than sunny places. It may be less evident, however, that the hours of sunlight over a year are the same for every point on the earth, provided that only the hours between sunrise and sunset are counted, regardless of cloud cover. Those regions of the earth closer to the poles that have long winter nights also have long summer days. However, since the sun, on average, is lower in the sky in the polar regions than in the tropics, sunlight must traverse greater air mass in the polar regions than in the tropics. As a result, polar sunlight carries less energy to the surface than tropical sunlight. Not surprisingly, the polar regions are colder than the tropics.

In this section, quantitative formulas will be presented that will enable the reader to determine exactly how long the sun shines in any particular place on any particular day and to determine how much sunlight can be expected, on average, during any month at various locations. Means will also be presented for determining the position of the sun at any time on any day at any location. Finally, the effects of varying the orientation of a PV array on the power and energy produced by the array will be discussed.

2.4.2 Definitions

Irradiance is the measure of the power density of sunlight and is measured in W/m^2. Irradiance is thus an instantaneous quantity and is often identified as the *intensity* of sunlight. The solar constant for the earth is the irradiance received by the earth from the sun at the top of the atmosphere, that is, at AM0, and is equal to 1367 W/m^2. After passing through the atmosphere with a path length of AM1, the irradiance is reduced to approximately 1000 W/m^2 and has a modified spectral content due to atmospheric absorption. The irradiance for AM1.5 is accepted as the standard calibration spectrum for PV cells.

Irradiation is the measure of energy density of sunlight and is measured in kWh/m^2. Since energy is power integrated over time, irradiation is the integral of irradiance. Normally, the time frame for integration is one day, which, of course, means during daylight hours.

Historically, irradiation was often expressed as *peak sun hours* (psh). The psh is simply the length of time in hours at an irradiance level of 1 kW/m^2 needed to produce the daily irradiation obtained from the integration of irradiance over all daylight hours. Figure 2.2 illustrates the result of this integration for three successive summer days in Florida. Note that the figure plots irradiance versus time to determine irradiation. If irradiation is measured in kWh, which is proportional to psh, then the relative irradiation on Day 1 is 54.7 kWh, Day 2 is 16.9 kWh and Day 3 is 50.5 kWh. This illustrates that, even on a very cloudy day, the psh may still be 30% of that of a sunny day.

Irradiance and irradiation both apply to all components of sunlight. Hence, at a given time, or for a given day, these quantities will depend on location, weather conditions and time of year. They will also depend on whether the surface of interest is shaded by trees or buildings and whether the surface is horizontal or inclined. The daily irradiation is numerically equal to the daily psh.

Over the years, a number of computer programs have been developed to estimate the performance of PV systems. These programs take into account the position of the sun, which will be discussed in the next few sections, as well as the specific

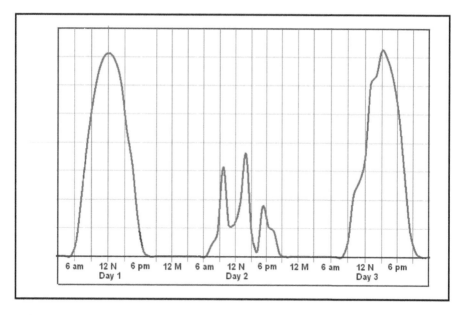

FIGURE 2.2 Comparison of hourly and daily radiation for three successive days showing weather effects.

equipment in use and the average weather, generally defined as a typical meteorological year. At least one of these programs is free to the public and will be used in later chapters. Since being able to determine the position of the sun along with the irradiance values at a given time in a given place is essential to the optimal design of PV systems, this chapter will deal with the sun, and equipment will be added in later chapters.

To determine the amount of irradiation available at a given location at a given time for conversion to electricity, it is useful to develop several expressions for the irradiance on surfaces, depending on the angle between the surface and the incident beam. It is also interesting to be able to determine the number of hours of sunlight on a given day at a given location.

2.4.3 THE ORBIT AND ROTATION OF THE EARTH

The earth revolves around the sun once per year in an elliptical orbit with the sun at one of the foci, such that the distance from the sun to the earth is given by [1]

$$d = 1.5 \times 10^{11} \left\{ 1 + 0.017 \sin\left[\frac{360(n-93)}{365} \right] \right\} m, \qquad (2.4)$$

where n represents the day of the year with January 1 as day 1. Since the deviation of the orbit from circular is so small, it is normally adequate to express this distance in terms of its mean value.

The Sun

The earth also rotates about its own polar axis once per day. The polar axis of the earth is inclined by an angle of 23.45° to the plane of the earth's orbit about the sun. This inclination is what causes the sun to be higher in the sky in the summer than in the winter in the northern hemisphere. It is also the cause of longer summer sunlight hours and shorter winter sunlight hours. Figure 2.3 shows the earth's orbit around the sun with the inclined polar axis. Note that on the first day of Northern Hemisphere summer, the sun appears vertically above the Tropic of Cancer, which is latitude 23.45° N of the equator. On the first day of winter, the sun appears vertically above the Tropic of Capricorn, which is latitude 23.45° S of the equator. On the first day of spring and the first day of fall, the sun is directly above the equator. From fall to spring, the sun is south of the equator and from spring to fall the sun is north of the equator.

The angle of deviation of the sun from directly above the equator is called the *declination*, δ. If angles north of the equator are considered positive and angles south of the equator are considered negative, then at any given day of the year, n, the declination can be found from

$$\delta = 23.45° \sin\left[\frac{360(n-80)}{365}\right]. \qquad (2.5)$$

This formula, of course, is only a good approximation, since the year is not exactly 365 days long and the first day of spring is not always the 80th day of the year. In any case, to determine the location of the sun in the sky at any time of day at any time of year at any location on the planet, the declination is an important parameter.

It is also important to be able to determine the clock time at which solar noon occurs. Solar noon occurs at 12 noon clock time at only one longitude, L_1, within any time zone. At longitudes east of L_1, solar noon will occur before 12 noon, and at longitudes west of L_1, solar noon will occur after 12 noon. On a sunny day, solar noon can be determined as that time when a shadow points directly north or directly south, depending on the latitude. Note that in the tropics, part of the year the shadow will point north and the rest of the year the shadow will point south.

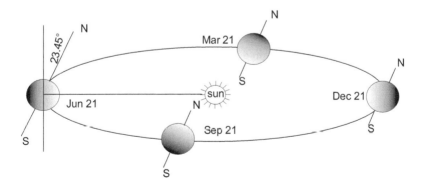

FIGURE 2.3 The orbit of the earth and the declination at different times of the year.

Fortunately, if the longitude is known, it is straightforward to determine the *approximate* relationship between clock noon and solar noon. Since there are 24 hours in a day and since the earth rotates 360° during this period, this means that the earth rotates at the rate of 15°/hr. It is also convenient that longitude zero corresponds to clock noon at solar noon. As a result, solar noon occurs at clock noon at multiples of 15° east or west longitude. Furthermore, since it takes the earth 60 minutes to rotate 15°, it is straightforward to interpolate to find solar noon at intermediate longitudes.

For example, at a longitude of 80° west, solar noon can be found by noting that 80° is between 75° and 90°, where solar noon occurs at clock noon *standard time*. Since 80° is west of 75°, when the sun is directly south at 75°, the sun will be east of south at 80°. The clock time at which the sun will be south at 80° (solar noon for 80°) is thus found by interpolation to be

$$t = 12 + \frac{80-75}{15} \times 60 = 12 + 20\,\text{min} = 12:20\,\text{p.m.}$$

Note that this time is standard time relative to the time zone for which solar noon occurs at 75° W. If 90° W is used as the solar noon reference, then solar noon at 80° will occur 40 minutes before solar noon occurs at 90°. Note that the answer is still the same. At 80° W, solar noon occurs at 12:20 p.m. Eastern Standard Time or at 11:20 a.m. Central Standard Time.

One glitch in the solar noon argument involves those unique locations on the earth such as Newfoundland, Canada or India, where there is only half an hour shift between adjacent time zones, or Alaska, where a single clock time zone covers nearly 30° of longitude.

Another glitch in determining solar noon results from the combination of the declination and the elliptical orbit of the earth. During half the year, the earth is moving closer to the sun and during half the year the earth is moving away from the sun. This results in a variation in solar noon called the analemma [3]. One way of describing the analemma is to plot the position of the tip of a shadow from a fixed object at a specific time each day, each week or each month. Whichever time period is used, the resulting plot over a year will look like the number 8. The reader familiar with globes may recall seeing the analemma on the globe.

Another convenient way of describing the analemma is to plot the clock time at which solar noon occurs over the period of a year. This plot will refine the solar noon value determined from the longitude correction. Solar noon data can be obtained from Equation of time—Wikipedia [4]. Figure 2.4 shows a plot of solar noon versus clock time for Boca Raton, FL, which is located at a longitude of approximately 80° west. Note that the greatest variation in solar noon is approximately ± 16 minutes from 12:20 p.m., the value calculated from the longitude correction formula.

2.4.4 TRACKING THE SUN

To completely specify the position of the sun it is necessary to specify three coordinates. If one assumes the distance from the sun to the earth to be constant, then the

The Sun 25

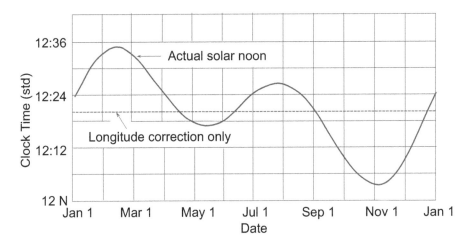

FIGURE 2.4 Daily variation of solar noon versus clock time in Boca Raton, FL due to the analemma effect.

position of the sun can be specified using two angles. Two common choices are the solar altitude and the azimuth.

The *solar altitude*, α, is the angle between the horizon and the incident solar beam in a plane determined by the zenith and the sun, as shown in Figure 2.5.

The angular deviation of the sun from directly south can be described by the *azimuth angle*, ψ, which measures the sun's angular position east or west of south. The azimuth angle is zero at solar noon and increases toward the east. It is the angle between the intersection of the vertical plane determined by the observer and the sun with the horizontal and the horizontal line facing directly south from the observer, assuming the path of the sun to be south of the observer. *The reader should note that, in many publications, the azimuth angle is referenced to north, such that solar noon appears at ψ = 180°*. In fact, compass angles are azimuth angles referenced to magnetic north.

Another useful, albeit redundant, angle in describing the position of the sun is the angular displacement of the sun from solar noon in the plane of apparent travel of the sun. The *hour angle* is the difference between noon and the desired time of day in terms of a 360° rotation in 24 hours. In other words,

$$\omega = \frac{12-T}{24} \times 360° = 15(12-T)°, \qquad (2.6)$$

where T is the time of day expressed with respect to solar midnight, on a 24-hour clock. For example, for T = 0 or 24 (midnight), ω = ± 180° and for T = 9 a.m., ω = 45°. By relating ω to the other angles previously discussed, it is possible to show [1] that the sunrise angle is given by

$$\omega_s = \cos^{-1}(-\tan\phi\tan\delta), \qquad (2.7)$$

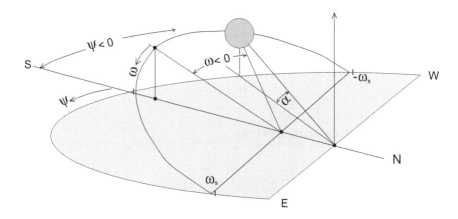

FIGURE 2.5 Sun angles, showing altitude, azimuth and hour angle.

which, in turn, implies that the sunset angle is given by $-\omega_s$. This formula is useful because it enables one to determine the number of hours on a specific day at a specific latitude when the sun is above the horizon. Converting the sunrise angle to hours from sunrise to solar noon, and then multiplying by 2 to include the hours from solar noon to sunset, yields the number of hours of daylight to be

$$\text{DH} = \frac{48}{360} \times \omega_s = \frac{\cos^{-1}(-\tan\phi\tan\delta)}{7.5} \text{ hr.} \quad (2.8)$$

Two very important relationships between α and ψ can be determined by the reader who enjoys trigonometry. If δ, ϕ and ω are known, then the position of the sun, in terms of α and ψ at this location at this date and time, can be determined from [1]:

$$\sin\alpha = \sin\delta\sin\phi + \cos\delta\cos\phi\cos\omega \quad (2.9)$$

and

$$\cos\psi = \frac{\sin\alpha\sin\phi - \sin\delta}{\cos\alpha\cos\phi} \quad (2.10)$$

Note that in all the previous expressions, angles are measured in degrees. Be careful as to whether your computer wants you to express angles in degrees or in radians when it evaluates trigonometric functions.

It is interesting to look at the values of (2.9) and (2.10) at solar noon, when $\omega = 0$. The results of Homework Problem 2.9 show that at solar noon, $\theta_z = \phi - \delta$, where θ_z, the zenith angle, is the angle between the vertical and the sun position. Thus, θ_z is the complement of α. During the year, the highest point of the sun in the sky will be at $\theta_z = \phi - 23.45°$ and the lowest point of the solar noon sun in the sky will be at $\theta_z = \phi + 23.45°$, provided that $\phi > 23.45°$. It is particularly interesting to note that if $\phi > 90° - 23.45° =$

66.55°, then the lowest point of the sun in the sky is below the horizon, meaning that the sun does not rise or set that day. This, of course, is the situation in polar regions, which are subject to periods of 24 hours of darkness. These same regions, of course, are also subject to equal periods of 24 hours of sun six months later. If $\phi < 23.45°$, θ_z will at some time during the summer be negative. This simply means that the sun will appear north of directly overhead at solar noon. The relationships among θ_z, ϕ and δ at solar noon are illustrated in Figure 2.6.

Since (2.9) and (2.10) are somewhat difficult to visualize, it is convenient to plot α versus ψ for specific latitudes and days of the year. Figure 2.7 shows a series of plots of altitude versus azimuth at a latitude of 30° N. The curves show approximately how high in the sky the sun will be at a certain time of day during a particular month, with the azimuth angle determined by the time of day.

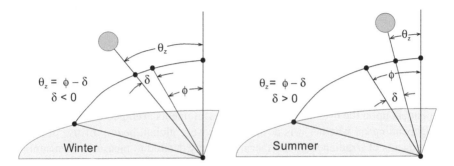

FIGURE 2.6 Relationships among zenith angle, latitude and declination at solar noon in winter and summer.

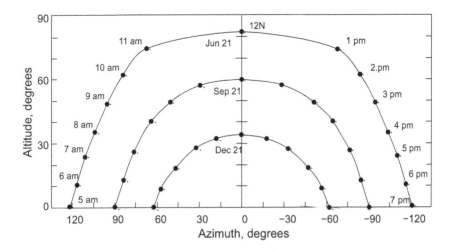

FIGURE 2.7 Plot of solar altitude versus azimuth for different months of the year at a latitude of 30° north.

Markvart [1] gives detailed formulas for the sun position and the components of global irradiance. The interested reader is encouraged to review them. Interestingly enough, when all of the mathematical work is finished, one ends up with an answer that does not account for cloud cover. The most reliable means of accounting for cloud cover is to make measurements over a long period of time to determine average figures. In fact, accurate prediction of the performance of PV systems also depends upon temperature.

Thus, if sunlight data is included as a result of measured meteorological data, which includes irradiance and temperature over an hourly basis for a given location for the 8760 hours per year, reasonably reliable estimates can be made of PV system performance. Of course, measuring hourly data for a given year, say 2023, will not necessarily be a good predictor for other years. To achieve a better predictor for annual performance, experts have defined the *typical meteorological year* (TMY) by selecting data for each month from a different year. In later chapters, various performance models based on TMY data will be discussed.

2.4.5 MEASURING SUNLIGHT

Since all the calculations and approximations in the world cannot yield exact predictions of the amount of sunlight that will fall on a given surface at a given angle at a given time on a given day in a given place, the design of PV power systems is dependent upon the use of data based on measurements averaged over a long time.

Depending on whether it is desired to measure global, beam, diffuse or albedo components of irradiance or whether it is simply desired to measure when sunlight exceeds a certain brightness level, different types of instruments are used.

2.4.5.1 Precision Measurements

The *pyranometer* is designed to measure global radiation. It is normally mounted horizontally to collect general data for global radiation on a horizontal surface. However, it is also often mounted in the plane of a PV collector to measure the global radiation incident on the inclined surface, which is essential when commissioning a system, that is, comparing actual system performance to predicted system performance.

The pyranometer is designed to respond to all wavelengths and, hence, it responds accurately to the total power in any incident spectrum.

The *black and white pyranometer* operates on the principle of differential heating of a series of black and white wedges. The temperatures of each wedge are measured with thermocouples that yield voltage differences dependent on the temperature differences. This instrument is somewhat less accurate than the precision unit.

The *normal incidence pyrheliometer* uses a long, narrow tube to collect beam radiation over a narrow beam solid angle, generally about 5.5°. Since the instrument is only sensitive to beam radiation, a tracker is needed if continuous readings are desired.

The pyranometer can be mounted on a *shadow band stand* to block out beam radiation so that it will respond only to the diffuse component. The stand is mounted so the path of the sun will be directly above the band during daylight hours. Because

The Sun

δ changes from day to day and because the band blocks out only a few degrees, it is necessary to readjust the band every few days.

2.4.5.2 Less Precise Measurements

Many inexpensive instruments are also available for measuring light intensity, including instruments based on cadmium sulfide photocells and silicon photodiodes. These devices give good indications of relative intensity but are not sensitive to the total solar spectrum and thus cannot be accurately calibrated to measure total energy. These devices also do not normally have lenses that capture incident radiation from all directions. Devices that capture solid angles from a few degrees to upward of 90° are available.

2.5 CAPTURING SUNLIGHT

2.5.1 Maximizing Irradiation on the Collector

The designer of any system that collects sunlight must decide on a means of mounting the system. Perhaps the easiest mounting of most systems is to mount them horizontally. This orientation, of course, does not optimize collection, since the beam radiation component collected is proportional to the cosine of the angle between the incident beam and normal to the plane of the collector, as shown in Figure 2.8. Depending on the ratio of diffuse to beam irradiance components, the fraction of available energy collected will be between cosγ and unity. Of course, in a highly diffuse environment, the beam irradiance will be only a small fraction of the global irradiance. Several alternatives to horizontal mounting exist. Since $\theta_z = \phi - \delta$ defines the position of the sun at solar noon, if a collector plane is perpendicular to this angle, it will be perpendicular to the sun at solar noon. This is the point at which the sun is highest in the sky, resulting in its minimum path through the atmosphere and corresponding lowest air mass for the day. Since the sun travels through an angle of 15° per hour, it will be close to perpendicular to the collector for a period of approximately two hours. Beyond this time, the intensity of the sunlight decreases due to the

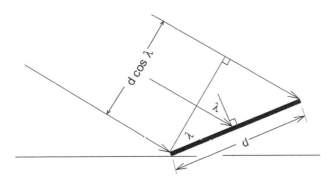

FIGURE 2.8 Two-dimensional illustration of the effect of collector tilt on effective area presented to beam component of radiation.

increase in air mass, and the angle between incident sunlight and the normal to the collector increases.

These two factors cause the energy collected by the collector to decrease relatively rapidly during the hours before 10 a.m. and after 2 p.m. Figure 2.9 shows the approximate cumulative irradiation received by a fixed, south-facing collector tilted at the latitude angle in summer and winter. Note that the curves essentially represent the daily integration of the array output shown in Figure 2.3 on a clear day. *In this location*, it can be seen that the summer and winter cumulative collection is about the same.

If the collector is mounted on a single north–south horizontal axis so it can track the sun in an east–west direction, then the incident irradiance is affected by the increasing air mass as the sun approaches the horizon, but the angle between the direct beam and collector surface remains more constant as the tracker rotates. Figure 2.9 also shows the additional cumulative irradiation received by a backtracking collector. Note that, for this location, the backtracker collects about 30% more per day in the summer and in the winter. In Seattle, WA, which receives somewhat more diffused sunlight than Phoenix, a tracking collector will collect about 35% more in the summer but only 9% more energy compared to an optimized fixed collector in the winter [5]. Whether to use a tracking collector then becomes an important economic decision for the engineer, since a tracking mount is more costly than a fixed mount.

To make the mounting selection even more interesting, one can consider a single-axis tracker, which rotates about an axis perpendicular to θ_z. One can then also consider a two-axis tracker or mountings that can be adjusted manually several times per

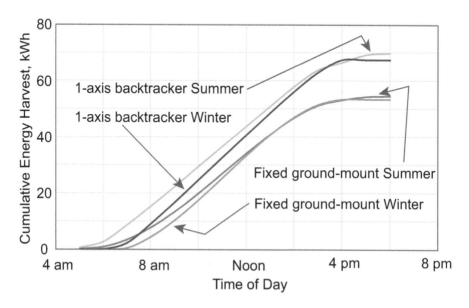

FIGURE 2.9 Cumulative daily irradiation received by fixed and single-axis backtracking collectors for summer and winter sunny days.

day or, perhaps, several times per year. Each of these options will enable the collection of more annual energy than a fixed collector. However, these options may also result in a larger area necessary to avoid the array shading itself, and, if the cost of land is included in the economic evaluation of the array, the cost of land may become an important factor as well. On the other hand, if the land is designed as multi-use, then the extra space between collectors may benefit the other use(s).

It is interesting to take a closer look at backtracking, since it is now becoming more popular to track from east to west such that the collector is flat at solar noon. Depending on the separation of adjacent rows and, perhaps, the slope of the terrain, when the sun is below a certain altitude, one row will begin to shade the adjacent row. Some trackers then employ "backtracking," during which the rows adjust back toward level to eliminate shading, even though the incident sunlight angle on the array is increased. Figure 2.10 shows an example of backtracking. Note that the rows run north–south and the arrays rotate from east-facing in the morning to west-facing in the afternoon. The numbers represent time, with 1 being earlier than 3 and 2 representing the time at which the row to the east begins shading the row to the west. The arrows pointed at the array represent the direction of the beam component at the corresponding time. Note that, at time 1, the E–W component of the direct beam is perpendicular to the array and no shading occurs. At time 2, the west row begins shading the adjacent row to the east. At this point, the trackers reverse direction (backtracking) and begin rotating toward a 0° tilt. Rotation continues until shading is eliminated. This rotation is represented by time 3. Note that the difference between times up to time 2 is that after time 2, the E–W component of the direct beam is no longer perpendicular to the array, but at least no shading occurs and the collection of the array is optimized. As the rows are spaced further apart, the time that backtracking begins is later in the day. At the end of the day, the array returns to east-facing to wait for the next day to begin.

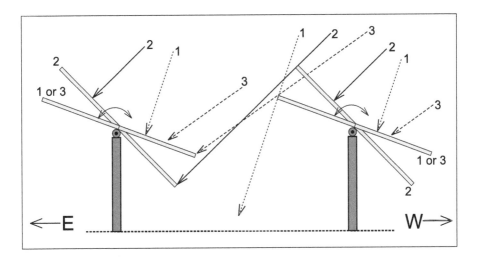

FIGURE 2.10 Illustration of operation of a backtracking version of a single-axis tracker with a horizontal N–S axis.

Collector orientation may also be seasonally dependent. For example, a remote cabin, used only during summer months, will need its collector oriented for optimal summer collection. However, if the cabin is used in the winter as a ski lodge or in the fall as a hunting base, then the collector may need to be optimized for one of these seasons. For optimal performance on any given day, a fixed collector should be mounted with its plane at an angle of $\phi - \delta$ with respect to the horizontal, as shown in Figure 2.11. This will cause the plane of the collector to be perpendicular to the sun at solar noon.

For optimal seasonal performance, then, one simply chooses the average value of δ for the season. For summer in the Northern Hemisphere, δ varies sinusoidally between 0° on March 21 to 23.45° on June 21 and back to 0° on September 21. If this variation is plotted as half a sine wave with an amplitude of 23.45°, those who have evaluated the average value of a sine wave in conjunction with the output of a rectifier circuit may recall that the average value of half a sine wave having amplitude A is $2A/\pi$. Hence, the average declination between March 21 and September 21 is 14.93°. Similarly, the average declination for the period from September 21 to March 21 is −14.93°. Hence, for the best average summer performance, a collector should be mounted at approximately $\phi - 15°$, and, for the best average winter performance, it should be mounted at $\phi + 15°$. Typically, only a small percentage of optimal annual energy collection for a fixed collector is lost if the collector tilt is within ±15° of latitude and within ±45° of south-facing (north-facing in the southern hemisphere). Figure 2.12 shows how monthly performance depends on the collector tilt angle. Different locations will show different relative monthly performances, depending on local weather seasonal behavior.

2.5.2 Shading

Even a small amount of shade on a PV module can significantly reduce the module output current. It is thus of paramount importance to select a site for a PV system where the PV array will remain unshaded for as much of the day as possible. This is easy if there are no objects that might shade the array, but it is probably more likely

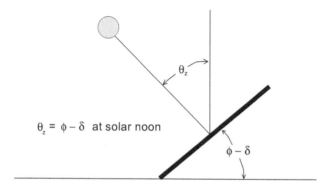

FIGURE 2.11 Optimizing the mount angle of a fixed collector.

The Sun

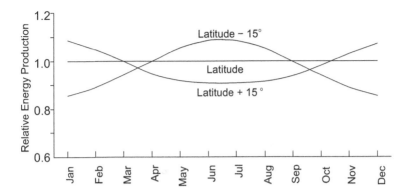

FIGURE 2.12 Monthly collector performance as a function of collector tilt angle.

that a site will have objects nearby that shade the array at some time of the day on some day of the year. The PV system designer must thus be able to use her knowledge of sun position to determine the times at which a PV array might be shaded. In later chapters a few systems will be discussed that incorporate electronic means of minimizing overall shading effects, but in all cases, system performance is optimized by avoiding as much shading as possible.

2.5.2.1 Field Measurement of Shading Objects

Figure 2.13 shows a device that incorporates plots of altitude versus azimuth for selected latitudes. The device is used at the proposed site to determine when the array will be shaded by observing the position of reflected objects on the screen of the device. By sketching the outlines of shading objects on the screen with the device at the proposed location of the collector, the user can then determine when the collector will be shaded [6].

Up to a few years ago, this method was used commonly. Recently, however, computational methods have pretty much replaced manual measurements, some of which involved measuring from precarious positions on rooftops.

In addition, new technology that will be presented in later chapters has enabled the system designer to minimize the effect of shading a part of the array on the performance of the remainder of the array. However, it is highly likely that a practicing PV system designer will encounter situations where it is necessary to limit the height of surrounding objects to maximize the performance of the array. The next section illustrates how to do this.

2.5.2.2 Computational Methods of Determining Shading

Computational methods of determining shading range from solving (2.9) and (2.10) for a specific site to using a variety of computer programs. Google Sketch-Up is a popular freeware program that can be downloaded from the web [7]. It can be used with Google Earth to determine shading at existing sites. Autocad has a plug-in that

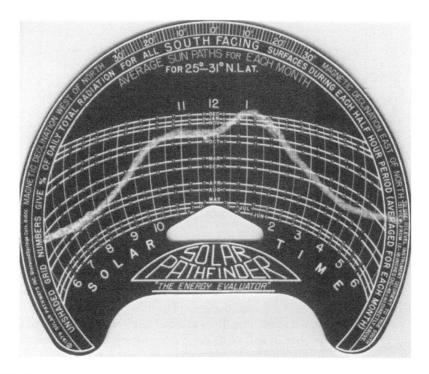

FIGURE 2.13 Solar Pathfinder showing the region of shading. Shading occurs at points above the white line on the pattern. (Florida Solar Energy Center Photo.)

does a very nice job of shading analysis on planned new construction drawings. Any of these methods are useful for any mounting configuration.

As an example, consider a ground-mounted array as shown in Figure 2.14. Assume the array is located at 30° north latitude (ϕ) and that it is desired to space the rows such that rows do not shade each other between the hours of 8:30 a.m. to 3:30 p.m. sun time at any time of the year.

If a shading program is not available, then the first step is to solve (2.9) and (2.10) for the sun position. Since the worst case will be on December 21, and since the angles will be symmetrical about south, one only needs to know δ, ϕ and ω for 8:30 a.m. sun time to solve for α and ψ. Solving for δ and ω yields

$$\delta = 23.45°\sin\left[\frac{360(n-80)}{365}\right] = 23.45°\sin\left[\frac{360(355-80)}{365}\right] = -23.45°$$

as expected, and

$$\omega = 15(12-T)° = 15(12-8.5) = 52.5°.$$

Thus, from (2.9),

$$\sin\alpha = \sin(-23.45°)\sin 30° + \cos(-23.45°)\cos 30°\cos 52.5° = 0.2847$$

The Sun

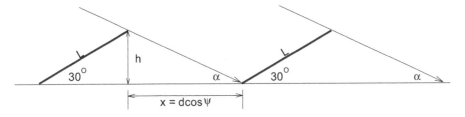

FIGURE 2.14 Determining the spacing between rows of PV modules.

and, from (2.10),

$$\cos\psi = \frac{0.2847 \times \sin 30° - \sin(-23.45°)}{\cos(\sin^{-1} 0.2847)\cos 30°} = 0.6508.$$

Thus, $\alpha = 16.54°$ and $\psi = 49.4°$ east of south.

Next, the length of a shadow, d, can be found using a bit of trigonometry. If the length of the PV modules is L, then $h = L\sin 30°$, and $d = h/\tan\alpha = L\sin 30°/\tan\alpha$. So, for example, if L = 62 in (157.4 cm), then d = 104 in (264.2 cm). Since this shadow points 49.4° west of north, the projection of the shadow to the north will give a distance of $104\cos(49.4°) = 68$ in (172.7 cm), which will thus be the minimum row spacing.

This method can easily be adapted to an Excel spreadsheet by including the formulas for sun position and then using the sun position coordinates at specified hours on specified days along with known shading object heights and positions, to calculate the corresponding shadow length of the object.

2.5.3 Special Orientation Considerations

Sometimes it is not possible or convenient to install a collector facing directly south. If south-facing is the preferred orientation, simulations show that the collector can be facing up to 22.5° away from south with less than a 2% reduction in annual collection at latitudes up to 45°. In fact, the loss in monthly or annual solar harvest can be determined quite readily for any location and array orientation by using NREL SAM [5].

In other cases, maximum PV system output may be desirable at a time of day other than solar noon. For example, in many regions, peak utility electrical generation occurs between the hours of 3 and 6 p.m. If a PV system is connected to the utility grid, it may be desirable to maximize system output during the utility peak. For a fixed mounting, this would require having the collector face the sun at the midpoint of this time period on a date near the middle of the period during which maximum collector output is desired. This orientation will produce peak output when the sun path is at a higher air mass than at solar noon, resulting in a slightly lower peak output than would be obtained with a south-facing collector.

Figure 2.15 shows a comparison of PV output for northeast, south and northwest-facing PV systems for Denver, CO, on the first day of summer. Each system is a 5 kW

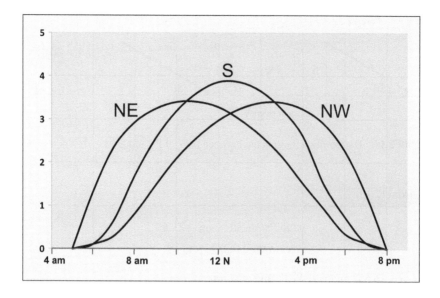

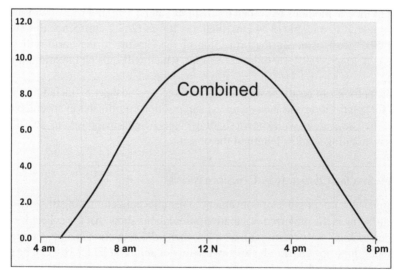

FIGURE 2.15 Comparison of PV output for northeast, south and northwest-facing PV systems for Denver, CO on the first day of summer.

system on a roof with a slope of 27°. Note the appreciable increase in system output later in the afternoon for the northwest-facing system and also early in the morning for the northeast-facing system, as might be expected. The daily and annual kWh produced by each of the 3 arrays are shown in Table 2.1.

Although the total daily and annual energy output of the northeast-facing and northwest-facing arrays is lower than that of the south-facing array, the loss of total energy may be acceptable if the added power output during utility peak hours will

TABLE 2.1
Daily and annual energy production of NE, S and NW PV arrays.

Array	Example day kWh production	Annual kWh production
5-kW NE-facing array	30.63	5520
5-kW S-facing array	31.45	8560
5-kW NW-facing array	30.05	5248
15-kW Total array	92.13	19,328

offset the need to use more expensive peaking generation. This observation is especially important if it will offset the need to install additional peaking generation in the first place. In addition, at 2023 installation prices, east, west, and north-facing arrays may possibly produce electricity at a cost to the owner that is less than the local utility rate. This possibility will be explored in more detail in Chapter 9.

Computation of the desired orientation for maximum output in a direction other than south is straightforward:

1. Determine the latitude of the location.
2. Calculate the declination for the design day, using (2.5).
3. Determine the time of day (local time) for peak PV system output.
4. Convert the local time to solar time, based upon site longitude.
5. Calculate the hour angle using (2.6).
6. Use (2.9) and (2.10) to determine α and ψ.

Example: Suppose it is desired to determine the position of the sun over Atlanta, GA, at 4 p.m. on July 1. Starting with step 1 of the 6 steps outlined earlier, the latitude of Atlanta is found to be $33°39' = 33.65°$ N. Then (2.5) is used to determine the declination for July 1. Since July 1 is the 181st day of the year, (2.5) gives

$$\delta = 23.45 \sin\left[\frac{360(181-80)}{365}\right] = 23.12°.$$

With 4 p.m. local time (Eastern Daylight Time) at the longitude of Atlanta ($84°36'$ W = $84.60°$W) as the time of day for peak system output, the solar time is found to be 3 p.m. (Eastern Standard Time) plus an additional

$$t = \frac{84.6 - 75}{15} \times 60 = 38.4 \text{ minutes, or 3:38 p.m.}$$

Next, substituting the solar time in (2.8) yields the hour angle as

$$= \frac{12 - 15.64}{24}(360°) = -54.60°.$$

The solar altitude is next found from (2.9) with the result

$$\sin\alpha = \sin 23.12° \sin 33.65° + \cos 23.12° \cos 33.65° \cos(-54.6°) = 0.6611.$$

Next, the azimuth is found from (2.10) to be

$$\cos = \frac{0.6611\sin 33.65° - \sin 23.12°}{\cos\left[\sin^{-1} 0.6611\right]\cos 33.65°} = -0.04216.$$

So, finally, $\alpha = \sin^{-1} 0.6611 = 41.38°$ and $\psi = \cos^{-1}(-0.04216) = -92.42°$. Be sure to note that $\cos^{-1}(-0.04216) = \pm 92.42°$. The minus sign is chosen because at 3:38 p.m. solar time the sun is west of south.

Finally, note that if a bit more precision is desired, an analemma check can be made and the hour angle corrected accordingly before proceeding with the calculation of altitude and azimuth.

Homework Problems

Chapter 2 Homework Problems

2.1 Show that, for a surface temperature of 5800 K, the sun will deliver 1367 W/m² to the earth. This requires integration of the blackbody radiation formula over all wavelengths to determine the total available energy at the surface of the sun in W/m². Numerical integration is recommended. Be careful to note the range of wavelengths that contribute the most to the spectrum. Then note that the energy density decreases as the square of the distance from the source, similar to the behavior of an electric field emanating from a point source. The diameter of the sun is 1.393×10^9 m, and the mean distance from the sun to the earth is 1.5×10^{11} m.

2.2 If the diameter of the sun is 1.393×10^9 m, and if the average density of the sun is approximately 1.4× the density of water, and if 2×10^{19} kg/yr of hydrogen is consumed by fusion, how long will it take for the sun to consume 25% of its mass in the fusion process?

2.3 Calculate the zenith angles needed to produce AM 1.5 and AM 2.0 if AM 1.0 occurs at zero degrees.

2.4 Calculate the zenith angle at solar noon at a latitude of 40° north on May 1.

2.5 Calculate the number of hours the sun was above the horizon on your birthday at your birthplace.

2.6 Calculate the irradiance of sunlight for AM 1.5 and for AM 2.0, assuming no cloud cover, using (2.2) and (2.3). Then write a computer program that will plot irradiance versus AM for $1 \leq AM \leq 10$ for each equation.

2.7 Write a computer program using Excel or something similar that will plot solar altitude versus azimuth. Plot sets of curves similar to Figure 2.8 for the months of March, June, September and December for Denver, CO, Mexico City, Mexico, and Fairbanks, AK.

2.8 Using (2.9) and (2.10), show that at solar noon, $\theta_z = \phi - \delta$ and $\psi = 0$.

The Sun   39

2.9 Write a computer program that will generate a plot of solar altitude versus azimuth for the 12 months of the year for your hometown. It would be nice if each curve would have time-of-day indicators.

2.10 Calculate the time of day at which solar noon occurs at your longitude. Then compare the north indicated by a compass with the north indicated by a shadow at solar noon and estimate the error in the compass reading or the error due to the analemma effect or both.

2.11 Calculate the collector orientation that will produce maximum summer output between 2 p.m. and 5 p.m. in Tucson, AZ. Use July 21 as the assumed midpoint of summer and correct for the longitude of the site.

2.12 Calculate the collector orientation that will produce maximum summer output at 9 a.m. Daylight Savings Time in Minneapolis, MN, using July 21 as the assumed midpoint of summer and using a longitude correction.

2.13 A collector in Boca Raton, FL ($\phi = 26.4°$ N, longitude = 80.1° W), is mounted on a roof with a 5:12 pitch, facing 30° S of W. Determine the time of day and days of the year that the direct beam radiation component is normal to the array.

2.14 Determine the location (latitude and longitude) where, on May 29, sunrise is at 6:30 a.m. and sunset is at 8:08 p.m. EDT. At what time does solar noon occur at this location?

2.15 Using Figure 2.14, make a table of unshaded collector times for each month of the year.

2.16 An array of collectors consists of three rows of south-facing collectors as shown in Figure p2.1. If the array is located at 40° N latitude, determine the spacing, d, between the rows needed to prevent shading of one row by another row between the hours of 8:30 a.m. and 3:30 p.m. sun time on the first day of winter.

2.17 A commercial building has a flat roof as shown in Figure p2.2. The roof has a parapet around it that is 3 ft high, and an air conditioner is located as shown that is 4 ft above the roof. The building is located at 40 degrees north. A PV system is to be installed using modules that measure 3 ft by 5 ft, in rows that will remain unshaded for at least 6 hours every day of the year. The modules are to be mounted at a tilt of 25 degrees, in portrait mode, facing south. The bottom edges of the modules will be 6 inches above the roof. Determine the maximum number of modules that can be

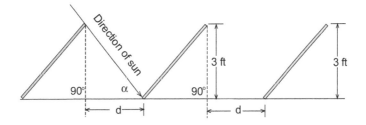

FIGURE P2.1 Rows of rack-mounted modules.

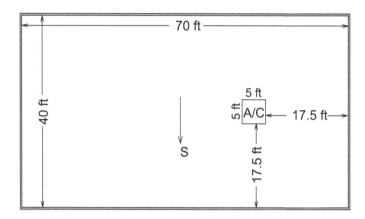

FIGURE P2.2 Roof layout for Problem 2.22.

installed on the roof so that no module will be shaded during the six-hour period. Pay attention to module rows shading other rows, shading from the air conditioner and shading from the parapet. Draw a diagram that shows your result.

2.18 Using NREL SAM [5] for Miami, FL; Sacramento, CA; and Boston, MA; create a table that shows the annual collection losses, in percent, for fixed-tilt collectors, compared to the maximum collection when the collector is tilted at latitude. Include tilt angles of latitude and latitude ±15°. If all that counts is total annual kWh, what generalizations can you offer with regard to array tilts? Assume a 1-kW array.

REFERENCES

[1] Markvart, T., ed., *Solar Electricity*, John Wiley & Sons, Chichester, UK, 1994.
[2] Meinel, A. B. and Meinel, M. P., *Applied Solar Energy, an Introduction*, Addison-Wesley, Reading, MA, 1976.
[3] Discussion and Explanation of Analemma, Complete with Formulas and Diagrams. Equation of Time—Wikipedia.
[4] "Sun or Moon Rise/Set Table for One Year" Tabulates Sunrise, Sunset, Moonrise and Moonset on a Daily Basis for a Year for a Large Number of Cities: https://aa.usno.navy.mil/data/RS_OneYear
[5] National Renewable Energy Laboratory, System Advisory Model, System Advisor 2022: https://sam.nrel.gov/download.html
[6] Solar Pathfinder Assistant—Latest Version/Updates: www.solarpathfinder.com
[7] A Web Search for Shading Analysis Programs Will Yield Hundreds of Results.

SUGGESTED READING

Nye, Bill, *Unstoppable: Harnessing Science to Change the World*, St. Martin's Press, New York, 2015.

3 Introduction to PV Systems

3.1 INTRODUCTION

Photovoltaic systems are designed around the photovoltaic cell. Since a typical photovoltaic cell produces about 6 watts at approximately 0.5 volt DC, cells must be connected in series-parallel configurations to produce enough power for high-power applications. Figure 3.1 shows how cells are configured into modules and how modules are connected as arrays. Modules may have peak output powers ranging from a few watts, depending on the intended application, to more than 500 watts. Typical array output power is in the kilowatt range, although megawatt and gigawatt arrays are now becoming more commonplace.

Since PV arrays produce power only when illuminated, PV systems sometimes employ an energy storage mechanism so the captured electrical energy may be made available at a later time. Most commonly, the storage mechanism consists of rechargeable batteries, but it is also possible to employ more exotic storage mechanisms. In addition to energy storage, storage batteries can also provide transient suppression, system voltage regulation and a source of current that can exceed PV array capabilities.

When a battery storage mechanism is employed, it is common to also incorporate some sort of charge control into the system, so the batteries can be prevented from reaching either an overcharged or overdischarged condition. It is also possible that some or all of the loads to be served by the system may be AC loads. If this is the case, an inverter will be needed to convert the DC from the PV array to AC. If a system incorporates a backup system to take over if the PV system does not produce adequate energy, then the system will need a controller to operate the backup system.

It is also possible that the PV system will be interconnected with the utility grid. Such systems may deliver excess PV energy to the grid or use the grid as a backup system in case of insufficient PV generation. These grid-interconnected systems need to incorporate suitable interfacing circuitry so the PV system will be disconnected from the grid in the event of grid failure. Figure 3.2 shows the components of several types of photovoltaic systems. This chapter will emphasize the characteristics of PV system components to pave the way for designing systems in the following chapters. The physics of PV cells, with an emphasis on the challenges to the cell innovator, are covered in Chapter 10, and current specific cell technologies are discussed in Chapter 11.

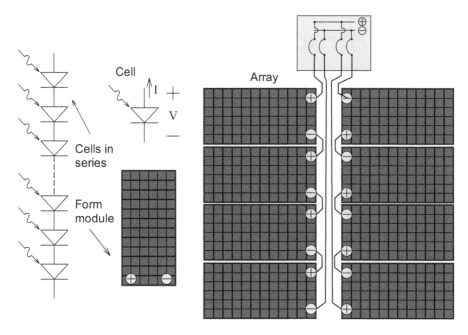

FIGURE 3.1 Cells, modules and arrays.

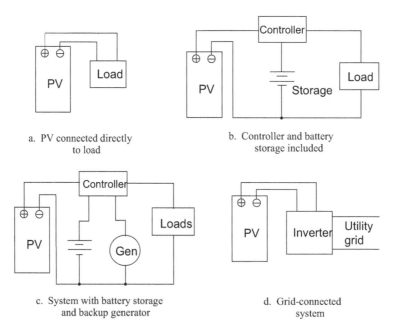

FIGURE 3.2 Examples of PV systems.

Introduction to PV Systems

3.2 THE PV CELL

The PV cell is a specially designed pn junction or Schottky barrier device. The well-known diode equation describes the operation of the shaded PV cell. When the cell is illuminated, electron–hole (EHP) pairs are produced by the interaction of the incident photons with the atoms of the cell. The electric field created by the cell junction causes the photon-generated EHPs to separate, with the electrons drifting into the n-region of the cell and the holes drifting into the p-region, provided that the EHP are generated sufficiently close to the pn junction. This process will be discussed in detail in Chapter 10. For this chapter, knowledge of the terminal properties of the PV cell is all that is needed.

Figure 3.3 shows the ideal I–V characteristics of a typical PV cell. Note that the amounts of current and voltage available from the cell depend upon the cell illumination level. In the ideal case, the I–V characteristic equation is

$$I = I_\ell - I_o \left(e^{\frac{qV}{kT}} - 1 \right) \qquad (3.1)$$

where I_ℓ is the component of cell current due to photons, $q = 1.6 \times 10^{-19}$ coul, $k = 1.38 \times 10^{-23}$ j/K and T is the cell temperature in K (0 K = −273.16°C). While the I–V characteristics of actual PV cells differ somewhat from this ideal version, (3.1) provides a means of determining the ideal performance limits of PV cells.

Figure 3.3 shows that the PV cell has both a limiting voltage and a limiting current. Hence, the cell is not damaged by operating it under either open circuit or short circuit conditions. To determine the short circuit current of a PV cell, simply set V = 0 in the exponent. This leads to $I_{sc} = I_\ell$. To a very good approximation, the

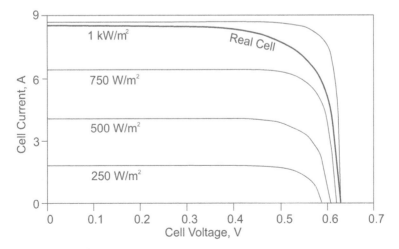

FIGURE 3.3 I–V characteristics of real and ideal PV cells under different illumination levels.

cell current is directly proportional to the cell irradiance. Thus, if the cell current is known under standard test conditions, $G_o = 1$ kW/m² at AM 1.5, then the cell current at any other irradiance, G, is given by

$$I_\ell(G) = (G/G_o)I_\ell(G_o). \qquad (3.2)$$

To determine the open circuit voltage of the cell, the cell current is set to zero and (3.1) is solved for V_{OC}, yielding the result

$$V_{OC} = \frac{kT}{q}\ln\frac{I_\ell + I_o}{I_o} \cong \frac{kT}{q}\ln\frac{I_\ell}{I_o} \qquad (3.3)$$

since normally $I_\ell \gg I_o$. For example, if the ratio of photocurrent to reverse saturation current is 10^{11}, using a thermal voltage (kT/q) of 26 mV yields $V_{OC} = 0.66$ V. Note that the open circuit voltage is logarithmically dependent only on the cell illumination, while the short circuit current is directly proportional to cell illumination.

Multiplying the cell current by the cell voltage yields the cell power, as shown in Figure 3.4. Note that, at any given illumination level, there is one point on the cell I–V characteristic where the cell produces maximum power. Note also that the voltage at which the maximum power points occur is minimally dependent upon the cell illumination level.

If I_m represents the cell current at maximum power, and if V_m represents the cell voltage at maximum power, then the cell maximum power can be expressed as

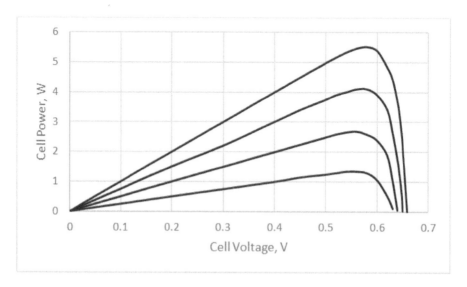

FIGURE 3.4 Power versus voltage for a PV cell at 250, 500, 750 and 1000 W/m² illumination levels.

Introduction to PV Systems

$$P_{max} = I_m V_m = FF I_{SC} V_{OC}, \tag{3.4}$$

where FF is defined as the cell *fill factor*. The fill factor is a measure of the quality of the cell. Cells with large internal resistance will have smaller fill factors, while the ideal cell will have a fill factor of unity. Note that a unity fill factor suggests a rectangular cell I–V characteristic. Such a characteristic implies that the cell operates as either an ideal voltage source or as an ideal current source. Although a real cell does not have a rectangular characteristic, it is clear that it has a region where its operation approximates that of an ideal voltage source and another region where its operation approximates that of an ideal current source.

For the cell having an ideal I–V characteristic governed by (3.1), with $V_{OC} = 0.596$ V and $I_{SC} = 9.0$ A, the fill factor will be approximately 0.825801. Typical fill factors for real PV cells, depending on the technology, may vary from 0.5 to 0.81. The secret to maximizing the fill factor is to maximize the ratio of photocurrent to reverse saturation current while minimizing series resistance and maximizing shunt resistance within the cell.

The cell power versus cell voltage curve is especially important when considering maximizing power transferred to the cell load. This topic will be investigated in more detail in Section 3.6, where methods of matching the load to the source are discussed.

The PV cell I–V curve is also temperature sensitive. A quick look at (3.3) might suggest that the open circuit voltage is directly proportional to the absolute temperature of the cell. A longer look, however, will reveal that the reverse saturation current is highly temperature dependent also. The net result, which will be covered in detail in Chapter 10, is that the open circuit voltage of an *ideal* crystalline silicon PV cell decreases by approximately 2.3 mV/°C increase in temperature, which amounts to approximately 0.4%/°C. The short circuit current, on the other hand, increases only slightly with temperature. As a result, the cell power also decreases

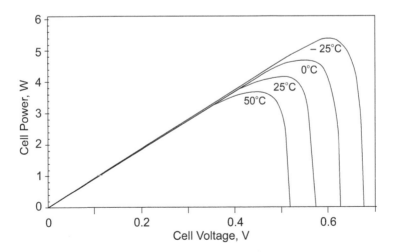

FIGURE 3.5 Temperature dependence of the power versus voltage curve for a PV cell.

by approximately 0.4%/°C. Figure 3.5 shows the temperature dependence of the PV cell power versus voltage characteristic.

It is important to remember that, when a cell is illuminated, it will generally convert less than 22% of irradiance into electricity. The balance is converted to heat, resulting in the heating of the cell. As a result, the cell can be expected to operate above ambient temperature. If the cell is part of a concentrating system, then it will heat even more, resulting in additional temperature degradation of cell performance.

The photocurrent developed in a PV cell is dependent on the intensity of the light incident on the cell. The photocurrent is also highly dependent on the wavelength of the incident light. In Chapter 2 it was noted that terrestrial sunlight approximates the spectrum of a 5800 K blackbody source. PV cells are made of materials for which conversion to electricity of this spectrum is as efficient as possible. Depending on the cell technology, some cells must be thicker than others to maximize absorption. Cells often have textured surfaces and may also be coated with an antireflective coating to minimize the reflection of sunlight away from the cells.

3.3 THE PV MODULE—ESSENTIALS AND IMPROVEMENTS

To obtain adequate output voltage, PV cells are generally connected in series to form a PV module. With silicon, single-cell open-circuit voltages are typically close to 0.65 V and maximum power voltages are close to 0.5 V at 25°C. Historically, when the majority of PV systems were stand-alone systems with battery backup, it was desirable to combine a sufficient number of cells in series to provide effective battery charging. Since nominal 12-V lead-acid battery charging voltages range from 14 to 16 volts, allowing for the operation of modules at elevated temperatures, 36-cell modules were the norm for many years. These modules operated close to their maximum power points when charging 12-V lead-acid batteries.

Recently, however, three things have changed the design parameters for PV modules. First of all, since 1999, more PV modules have been used in grid-connected systems worldwide than in stand-alone systems [1]. In a straight grid-connected system with no battery backup, it is now common to design the PV array such that the maximum open circuit voltage is just under 600 V DC for smaller systems and up to 1500 V DC for larger systems. These arrays are connected directly to maximum power point tracking (MPPT) inverter inputs. Thus, many modern modules have 54 to 96 cells and sometimes even more, with correspondingly higher open circuit voltages and higher module power ratings.

The other change has been in the technology of charge controllers. When the first edition of this book was published, charge controllers connected the array directly to the batteries at the battery voltage. Thus, the array was designed to operate close to the normal charging voltage of the batteries to approximate maximum power operation. Modern charge controllers, however, incorporate MPPT input circuitry, so the array can operate at a maximum power voltage that exceeds the battery voltage and still, via the charge controller, deliver the correct charging voltage to the batteries. The third factor is cost. Over the past 20 years, the cost of PV modules has dropped by a factor of more than 20, such that much larger systems are now dominating the market. Larger cells and larger modules simplify the installation.

Introduction to PV Systems

An important observation relating to the series connection of PV cells relates to the shading of individual cells. If any one of the cells in a module should be shaded, the performance of that cell will be degraded. Since the cells are in series, this means that the cell may become forward biased if other unshaded modules are connected in parallel, resulting in the heating of the cell. This phenomenon can cause premature cell failure. To protect the system against such failure, modules are generally protected with *bypass diodes*, as shown in Figure 3.6. If PV current cannot flow through one or more of the PV cells in the module, it will flow through the bypass diode instead. For lesser amounts of shading, the remaining cells simply adjust their current down to the current of the lowest-current-producing cell.

Another clever adaptation of technology has been to use half cells instead of whole cells. Thus, a 120 half-cell or a 144 half-cell module is equivalent to a 60 full-cell or 72 full-cell module. The two sets of half-cells are each connected in series sets of either 60 or 72 and each set fills half the module. The outputs of the two sets are then connected in parallel with the advantage that each set carries only half the overall module current and thus the I^2R losses due to internal resistance are reduced by a factor of 4. In reality, however, the half-cell technique also allows for smaller cross-section interconnects of the cells, thus reducing the amount of interconnect material and exposing a larger fraction of the cell surface to the incident sunlight. Finally, if one group of half-cells is shaded and the other is not shaded, the unshaded group is not significantly affected by the shaded group as a result of the parallel connection. One can imagine the optimization process must be very interesting when all parameters, including material cost, are taken into account.

When cells are mounted into modules, they are often covered with antireflective coating and then with a special laminate to prevent degradation of the cell contacts. The module housing is generally metal, which provides physical strength to the module. When the PV cells are mounted in the module, they can be characterized as having a *nominal operating cell temperature* (NOCT). The NOCT is the temperature

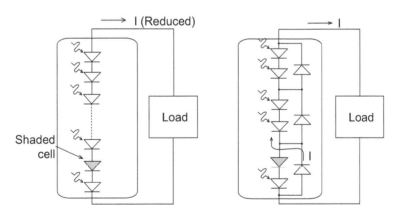

a. Module without bypass diodes b. Module with bypass diodes

FIGURE 3.6 Use of bypass diodes to protect shaded cells.

the cells will reach when operated at an open circuit in an ambient temperature of 20°C at AM 1.5 irradiance conditions, G = 0.8 kW/m² and a wind speed less than 1 m/s (3.28 ft/sec, 2.24 mph). For variations in ambient temperature and irradiance, the cell temperature (in °C) can be estimated quite accurately with the linear approximation that

$$T_C = T_A + \left(\frac{NOCT - 20}{0.8}\right)G \tag{3.5}$$

The combined effects of irradiance and ambient temperature on cell performance merit careful consideration. Since the open circuit voltage of a silicon cell decreases by approximately 2.3 mV/°C, the open circuit voltage of a module will decrease by 2.3n mV/°C, where n is the number of series cells in the module.

Hence, for example, if a 60-cell module has a NOCT of 45°C with V_{OC} = 41.3 V, when G = 0.8 kW/m², then the cell temperature will rise to 61.25°C when the ambient temperature rises to 30°C and G increases to 1 kW/m². This 31°C increase in cell temperature will result in a decrease of the open circuit voltage to 37.0 V, a 10.4% decrease. Furthermore, excessive temperature elevation may cause the cell to fail prematurely.

Finally, a word about module efficiency. It is important to note that the efficiency of a module will be determined by its weakest link. Since the cells are series connected, it is important that cells in the module be matched as closely as possible. If this is not the case, while some cells are operating at peak efficiency, others may not be optimized. As a result, the power output from the module will be less than the product of the number of cells and the maximum power of a single cell.

Another effect that results from series-connected cells that do not have identical I–V curves, is when the module is short-circuited, some of the cells will be generating power while others will be dissipating power. The greater the mismatch among cells in the module, the greater the level of power dissipated in the weaker cells. If all cells are perfectly matched, no power is dissipated within the module under short circuit conditions and the overall efficiency of the module will be the same as the efficiency of individual cells. Fortunately, modern modules have very closely matched cells, but for their predecessors this was not necessarily the case. Historic modules were guaranteed to be within ±10% of the rated value, whereas modern modules are rated within approximately 1.2% of the rated value as long as they are manufactured by the same manufacturer and are the same model.

3.4 THE PV ARRAY

If higher voltages or currents than are available from a single module are required, modules must be connected into arrays. Series connections result in higher voltages, while parallel connections result in higher currents. Just as in the case of cells, when modules are connected in series, it is desirable to have each module's maximum power production occur at the same current. When modules are connected in parallel, it is desirable to have each module's maximum power production occur at the same voltage. Fortunately, modern quality control, along with individual module testing as the last manufacturing step, pretty much guarantees that all modules of the same module model will have nearly identical I–V characteristics over a wide temperature range. Thus, as long as module models are not mixed, module voltages, currents and powers will generally be well-balanced in an unshaded array. In later chapters, criteria will be developed for deciding how many modules to connect in series and how many modules to connect in parallel to optimize the performance of a PV array that provides power to the designer's choice of inverter or charge controller. Which, if any, of the positive or negative output leads of an array should be grounded will also be discussed later.

3.5 ENERGY STORAGE

Historically, energy storage and PV have been emerging together. Early PV systems typically incorporated battery storage to provide power for nighttime use. Recently, as a result of a significant increase in the use of PV, wind and other less predictable renewable energy sources, a major effort has gone into the development of improved energy storage methods. Chapter 6 discusses a wide range of energy storage options that are becoming candidates for meeting the many gigawatts and gigawatt-hours of storage that are anticipated to be needed as the transition to renewable energy sources continues.

3.6 PV SYSTEM LOADS AND MPPT

The importance of operating PV modules near their maximum power points has already been discussed. Maximum power operation is a challenging problem, since it requires that the system load be capable of using all power available from the PV system at all times. Not only does this mean that the system load needs to be maximum when PV system output is maximum, but it also means the system load must adjust itself rather quickly at the onset or dispersion of cloud cover. The opposite challenge occurs when the load is satisfied and there is still additional PV power available. If there is no use for the power, the system power output needs to be curtailed or diverted to the point where the source and load are balanced.

Nearly all PV systems employ maximum power point trackers to ensure maximum power delivery by the PV array. Maximum power point trackers can be employed at the module level, at the string level or at the array level. In later chapters, they will be employed extensively. Because of this, it makes sense for the reader to understand how they work.

MPPTs generally employ pulse-width modulation techniques to switch from an input DC voltage to an output DC voltage at a different level, similar to a switching DC power supply. The MPPT employs a feedback loop to sense the PV array output power and change the array output voltage accordingly until the output power is maximized. In some cases, when all loads, including storage, have been satisfied, the MPPT will need to recognize this, or, at the recognition of this by other system controller components, the MPPT will need to back off the power production of the array to maintain system stability. Figure 3.7 shows how the MPPT electronically

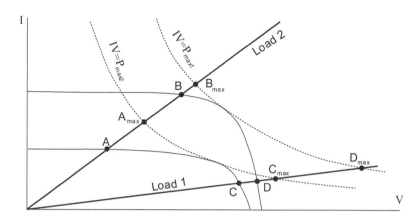

FIGURE 3.7 Operation of the MPPT.

moves the operating point along the maximum power hyperbola (IV = constant) associated with the PV array until it intersects the load I–V characteristic.

Another consideration for loads on PV systems is the trade-off between the lifetime operating cost of a load and the initial cost of the load. Sometimes a more efficient load is available at a higher initial cost and sometimes a more efficient load is available at a lower initial cost. If the initial cost of a load is lower, and there are no hidden maintenance costs, then it is a no-brainer to use the more efficient load, unless aesthetic or functional considerations should rule it out.

On the other hand, if the initial cost of a more efficient load is higher, then it is important for the PV system designer to perform an economic analysis of both system options, to determine which load is a better choice. In particular, the choice between relatively inexpensive first-cost incandescent lamps and more expensive but more efficient LED lamps involves an interesting cost analysis. Now that LED lamps are becoming commonplace, but at somewhat higher initial cost and correspondingly longer lifetimes, these devices add yet another interesting dimension to the choice of lighting sources. These considerations will be accounted for in Chapter 8.

3.7 PV SYSTEM AVAILABILITY—TRADITIONAL CONCERNS AND NEW CONCERNS

3.7.1 Traditional Concerns as Applied to Stand-Alone Systems

Critical loads are defined as loads for which power is required at least 99% of the time. Noncritical loads require power for at least 95% of the time. It is important to recognize that these definitions involve a statistical distribution of downtime over the expected lifetime of the system. In other words, a critical system may have availability during any given month or year of its operation that may be less than 99%, while it will have availability during other years in excess of 99%. The 99% figure is considered to be an average over the lifetime of the system.

Introduction to PV Systems

The reader is probably familiar with the causes of downtime on conventional grid-connected systems. Sometimes, it is a failure of an electrical generator. Sometimes, it is a system overload that requires parts of the system to be shut down selectively to prevent generator overload. Sometimes, a tree falls and breaks a power line. About every 100 years, an ice storm of the century causes long-term power outages for millions of utility customers, and sometimes a short circuit occurs and causes a circuit breaker to trip. Even hurricanes have been known to cause an occasional power outage. Clearly, many other situations may occur that will result in the loss of power.

Photovoltaic systems are subject to similar failure modes. Loose or corroded connections, battery failure, controller failure and module failure represent a few of the things that might go wrong in a PV system. However, a good preventive maintenance program can keep these failure modes at a minimum. The warranties on most major PV system equipment are now 25 years or longer.

Photovoltaic systems do have a factor that affects system performance to which conventional systems are not subjected—unpredictable cloud cover. Unless a PV system is grid-connected, so grid power will be available at night and when it is cloudy, it is necessary to provide battery backup not only for hours of nighttime operation but for those days when the sun shines too little to make sufficient electricity to meet the required daily need. Most readers are aware that different geographical locations have different seasons when cloudy days are more common. The reader in the north will be familiar with the winter weeks when the sun almost never appears through the cloud cover. The reader in the south will be familiar with the rainy summer days when little sunshine is available. The bottom line is that for different geographical locations, different amounts of battery backup are required for critical and noncritical loads if the PV system is not connected to the grid.

Figure 3.8 [3] shows the statistical distribution of downtimes due to weather over a 23-year system lifetime for critical and noncritical systems. Note that a 95%

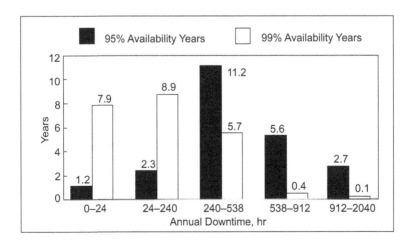

FIGURE 3.8 Statistical distribution of annual downtime for critical and noncritical PV systems over a 23-year system lifetime (*From Stand-Alone Photovoltaic Systems: A Handbook of Recommended Design Practices*, Sandia National Laboratories, Albuquerque, NM, 1995) [3].

available system is allowed 438 hours of downtime per year but, that for 3.5 years of the 23 years, the system will be down for less than 240 hours and for 8.3 years the system will be down for somewhere between 538 and 2040 hours per year. The critical system is allowed to be down for 88 hours per year but will be down for less than 24 hours per year for 7.9 years and will be down for more than 240 hours per year for 6.2 years out of the 23-year lifetime.

Figure 3.9 [3] shows a comparison of necessary days of battery backup for critical and *noncritical* operations as a function of available peak sun hours (psh), assuming a stand-alone system. Peak sun hours represent the total irradiation in a day expressed as the number of hours necessary at peak sun to account for the total irradiation between sunrise and sunset. An internet search will generally result in tables of psh for various geographic locations, months and array tilts and orientations. These tables, however, do not yield psh for all possible tilts and orientations, so an alternative method of determining psh is to use the PVWatts model found in NREL SAM for the desired location and array tilt and orientation.

By selecting an array tilt and orientation and then running SAM [4] for a 1 kW PV array at that location, the simulation will present a monthly AC kWh summary for the array. A careful look at the hourly simulated output will show for a DC:AC ratio of 1.15, the system AC output will essentially be equal to the system DC output, especially if the system is microinverter or optimizer based. Thus daily kWh/kW can be estimated by simply adding up the average hourly kW system AC output values for the day of interest.

The next step is to recognize that, if the monthly psh is defined as the number of hours per month the sun would need to shine at peak intensity (1 kWh/m^2) to

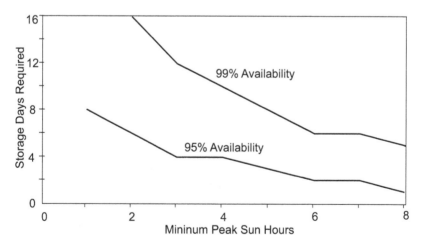

FIGURE 3.9 Necessary days of autonomy for critical and noncritical PV systems operation versus minimum available peak sun hours (*From Stand-Alone Photovoltaic Systems: A Handbook of Recommended Design Practices*, Sandia National Laboratories, Albuquerque, NM, 1995) [3].

generate the simulated DC monthly kWh output, then simply dividing by the number of days in the month will give the average daily psh for that month at the specific simulation conditions. As an example, suppose the SAM PVWatts simulation results in an estimated 110 kWh (AC)/kW production for January. The equivalent daily psh will thus be 110 ÷ 31 = 3.55.

Daily estimated system output can also be simulated, as long as the simulation is recognized as an estimated performance for a particular year but not to be used for a future prediction. For example, the daily simulated kWh outputs for an array facing south with a 26-degree tilt assuming total system losses of 10% in Dallas, TX, on December 12, 13 and 14 are found to be 4.95 kWh, 0.494 kWh, and 1.246 kWh. The average daily kWh for December is found to be 3.48 kWh. The point here is simply that on some days that for only one day of storage, there will obviously be some days where the storage does not meet the system needs. In this particular case, for enough days of storage, one comes closer to being able to expect the average daily performance for use in calculating the number of days of storage needed for the desired system availability. For more precision, it is probably better to look up the psh numbers, since even though psh and kWh/kW/day are pretty much proportional, the constant of proportionality is dependent upon what assumptions are made in the simulation of kWh/kW.

Days of battery backup are also known as storage days or *days of autonomy*. Using Excel, second-order polynomial curve fits can be obtained for the data in Figure 3.9. For those familiar with least mean square curve fitting to data, it will be recognized that an R^2 value of 1.0 indicates a perfect fit of the curve to the data, while an R^2 value that approaches zero indicates random scatter of the data. The best fits the data of Figure 3.9 yield the following equations for estimating necessary storage days, based on minimum average peak sun hours over the year, T_{min}, as determined from insolation data for the listed sites.

$$D_{crit} = 0.2976 T_{min}^2 - 4.7262 T_{min} + 24 \quad (R^2 = 0.9914) \quad (3.6a)$$

for critical applications, and

$$D_{non} = 0.1071\, T_{min}^2 - 1.869 T_{min} + 9.4286 \quad (R^2 = 0.9683) \quad (3.6b)$$

for *noncritical* applications, provided that $T_{min} > 1$ hr. These equations should be used only if the critical and *noncritical* storage times have not already been determined for a site, since site-specific cloud cover or sunlight availability may differ from the averages assumed in arriving at (3.6).

Extending the availability of a PV system from 95% to 99% may at first appear to be a simple, linear extension of the 95% system. Such a linear extension would involve only an additional 4% in cost. However, this is far from the case, and should obviously be so, considering that essentially no systems can provide 100% availability at any cost. Figure 3.10 shows the sharp increase in system cost as the availability of the system approaches 100%.

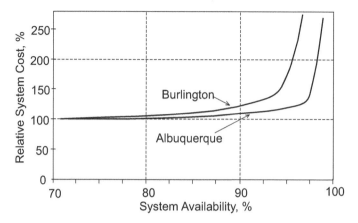

FIGURE 3.10 Relative costs versus availability for PV systems in Burlington, VT, and Albuquerque, NM (*From Stand-Alone Photovoltaic Systems: A Handbook of Recommended Design Practices*, Sandia National Laboratories, Albuquerque, NM, 1995) [3].

The cost of extending the availability of a system toward 100% depends on the ratio of the available sun during the worst time of the year to the available sun during the best time of the year. The extreme of this situation is represented by the polar latitudes, where no sun at all is available during the winter. Thus, near-100% availability would require enough batteries to store enough energy to meet all the no-sun load needs, plus a sufficient number of PV modules to charge up the batteries. This situation is further complicated by the likelihood that the winter loads would be larger than the summer loads. Clearly, a system with 180 days of autonomy should be more costly than a system with 10 days.

In the less extreme situation, such as found in Seattle, to provide the winter system needs, the system must be overdesigned with respect to the summer needs, resulting in excess power from the array during the summer. There is a good chance that this excess availability will be wasted unless a creative engineer figures out a way to put it to use. One way to do so, of course, is to connect it to the local electrical grid if the grid is available. However, in remote locations, the system may be stand-alone and the creativity of the engineer will be challenged.

3.7.2 New Concerns About PV System Availability

In the United States, for the past few years, more grid-connected renewable capacity has been installed annually than any other generation technology. In 2021, new wind capacity (12.2 GW) accounted for 30.7% of total new energy capacity, new utility-scale solar (15.4 GW) accounted for 38.7% of new capacity and natural gas (6.6 GW) contributed 16.6%. The remaining 14.1% included batteries (4.3 GW), nuclear (1.1 GW) and 0.2 GW from other sources [5].

Because of this rapid growth in small, medium and large systems, some utilities are now expressing concern over how PV generation may impact the rest of the grid. For example, PV produces power only during the day, but, for the most part, nighttime electrical needs must be provided by nonrenewable utility sources, except for nighttime wind and storage of renewable energy.

A somewhat less obvious concern regards the potential rapid variation of PV output during the day from any one PV system. If a 10-kW residential grid-connected PV system is clouded over in a matter of five seconds, reducing its output from, say, 8 to 3 kW, this will have about the same effect on the grid as an electric water heater or an electric clothes dryer turning on. When the sun returns in a few minutes, if the PV output returns to 8 kW, this will be equivalent to the appliance turning off. For a single appliance or PV system, the grid can readily handle the disturbance. But, for hundreds of similar systems in the same neighborhood, the grid impact of changing daytime cloud cover presents an interesting transient power flow analysis challenge, especially if the clouds are small and not all systems change power output simultaneously.

Another concern is the effects of cloud cover changes on utility-scale PV systems. An increase or decrease of 10 MW over a few seconds resulting from changing cloud cover presents an even more serious challenge to the utility in terms of transient power flows. Major utility disruptions have generally involved the reaction of the grid to the loss of feeder capacity. If the loss of major PV generation occurs, can the remaining utility generation make up for the PV loss?

And what if the utility is operating near peak capacity and a conventional generation facility goes offline? In Chapter 4, IEEE 1547 will be discussed in detail. This standard initially required a PV system to shut down if utility power was lost. Recent versions of IEEE 1547 now allow PV to be allowed to continue to operate to help stabilize the grid if utility generation is lost or if power factor correction is needed. All of the preceding concerns will become of greater interest as PV generation continues to become more cost-effective. Predictably, massive energy storage will play an important role in smoothing out the bumps on this road. In fact, these and other efficiency considerations will be involved in the continued evolution of the smart grid.

3.8 INVERTERS—CONVERSION OF DC TO AC

3.8.1 INTRODUCTION

Depending on the requirements of the load, a number of different types of inverters are available. Selection of the proper inverter for a particular application depends on the waveform requirements of the load and on the efficiency of the inverter. Inverter selection will also depend on whether the inverter will be a part of a grid-connected system or a stand-alone system. Even though warranties are now up to 25 years on some PV inverters, many opportunities still exist for the design engineer to improve on inverters.

Table 3.1 [6, 7] summarizes inverters presently available. Inverter performance is generally characterized in terms of the rated power output, the surge capacity, the efficiency and the harmonic distortion. Since maximum efficiency may be achieved near rated output, it is important to consider the efficiency versus output power curve for the inverter when selecting the inverter. Certain loads have significant starting currents, so it is important to provide adequate surge current capacity in the inverter

TABLE 3.1
Summary of Inverter Performance Parameters [6, 7]

Parameter	Square Wave	Modified Sine Wave	Pulse Width Modulated	Sine Wave*
Output power range (watts)	Up to 1,000,000	Up to 6000	Up to 800,000	Up to 10,000,000
Surge capacity (multiple of rated output power)	Up to 20×	Up to 4×	Up to 2.5×	Up to 4×
Typical efficiency over the output range	70–98%	>95%	>90%	>97%
Harmonic distortion	Up to 40%	>5%	<5%	<5%

* Multilevel H-Bridge or similar technology to yield utility-grade sine wave output.

to meet the load surge requirements. Other loads will either overheat or introduce unwanted noise if the harmonic distortion of their power supply is not below a specific level.

In general, the square wave inverter is the least expensive and relatively efficient, but has limitations in its applications. It has the best surge capacity but the highest harmonic distortion. The modified sine inverter is more complicated, but still relatively efficient. The pulse width modulated (PWM) inverter has a higher cost, high efficiency and minimal distortion. The pure sine inverter, with digitally generated waveform, has the least distortion and can have efficiencies in excess of 97%.

3.8.2 Square Wave Inverters

The simplest inverters are square wave inverters. These inverters employ solid-state switches connected as either astable multivibrators or externally controlled switches. The use of astable multivibrators enables the inverter to be used in a self-contained PV system, whereas externally synchronized switches are used when the inverter is to be synchronized with an external AC power source. Figure 3.11 shows how a single source can be switched alternatively in a positive direction and then in a negative direction to produce a square wave. This configuration is called an "H-Bridge."

An important feature of Figure 3.11 is that *there is no common negative terminal between DC input and AC output*. If a common ground is desired between the DC input and the AC output, it is necessary to couple V_o via a transformer to the inverter output. Otherwise, the common ground connection will be a short circuit across Q_2. As long as the transformer's primary and secondary windings are isolated, this enables the connection of one side of the secondary winding to a common DC/AC ground. Furthermore, it allows the selection of a transformer to step the inverter bridge circuit output voltage either up or down. The transformer can also assist in filtering the output waveform to remove harmonics.

Introduction to PV Systems

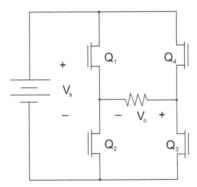

FIGURE 3.11 Converting a DC source to a square wave.

The disadvantage of transformer coupling is in the cost of the transformer, the weight of the transformer and the losses encountered in the transformer. In the last few years, with the continued improvement in the design and control of multi-level H-Bridges, inverter design has been transitioning away from transformer coupling. This requires that the DC input to the bridge must be floating, which is accomplished by using an ungrounded array. This practice will be further discussed in Chapter 4.

The key to efficient switching is to have current flowing from the array at all times while maintaining zero volts across the switching element when it passes current, and zero current through the switching element when it has voltage across it. This can be approximated reasonably well with insulated gate bipolar transistors, power MOSFETs or silicon-controlled rectifiers. Once the DC is converted into a square wave, its amplitude normally needs to be increased to produce a 120 V rms AC waveform. Since the rms value of a square wave is simply the amplitude of a square wave, a system with a 12-volt DC input will need a transformer with a 10:1 turn ratio.

It is important, however, to realize that the transformer must be designed with a sufficient number of turns so that the time constant determined by the magnetizing inductance of the transformer and the source resistance will be long enough to maintain the square wave. Too few turns on the transformer will cause the output waveform to droop as shown in Figure 3.12. When specifying the transformer for this application, it is easy to assume that transformers are ideal and to simply try to use a 120:12 volt transformer backward. The 120:12 volt transformer, however, probably would have been designed for use with a sinusoidal signal, and thus would not have enough turns to handle the square wave effectively. A square wave inverter with a good transformer and efficient switching can operate at efficiencies in the 90% range.

3.8.3 Multilevel H-Bridge Inverters

For a number of applications, a square wave is inadequate for meeting the harmonic distortion requirements of the load. For example, since square waves have significant harmonic content, and since hysteresis and eddy current losses in magnetic materials increase significantly with an increase in frequency, square wave

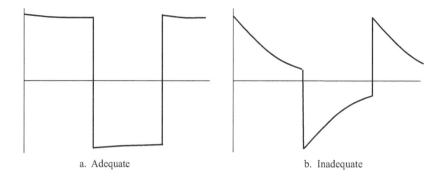

FIGURE 3.12 Output waveforms for a square wave inverter with adequate and with inadequate transformer turns.

excitation may cause some motors or other magnetic devices to overheat. Square wave harmonics can also introduce noise into a system. Thus, before selecting an inverter, it is important to verify that the proposed load will operate with square wave excitation.

If square wave excitation is not suitable for a load, it is possible that a modified sine wave will work. A number of methods are available to convert the output of a DC source into some approximation of a sine wave. One such method involves using a multilevel H-bridge as shown in Figure 3.13a to generate a waveform such as the one shown in Figure 3.13b. The idea is to time the individual voltage levels so that harmonic distortion will be minimized and that the rms value of the output voltage will remain constant in the event that the input DC voltage should vary.

The sequence for closing switches to obtain portions of the waveform in Figure 3.13b is shown on the waveform diagram. The actual switches may be bipolar transistors, MOS transistors, SCRs or insulated gate bipolar transistors. As an example of how the various voltage levels in v_o are obtained, consider the interval T_1, where, according to the output voltage waveform, $v_o = 0.5V_{dc}$. If S_1 is closed, diode D_1 becomes reverse biased and appears as an open circuit. If S_2 is closed, then the positive terminal of v_o is connected to $0.5V_{dc}$. Closing S_7 provides a path to ground from the negative terminal of v_o through forward-biased D_4. Hence, if the voltage drops across diodes and electronic switches are neglected, $v_o = 0.5V_{dc}$. Next, consider the interval where v_o is shown as $-V_{dc}$. During this interval, when S_3 and S_4 are closed, D_3 becomes reverse biased and appears as an open circuit, and the positive terminal of v_o is thus connected to $-0.5V_{dc}$. Closing S_5 and S_6 reverse biases D_2 and connects the negative terminal of v_o to $+0.5V_{dc}$. Hence, $v_o = -V_{dc}$. It is left as an exercise for the reader to verify the remaining switching combinations.

Noting the symmetry of v_o, and recognizing that $T_0 + T_1 + T_2 = 0.25T$, where T is the period of the waveform, it is possible to determine the rms value of v_o from the first quarter cycle by solving

$$V_{rms} = \left[\frac{4}{T}\int_0^{\frac{T}{4}} v_o^2(t)dt\right]^{\frac{1}{2}} = \left[\frac{4}{T}\left\{\int_0^{T_0} 0 dt + \int_{T_0}^{T_0+T_1}(0.5V_{dc})^2 dt + \int_{T_0+T_1}^{T_0+T_1+T_2} V_{dc}^2 dt\right\}\right]^{\frac{1}{2}}$$

$$= V_{dc}\left[\frac{T_1}{T} + \frac{4T_2}{T}\right]^{\frac{1}{2}} \tag{3.7}$$

As an example, assume that $V_{dc} = 160$ V, and it is desired to produce $V_{rms} = 120$ V by keeping $T_0 = T_1$ and solving for T_2. The result is that $T_0 = T_1 = T/16$ and $T_2 = T/8$. Problem 3.11 offers the reader a chance to solve for T_0, T_1 and T_2 for different values of V_{dc}. Note that since $T_1 + 4T_2 < T$, if $V_{dc} < V_{rms}$, there is no solution for T_0, T_1 and T_2.

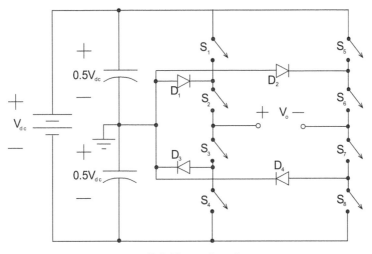

a. Switching configuration

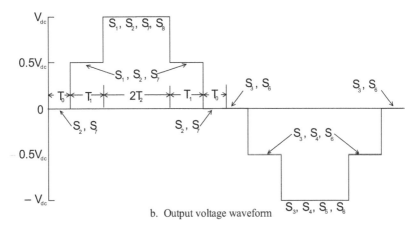

b. Output voltage waveform

FIGURE 3.13 Multilevel H-bridge modified sine wave inverter.

H-Bridges can be designed with 4n + 1 levels, where n is an integer. The more levels, the closer v_o can be made to approximate a sine wave. In addition, the voltage drop across any individual switch is reduced as the number of levels is increased. Problem 3.13 involves the design of a nine-level H-bridge.

The design challenge in modern inverters is to maximize the range of input voltages over which the output of the multilevel H-Bridge maintains a constant rms value with minimal distortion. The next design constraint, of course, is to ensure that the inverter input will track the PV array maximum power point and, finally, the inverter should have an efficiency close to 100%. Modern inverters generally have MPPT input voltage ranges of 300 V or more between maximum and minimum tracking voltages.

Many computer uninterruptible power supplies produce a five-level modified sine wave, so it is safe to say that this waveform is acceptable for use in providing backup AC power to a computer. However, IEEE Standard 1547–2018 [8] requires that any source connected to the utility line must have less than 5% total harmonic distortion (THD). For those who haven't memorized the formula for THD from their electronics textbook, recall that THD is the percentage ratio of the sum of the rms values of all the harmonics above the fundamental frequency to the rms value of the fundamental frequency. It is unlikely that a five-level modified sine wave inverter will meet the 5% THD rule, especially if V_{dc} is allowed to vary while keeping V_{rms} constant. This is the reason for the existence of pure sine inverters. Using higher-level H-Bridges to approximate pure sine waves is one method. Another method is to use pulse width modulation techniques.

3.8.4 PWM Inverters

The PWM inverter produces a waveform that has an average value at any instant equivalent to the level of a selected wave at that instant. PWM inverters are perhaps among the most versatile of the family of inverters. They are similar to the PWM described in the discussion of the MPPT with the exception that the MPPT PWM signal is designed to have a constant average value to produce a regulated DC output, while the inverter PWM signal is designed to have a time-dependent average value that can have any arbitrary waveform at any arbitrary frequency at any arbitrary amplitude. For use in PV applications, it is generally desirable to have a sinusoidal waveform with a predictable amplitude and frequency.

Figure 3.14 [9] shows how a PWM waveform can produce waveforms of differing amplitudes and frequencies by controlling the on-and-off time of a pulse waveform. The waveform is controlled by controlling the relative duty cycle of successive pulses. The amplitude is controlled by controlling the overall duty cycle and the frequency is determined by controlling the repetition time for the pulse sequence.

By switching the pulse between a positive level and a negative level, it is possible to construct waveforms with zero average value, which is particularly important when driving loads for which a DC component in the excitation may cause losses in the load. This includes applying DC to an AC motor or to a transformer. Application of DC to such loads can result in significant I^2R heating with possible failure of the winding insulation and subsequent catastrophic failure of the motor or

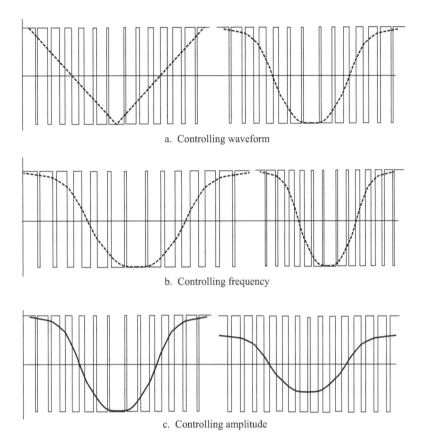

a. Controlling waveform

b. Controlling frequency

c. Controlling amplitude

FIGURE 3.14 PWM control of waveform, frequency and amplitude [9].

the transformer, including the possibility of fire. For this reason, the elimination of any DC component in the output of an inverter is important regardless of the type of inverter.

PWM inverters are particularly useful when used as AC motor controllers since the speed of an AC motor can be controlled by adjustment of the frequency of the motor excitation. It must be recalled, however, that the flux developed in the motor is inversely proportional to the excitation frequency and that every motor has a saturation flux above which the motor draws significant current without further flux increase. Hence, as a motor excitation frequency is decreased, the peak value of the applied voltage must also decrease proportionally to keep the motor out of saturation. A PWM controller coupled with MPPT capability can be a very efficient means of maximizing the efficiency of a PV pumping system.

Figure 3.15 shows one means of generating a PWM waveform. The process begins with a triangle wave at the frequency of the pulse waveform applied to one input of a voltage comparator. Next, the desired waveform is applied as a modulating waveform at the other input of the comparator. Whenever the modulating signal exceeds the triangle waveform, the comparator output goes high, and whenever the modulating signal is less than the triangle waveform, the comparator output goes low. The comparator output is then used to switch the output pulse on and off. The result is an output PWM waveform that has a moving average value proportional to the modulating signal. The average value of the PWM output is obtained in a manner similar to the detection process used to pick off the modulation signal of an amplitude-modulated communications signal. A filter needs to be incorporated that will not allow the voltage (or current) of the load to change instantaneously but will allow it to change quickly enough to follow the moving average of the PWM waveform. This can be achieved with inductive and/or capacitive filtering, depending on whether it is desired to smooth the current, the voltage or both.

The harmonic content of the two-level PWM waveform can be significantly reduced through the use of a three-level PWM waveform. A three-level PWM waveform can be generated by incorporating a second comparator into the circuit so that one comparator will control the positive-going pulse when the modulating signal is positive and the other comparator will control the negative-going pulse when the modulating signal is negative. The challenge here, however, is to switch the PV input so the array operates at maximum power voltage if the inverter is powered directly from the array. If a storage mechanism is used for array current, so the array is continuously supplying DC power to the storage mechanism, then the storage mechanism can be switched on and off to meet the load requirements without interrupting the power flow from the array. Alternatively, if the array voltage is switched between V_{OC} and some voltage less than V_{mp}, then the duty cycle of the connection to the array can be varied such that the average value of the array voltage seen by the inverter is V_{mp}. Figure 3.16 shows a three-level PWM inverter output signal derived from DC input from battery storage.

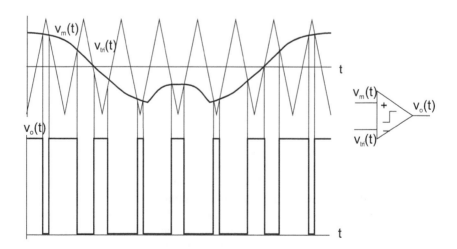

FIGURE 3.15 Generation of a PWM waveform.

Introduction to PV Systems 63

Figure 3.17 shows a PWM inverter circuit using a full-bridge switching network for the DC source, including MPPT. This is an adaptation of the full bridge that was shown for the generation of a square wave. The inverter can be fixed frequency, line commutated (synchronized) or variable frequency, depending on the source of the modulation signal. The controller generates the triangle waveform and the modulating waveform, v_m, with the amplitude and/or frequency of v_m determined by the PV array output voltage and current as sensed by the MPPT. Switches Q_1 and Q_3 of the bridge are switched on during the positive pulse excursions and Q_2 and Q_4 are switched on during the negative excursions. This provides a continuous power drain on the PV array, thus enabling MPPT.

Keeping the switching elements as close to ideal as possible, along with minimal loss in the filtering elements, can yield very high overall efficiencies for these units. When a switching element has either zero voltage or zero current, it does not dissipate power. During a switching transition, however, switching elements have both voltage and current and, hence, dissipate power. The tradeoff in a PWM inverter, then, is to have the frequency as high as possible to give adequate waveform reproduction, but

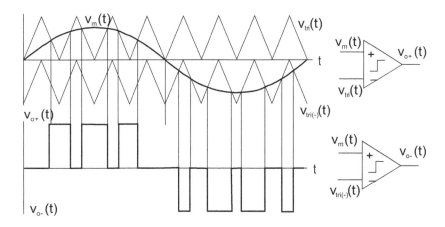

FIGURE 3.16 Three-level PWM inverter configuration and output signal.

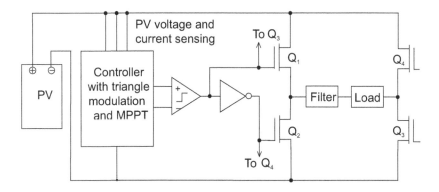

FIGURE 3.17 Full-bridge PWM converter showing controller and PV connections.

not so high as to increase switching losses to unacceptable levels. Most manufacturers now advertise PWM inverters with efficiencies above 95% over a wide range of inverter output power.

3.8.5 Transformerless Inverters

A closer look at the multi-level H-Bridge inverter schematic (Figure 3.13a) shows that there is no common connection between the input and output that can be grounded without shorting out at least two of the switches. This means that if a common ground is desired between input and output, the output must be isolated from the switching network output with a transformer connected across V_O. The transformer primary winding is connected across V_O and then either the transformer can have a grounded center-tapped winding to generate a dual-voltage output, such as 120/240 V, or the secondary can have just two terminals, one of which can share the ground connection with the input. Other convenient features of a transformer are that it tends to smooth out the high-frequency components of the inverter output and is capable of stepping the output voltage either up or down, depending upon the DC input voltage of the inverter and the corresponding rms value of the output of the multi-level H-bridge.

But there are inconveniences to having a transformer, not the least of which is the cost of the transformer as well as the inconvenience of the additional weight of the transformer. Perhaps even more important is the observation that the transformer losses are close to half the total losses of an inverter. So, for example, a 10,000-watt inverter with a transformer that operates at 96% efficiency, will lose 400 W at full output, whereas a transformerless inverter of the same size will lose only about 200 W. Thus, the challenge is to figure out how to eliminate the transformer.

Fortunately, *National Electrical Code (NEC)* 690.35 already allows for ungrounded PV source circuits and output circuits, provided that they meet the conditions of sections A–G. Section G allows PV arrays to be ungrounded provided that the inverter is listed for use with ungrounded DC input conductors. Note that even though the current-carrying conductors of the array are ungrounded, all the rest of the metal parts of the array (module frames, module mounts, etc.) must be separately grounded with equipment grounding conductors.

There remains a challenge for the transformerless inverter designer to provide an acceptable range of operating voltages and currents. Fortunately, this challenge has been met for inverters with input voltage ranges up to 1500 V and MPPT tracking range between 15% and 87% of maximum input DC voltage and output voltage ranging between 208 and 600 V, depending upon the specific inverter model.

3.8.6 Other Desirable Inverter Features

Modern inverters for PV applications often incorporate very sophisticated features to optimize inverter performance. In fact, the instruction and installation manuals for some of these inverters are nearly as long as this textbook [10, 11].

For stand-alone inverters and for grid-connected inverters with stand-alone options, a "search" mode is often incorporated into the design. In the search mode, the inverter uses minimal energy to keep its electronics operational when no loads are connected. In the search mode, approximately once every second, the inverter sends out an AC voltage pulse a few cycles long. If the inverter senses that no current flows as a result of the pulse, it concludes that no load is connected and waits to send another pulse. On the other hand, if current flows at a magnitude greater than a preprogrammed value, the inverter then recognizes that a load has been connected and supplies a continuous output voltage. If the inverter is not producing an output voltage on a continuous basis, power losses in the inverter are minimized.

In some cases, however, an inverter in the search mode can be confused. For example, if the inverter is set to provide power as long as more than 25 watts is connected, it can be confused with a 20-watt incandescent light bulb. The reason is that, when the lamp filament is cold, it has a lower resistance and a surge of current flows until the filament is heated. Hence, a 20-watt incandescent light bulb will appear to have a higher wattage load when it first turns on and then will drop back to 20 watts when the filament is hot. This, then, will cause the inverter to turn off and resume the search mode, where it will find the same light bulb and will turn it on again until it gets hot, resulting in a light that flashes on and off. Other confusing loads include the remote receivers in electronic equipment such as televisions, DVDs and Blu-ray players. When they are on, they draw very small amounts of power, so, when an inverter is in search mode, it is quite possible that the remote receivers will shut down.

For stand-alone inverters, it is necessary that the output appear as a voltage source, preferably as close to ideal as possible. For utility interactive inverters, however, it is more convenient to let the grid voltage fulfill the voltage source role and to have the inverter act as a current source that feeds current into the utility grid. Utility interactive inverters with battery backup that are capable of supplying power to standby loads if the utility grid is disconnected must be capable of switching over from grid-synchronized current source to an internally synchronized voltage source to power standby loads. These inverters are now generally referred to as multimode inverters.

Utility interactive inverters must be designed so that, if the utility goes down, the inverter output to the utility also shuts down. This safety feature is relatively easy to design into an inverter if no other distributed electrical source is connected to the utility. However, if another nonutility source is connected to the system, it is important that the PV system inverter will not recognize this other source as a utility source and continue to supply power to the utility. This is called "islanding" and will be discussed in detail in Chapter 4.

Many utility-interactive inverters that have a battery backup feature also incorporate a battery charger in the inverter so the utility can be used to charge the batteries in the event that the PV system has not fully charged the batteries. This feature only works, of course, when the utility grid is energized. Fortunately, this is generally most of the time.

Another convenient feature found in many stand-alone or grid-connected battery backup inverters is a generator start option. This allows for a separate AC generator to be used as a backup to the PV system either when the sun is down or when the battery level of charge drops to a prescribed level. If the inverter has a real-time clock, the time of day for the generator starting can also be controlled, so the generator will not come on just before the sun is about to charge the batteries.

Essentially all straight grid-connected inverters and recent versions of battery-backup or grid-forming inverters that are connected directly to a PV array incorporate MPPT circuitry at their DC input. This eliminates a separate MPPT charge controller external to the inverter and enables the PV array to deliver power at its maximum power level to the inverter. Ground fault detection and interruption (GFDI) and arc-fault circuit protection (AFCI) are now also often incorporated integral to the DC input circuitry of most inverters. Both will be discussed in more detail in Section 3.9.

Over the past decade or so, microinverters have played a significant role in smaller PV systems that may be subject to individual module shading or different module orientations, such as on a residential rooftop. In some cases, the microinverters are matched to the module by the manufacturer and sold as a single "AC Module" unit. Since the microinverters must comply with the same standards as larger inverters, since they are mounted beneath the modules, they conveniently serve as rapid shutdown devices if the utility is lost.

In addition to the rapid shutdown function, some microinverters are now classified as grid-forming. These inverters can be switched from the current source to the voltage source by an external controller, making possible "sunlight backup" for loads if the utility is lost during daytime hours. In later chapters, the distinction between AC-coupled battery backup and DC-coupled battery backup systems will be made. In the AC-coupled version, non-grid-forming microinverters can be used in the array and controlled by an external controller for operation during utility outages.

Finally, many modern inverters and associated energy storage systems have communication links with the utility that allow the utility to interact with the inverter for the purpose of grid stabilization. This can take the form of using energy storage and/or PV for power factor correction, demand management or power export to the utility. These inverters are often distinguished as grid-forming inverters. All of these features will be discussed in the design examples in the following chapters.

3.9 BOS COMPONENTS

3.9.1 Introduction

This section introduces the remaining basic electronic components of PV systems. These components include charge controllers, combiners, system controllers, optimizers, gateways, communications and rapid shutdown devices. All of these components handle relatively large amounts of power and are thus classified under the realm of power electronics. In each case, a simplistic explanation of the operation of the component will be given, with an emphasis on system performance requirements and how to best achieve them. Readers are encouraged to extend their understanding of these systems by consulting a power electronics text.

Introduction to PV Systems

3.9.2 Charge Controllers

3.9.2.1 Introduction

In nearly all systems with battery storage, a charge controller has been an essential component. The charge controller must shut down the PV array when the battery is fully charged and may also shut down the load when the battery reaches a prescribed state of discharge. When the "battery" is really a system of batteries connected in series and parallel as needed to meet system needs, the control process becomes somewhat more of a challenge. The controller should be adjustable to ensure optimal battery system performance under various charging, discharging and temperature conditions for lead-acid battery systems and will require more sophisticated individual cell charge management for lithium-ion battery systems.

Charge controllers still exist as individual devices. These are mostly found in smaller, grid-independent PV systems. In other systems, charge controllers are often incorporated into the overall function of a larger device, such as an inverter or system controller. Some charge controllers incorporate MPPT and some do not. For lithium battery systems, as will be discussed in Chapter 6, the charge control function is very complex and is generally managed by a microcontroller that monitors every cell in the battery, which could number in the thousands for a single "battery," or, perhaps more precisely, a single storage element.

Battery terminal voltage depends on various conditions of charge, discharge and temperature as will be shown in Chapter 6. The Thevenin equivalent circuit for the battery system is affected by all of these variables. The key is that, during charging, the battery terminal voltage, V_T, will exceed the battery cell voltage, V_B, since $V_T = V_B + IR_B$, where I is the charging current and R_B is the internal, or Thevenin equivalent, resistance of the battery. During discharge, $V_T < V_B$, since, under discharge conditions, $V_T = V_B - IR_B$ as the current direction is now reversed. The battery cell voltage is simply the battery open circuit voltage. Figure 3.18 shows the Thevenin equivalent circuits of the batteries under charging and discharging conditions.

The requirements for charging and discharging are made more complicated by the fact that the Thevenin equivalent circuit for the battery system is temperature dependent for both the open circuit voltage and the resistance. Researchgate.net shows a collection of internal resistance versus temperature curves for different battery types. In general, as temperature decreases, internal resistance increases, some more than others [12], and terminal voltage decreases. Furthermore, the Thevenin

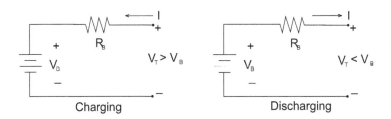

FIGURE 3.18 Thevenin equivalent circuit for a battery under charging and discharging conditions.

equivalent circuit for an old battery is different from that of a new battery of the same type. Hence, for a charge controller to handle all of these parameters, it should incorporate several important features. Depending on the specific application, it may be possible to omit one or more of the following features.

3.9.2.2 Charging Considerations

First, consider the charging part of the process. Assume that the battery is fully charged when the terminal voltage reaches 15 volts with a specific charging current. Assume also that, when the terminal voltage reaches 15 volts, the array will be disconnected somehow from the batteries and that when the terminal voltage falls below 15 volts, the array will be reconnected. Now note that, when the array is disconnected from the terminals, the terminal voltage will drop below 15 volts, since there is no further voltage drop across the battery's internal resistance. The controller thus assumes that the battery is not yet charged, and the battery is once again connected to the PV array, which causes the terminal voltage to exceed 15 volts, leading the array to be disconnected. This oscillatory process may continue until ultimately the battery becomes overcharged or until additional circuitry in the controller senses the oscillation and decreases the charging current or voltage. It should be noted that if the charging current pulses on and off; in fact, its average value will decrease to the duty cycle of the pulsing times the peak value of the current pulses.

An important difference between charge/discharge curves for lead-acid batteries and for lithium batteries is that *lithium battery charge must be limited to no more than 100%, whereas lead-acid batteries can be overcharged, but hydrogen is released when the charge level reaches 100%.* If C/x represents taking x hours to deliver full charge or discharge of the battery, then, for example, if a lead-acid battery is charged at a C/5 rate, full charge may be reached at a terminal voltage of 16 V, whereas if the battery is charged at C/20, then the battery may reach full charge at a terminal voltage of 14.1 V. If the charging current is then reduced to zero, the terminal voltage will drop to below 13 volts [13].

One way to eliminate overcharging resulting from the oscillatory process would be to reduce the turnoff set point of the controller. This, however, may result in insufficient charging of the battery. Another common method is to introduce hysteresis into the circuit, as shown in Figure 3.19, so that the array will not reconnect to the batteries until the batteries have discharged somewhat. The reader who has been wondering what to do with the regenerative comparator circuit that was presented in an electronics course now may have a better idea of a use for this circuit.

In fact, an even better charging algorithm might be to initially charge at a relatively high rate, such as C/5. When the terminal voltage reaches about 15 V, indicating approximately 85% of full charge, the charging rate is then decreased, taking temperature into account, until the battery ultimately reaches 100% charge at a very low charging rate and a correspondingly lower voltage. This method is employed in many of the charge controllers currently being marketed for use with PV systems.

Figure 3.20 shows the regions of charging associated with the algorithm suggested in the previous paragraph. Initially, the charge controller acts as a current

Introduction to PV Systems 69

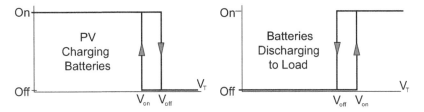

FIGURE 3.19 Hysteresis loops in charge controller using regenerative comparator for voltage sensing.

source. If the charging mechanism is a PV array, then presumably full array current will be used for charging. This is the bulk stage. When the charging voltage reaches a preset level, the *bulk voltage*, the charging mode is switched to constant voltage, during which the charging current decreases nearly linearly. This is called the *absorption* stage. Note that sometimes the bulk voltage is referred to as the absorption voltage, since it is the voltage at which constant voltage charging begins. The absorption mode is continued for a time preprogrammed into the controller, after which the charging voltage is decreased to the *float* voltage. The float voltage is then maintained by the charge controller. The float voltage must be set to a level that will not result in damage to the battery.

In fact, since battery temperature affects battery terminal voltage and state of charge, modern charge controllers incorporate battery temperature sensor probes that provide temperature information to the controller that results in automatic adjustment of charging set points for the charging modes. Since battery cell voltage drops slightly with temperature decrease, but battery internal resistance increases with temperature decrease, the battery charging voltage is increased as temperature decreases to overcome the increased resistance and maintain the desired charging current levels.

A further mode that is available in modern lead-acid chargers is the equalization mode. The equalization mode involves the application of a voltage higher than the bulk voltage for a relatively short time after the batteries are fully charged. This interval of overcharging causes gassing, which mixes the electrolyte as a result of the turbulence caused by the escaping gases. This mixing helps prevent sulfate buildup on the plates and brings all individual cells to a full state of charge.

Only unsealed or vented batteries need equalization. For specific equalization recommendations, manufacturers' literature on the battery should be consulted. Some charge controllers allow for automatic equalization every month or so, but often it is also possible to set the controllers for manual equalization. Electrolyte levels should be checked before and after equalization, since the gassing process results from the decomposition of the electrolyte.

The battery disconnect may result in the array being short-circuited, open-circuited or, perhaps, connected to an auxiliary load that will use excess array energy. If the array is short-circuited to disconnect it from the batteries, the controller is called a shunt controller. Open circuiting the array is done by a series controller.

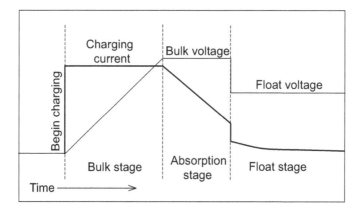

FIGURE 3.20 Three-stage battery charge control.

One advantage of the shunt controller is that it maintains a constant battery terminal voltage at an acceptable level by bypassing enough charging current to achieve this result. The disadvantage is the amount of power that must be dissipated by the shunt and the heat sinking necessary to remove this heat from the shunt device.

3.9.2.3 Discharging Considerations

Now consider the discharge part of the cycle. Assume the battery terminal voltage drops below the prescribed minimum level. If the controller disconnects the load, the battery terminal voltage will rise above the minimum and the load will turn on again, and once again an oscillatory condition may exist. Thus, once again an application for hysteresis is identified, and another regenerative comparator circuit is justified for the output of the controller.

At this point, all that remains is to make the set points of the charging regenerative comparator temperature sensitive with the correct temperature correction coefficient and the controller is complete. Of course, if the controller is designed to reduce charging current to bring the batteries up to exactly full charge before shutting down, and if the controller selectively shuts downloads to ensure the battery is optimally discharged, then the overall system efficiency will be improved over the strictly hysteresis-controlled system.

New designs continue to emerge for controllers as engineers continue their quest for the optimal design. Ideally, a charge controller will make full use of the output power of the PV array, charge the batteries completely and stop the discharge of the batteries at exactly the prescribed set point, without using any power itself.

3.9.3 MPPTs and Linear Current Boosters

Electronic MPPTs have already been mentioned in Section 3.6. Linear current boosters (LCB) are special-purpose maximum power point trackers designed for matching the PV array characteristic to the characteristic of DC motors designed for daytime

Introduction to PV Systems

operation, such as in pumping applications. In particular, a pump motor must overcome a relatively large starting torque. If a good match between the array characteristic and pump characteristic is not made, it may result in the pump operating under locked rotor conditions and may result in a shortening of the life of the pump motor due to input electrical energy being converted to heat rather than to mechanical output.

Figure 3.21 shows a typical pump I–V characteristic, along with a set of PV array I–V curves for different illumination levels. The fact that the pump characteristic is relatively far from the array characteristic maximum power point for lower illumination levels shows why an LCB can enable the pump to deliver up to 20% more fluid. The LCB input voltage and current track V_m and I_m of the PV array. The LCB output voltage and current levels maintain the same power level as the input, except for relatively small conversion losses, but at reduced voltage and increased current levels to satisfy the pump motor characteristic. The fact that the LCB increases current to the load accounts for the name of the device. In effect, the LCB acts as an electronic transformer for DC currents and voltages.

Maximum power point trackers and LCBs are generally adaptations of DC:DC switching voltage regulators, as indicated in Section 3.6. Coupling to the load for maximum power transfer may require providing either a higher voltage at a lower current or a lower voltage at a higher current. A buck, a boost or a buck-boost conversion scheme is commonly used in conjunction with load voltage and current sensors tied into a feedback loop using a microcontroller to vary the switching times on the switching device to produce optimal output voltage.

The LCB is used in special cases where only a boost of current is needed. This means a decrease in voltage will accompany the current boost to keep output power equal to input power. Since only a decrease in voltage is required, no boost is needed in the converter. Hence, a simple buck converter with associated tracking and control electronics will meet the design requirements of the device. In fact, this is also the case for most MPPTs. They typically convert a higher array voltage and lower array current to a lower voltage and a higher current, with minimal loss of power in the conversion process. The efficiency of good MPPT devices is typically 98% or better.

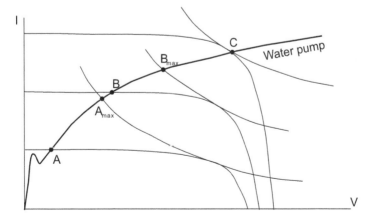

FIGURE 3.21 Pump and PV I–V characteristics, showing the need for the use of LCB.

Figure 3.22a shows a simplified diagram of a buck converter circuit. When the MOSFET is switched on, current from the PV array can only flow through the inductor into the parallel RC combination, where the capacitor voltage increases. When the MOSFET is off, current must remain flowing in the inductor, so the inductor current is now supplied by the capacitor through the diode, causing the capacitor to discharge. The extent to which the capacitor charges or discharges depends upon the duty cycle of the MOSFET. If the MOSFET is on continuously, the capacitor will charge to the array voltage. If the MOSFET is not on at all, the capacitor will not charge at all. In general, the output voltage and current of an ideal buck converter are given by

$$V_{out} = DV_{in}, \qquad (3.8a)$$
$$\text{and} \qquad I_{out} = I_{in}/D, \qquad (3.8b)$$

where D is the duty cycle of the MOSFET, expressed as a fraction ($0 < D < 1$). Note that the polarity of the output voltage is the same as the polarity of the input voltage, so there is no problem keeping the same grounded conductor at the input and the output of the converter.

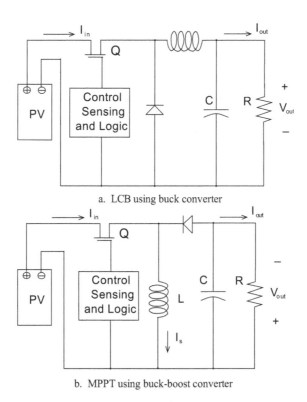

a. LCB using buck converter

b. MPPT using buck-boost converter

FIGURE 3.22 Linear current booster and maximum power point tracker.

Figure 3.22b shows the basic elements of a buck-boost converter that might be used in an MPPT. Note that, in this system, the polarity of the output voltage is opposite the polarity of the input voltage. This can be a problem if a common negative (grounded) conductor is required for the PV circuit and the output of the converter. In this case, the circuit needs to be somewhat more sophisticated. On the other hand, if, for example, the PV input requires a grounded positive conductor and the controller output requires a grounded negative conductor, this switch of grounds can be achieved with this design. However, presently, the preferred solution is to use an ungrounded PV array, as allowed by the *NEC*.

The output voltage and current of the ideal buck-boost converter are given by

$$V_{out} = \frac{D}{1-D} V_{in}, \tag{3.8c}$$

$$I_{out} = \frac{1-D}{D} I_{in}. \tag{3.8d}$$

where, again, D is the duty cycle of the MOSFET. Note that if $D < 0.5$, $V_{out} < V_{in}$, and if $D > 0.5$, $V_{out} > V_{in}$. Hence, the MPPT is capable of either increasing or decreasing its output voltage to track an array's maximum power point. Of course, if the MPPT output voltage decreases, then its output current will increase, and vice versa, such that, in the ideal case, the output power will equal the input power.

During the time Q is on, energy is stored in the inductor. When Q turns off, the inductor current must continue to flow, so it then flows through R and C and the diode, charging C to V_{out}, since the capacitor voltage equals the output voltage. When Q turns on again, the diode becomes reverse biased, and current is built up again in the inductor while the capacitor discharges through the resistor. Finally, when Q turns off, the cycle repeats itself. The values of L and C and the switching frequency determine the amount of ripple in the output voltage. Since no energy is lost in ideal inductors and capacitors, and since Q and the diode approximate ideal switches, essentially all power extracted from V_{in} must be transferred to the load. Of course, in reality, these components will have some losses, and the efficiency of the MPPT will be less than 100%. However, a well-designed MPPT will have an overall efficiency greater than 95%, with many units currently being marketed having advertised efficiencies that are close to 98% [14].

Another application of the MPPT is to ensure optimal charging of batteries. The MPPT charge controller electronically tracks the PV array maximum power point to ensure that maximum charging current is delivered to the battery bank. The result is a charge controller with high input voltage, low input current, lower output voltage and higher output current. Charge controllers without MPPT input circuitry connect the PV array directly to the battery voltage, while MPPT charge controllers operate the array at its maximum power point for charging the batteries at the proper charging voltage, as shown in Figure 3.23.

By designing the array to have a higher maximum power voltage, smaller wiring can be used between the array and charge controller, with lower power loss in the wiring and lower wiring cost. Note that with the PV array of Figure 3.23, the MPPT

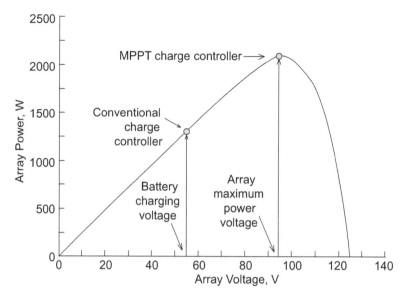

FIGURE 3.23 Comparison of PV array operating points for conventional and MPPT charge controllers.

charge controller will deliver about 2100 W to the batteries, while the conventional controller will deliver only about 1300 W to the batteries.

In terms of charging current, the MPPT charge controller delivers approximately 2100 ÷ 56 = 37.5 A, while the conventional charge controller delivers only 1300 ÷ 56 = 23.2 A. In fairness to the conventional charge controller, however, one should admit that normally with a conventional charge controller, the array would be designed with its maximum power point closer to 60 V so the battery charging voltage will be closer to the array's maximum power point voltage.

At the time of this writing, essentially all charge controllers and inverters used in grid-connected PV systems used MPPT. The array-to-battery efficiency of good MPPT charge controllers is typically close to 98%. Problem 3.17 explores this concept further.

3.9.4 Optimizers, Gateways, Communications and Rapid Shutdown

3.9.4.1 Optimizers

In Chapter 2, shading of modules was mentioned as one of the problems that can reduce overall array performance. Several methods for reducing shading effects through module design were introduced. When modules are connected in strings (source circuits) and strings are connected in parallel to produce arrays, and some of the modules in some of the strings may end up with different orientations, module-level compensation for nonuniform module voltage and/or current output can make a significant difference in overall array performance. Figure 3.24 shows the DC part of a PV system with three strings of modules. Each string has a different number of modules and some of the modules in some of the strings have different orientations,

Introduction to PV Systems 75

meaning different sun exposure. The three strings feed a single inverter with three inputs that are connected in parallel.

In this particular system, each module is equipped with an optimizer that has an input and an output. The DC output of the module is connected to the input of the optimizer. The optimizer has MPPT capability, and the optimizer itself acts as a DC:DC converter to enable its DC output voltage and current to be adjusted to values requested via a control loop with the inverter. Optimizers must be matched to the modules, since the optimizers will have maximum allowed input voltages and currents that need to be larger than the rated maximum module output voltages and currents. The optimizers for each string are connected in series, just as the modules would be in the absence of the optimizers. Optimizer system design details will be discussed in Chapter 4.

The inverter is now faced with the task of collecting maximum power from the array at acceptable voltage and current levels for the inverter. These levels will depend upon the output power and output voltage rating of the inverter. Since the optimizers are connected in series, each optimizer in a string will carry the same current but may end up with a different voltage than other modules in the string, depending upon the degree of shading experienced by its module. If all modules in a string are generating equal power, all optimizer output voltages will be the same, whereas if some of the modules are shaded, their output voltages will be less than the output voltages of the other modules in the string. The inverter itself does not do MPPT, but by acting as a controller for the optimizers, it adjusts their output voltages and currents such that none of the optimizer output voltages or currents exceeds the maximum rated voltage or current of the optimizers.

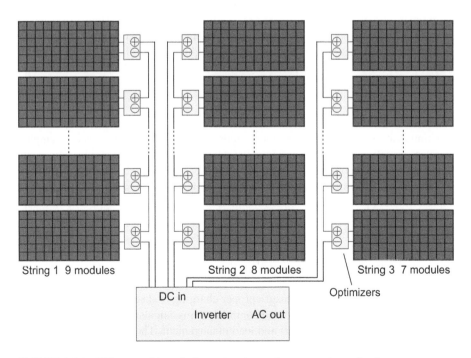

FIGURE 3.24 PV array with optimizers showing series connections of strings.

As an example, suppose the system is designed with 400 W modules. Then the maximum power from string 1 could be expected to be 9 × 400 = 3600 W, string 2 could be 3200 W and string 3 could be 2800 W. In fact, the likelihood that any of the modules will reach their rated power is small, simply because as the modules capture more sunlight, they get hotter and their power output decreases, as shown in Chapter 2.

In any case, now suppose the inverter sets the optimizers to establish string voltages of 400 V. In this optimal case, String 1 would generate 3600 W ÷ 400 V = 9 A, String 2 would generate 8 A and String 3 would generate 7 A, assuming all the optimizers have tracked the maximum power points of their associated modules. In the event that any of the strings is generating less than the maximum rated power, the string current will still be determined by the total power generated by all modules in the string divided by the established system voltage, regardless of whether all modules are generating identical power. Thus, the story of the optimizer. It enables its module to deliver maximum power to the system.

3.9.4.2 Gateways

Gateways will be encountered in PV systems with energy storage. In effect, they act as system controllers. Outputs of PV inverters and energy storage inverters are combined and connected to a common busbar along with utility power and backup loads. If utility power is lost, the gateway disconnects the system from the utility and maintains the disconnect until it recognizes that utility power has been restored and has been stable for a period of five minutes. In effect, the gateway is the brains of the overall system, since it also may be required to provide frequency and voltage amplitude stabilization for the PV components that normally will have synchronized with the utility. Not all gateways are necessarily called gateways. Some are listed as system controllers and some have their own specific product name and set of control and operational functions, such as possibly providing a neutral connection to create a split-phase (120/240 V) system from a 240 V system output. A number of different types of gateways will be discussed in Chapter 7.

3.9.4.3 Communications

Communications are essential to most modern PV systems. Some systems employ wireless or cell communications. Some systems employ wired communications via Cat 5 cabling, and some systems use powerline carrier communications. Both control and monitoring are achieved via the communication systems. Some systems can monitor system performance at the module level, some at the string level and others at the inverter level and transmit the data to locations convenient to the system owner. The rapid shutdown function, to be discussed in the next section, is generally achieved by a communication signal to switching devices located at or near the PV modules.

Control functions performed within the realm of the communication system include control of optimizers as discussed in Section 3.9.4.1 and control of inverter and/or microinverter output to ensure proper charging or discharging of storage elements. Recently, communication and control systems can also provide the necessary system functions for grid support and load management. These systems will be discussed in specific system design examples in the Chapters to follow.

Introduction to PV Systems

3.9.4.4 Rapid Shutdown

Once upon a time, when the PV inverter was disconnected from the utility, the open circuit DC output voltage of the PV strings was still present from the roof down to the inverter input. This meant that during an emergency, emergency responders could be exposed to voltages as high as 600 VDC if they were on the roof of the building or anywhere near where they could be exposed to the wiring between the roof and the ground-level inverter. The *NEC* [15] now requires that all PV systems installed on residences be equipped with rapid shutdown devices such that no PV voltage will be present outside the boundary of the PV array.

Because of large capacitors at inverter inputs, when a DC circuit is shut down, it may take a while for the capacitors to discharge. Thus, rapid shutdown has been defined to take into account the time it takes for the system voltage to drop below 30 V and the DC VA to drop below 240 VA. The VA requirements add an additional safety element, since, otherwise, at 30 V, if the system current is still in excess of 8 A, the VA requirement would not be met.

These rapid shutdown devices are allowed to take any form, as long as they can perform the required shutdown. For string inverters at the ground level, a number of options are available that incorporate remote switching contacts at the array that are controlled by a controller that is activated by a rapid shutdown switch at ground level. Two particularly convenient means of rapid shutdown are microinverters and optimizers. Each simply shuts down its output when signaled by a controller, an inverter or loss of the utility connection, causing the microinverter to shut down as required by UL 1741.

3.9.5 FOSSIL FUEL GENERATORS

In some cases, where either a large discrepancy between seasonal loads exists or seasonal sun availability varies greatly, a *stand-alone* system designed completely around PV components will result in the deployment of a large PV array to meet the needs of one season. Meanwhile, during other seasons, much of the energy available from the array is not used. This is similar to the problem of meeting critical system needs with PV, where the cost generally increases rapidly as the system availability exceeds 95%. In such cases, it has often been more cost effective to employ an alternate source of electricity to be available when the PV array is not meeting system needs. While it is conceivable that the backup source for a stand-alone system may be wind or other renewable source, it has been more common to employ a gasoline, diesel or propane generator as a system backup. In a grid-connected system, the utility grid provides any needed backup power, unless, of course, the grid goes down.

With relatively low acquisition costs of small and portable generators, it may appear that it would make better economic sense to simply use the generator without the PV array. However, life cycle cost analysis, as is discussed in Chapter 9, often shows the use of a mix of PV and conventional generation to be more economical than an engine-powered generator. Furthermore, the fact that PV generation is quiet and clean adds further appeal to using a maximum practical amount of PV generation in the system.

When it makes economic sense to use a generator or when a source of emergency power is required, it is important for the engineer to have some knowledge of generator options, provided that it also makes environmental sense. Normally, the generator will be supplied and installed by other than the PV contractor. The job of the PV engineer and PV contractor will be to recognize the various methods of safely incorporating fossil generators along with PV power in an overall electrical system.

Most readers probably already know that it is NOT a good idea to attempt to connect the output of a fossil generator directly to the utility grid. Properly installed generator systems include either an automatic or manual transfer switch that disconnects the load from the utility before connecting it to a generator. The automatic transfer switch will monitor the generator output voltage before switching over to the generator to be sure the voltage and frequency of the generator have stabilized. When a manual transfer switch is used, usually the generator has been manually started before the switch is transferred over to the generator while simultaneously disconnecting the load from the utility.

When a generator and a PV system are both a part of the building's electrical system, the important issue aside from making sure the generator is never connected directly to the grid is to be sure the generator is never connected to the PV system unless proper provisions have been made. If the PV system has no battery backup and thus shuts down when grid power is lost, it needs to be connected on the grid side of the generator transfer switch to be sure the PV system will not think the generator is the grid and continue to operate. The problem with this is if the PV system is making more power than is required by the loads, it will then try to export it to the generator. This is generally not a good way to keep a generator happy.

If the PV system includes energy storage for utility backup, some systems have provisions for connecting the generator in a manner that prevents the generator from connecting directly to the PV inverter output but does allow the generator to charge batteries or run loads when the PV system output is not available, such as at night.

Proper connections of generators along with PV systems will be discussed in Chapters 4 and 7 with several of the design examples. In general, it is important to realize that electricity from a fossil generator is relatively expensive due to the fuel cost and also an environmental issue.

For more detailed information on generators, a good reference is [16].

3.9.6 Miscellaneous Smaller But Important Components

3.9.6.1 Introduction

Aside from the major components of a PV system, there are numerous other smaller/less expensive components necessary to complete a PV system installation. They include overcurrent protection devices, disconnects, surge protectors, array mounts, receptacles, GFDI devices, arc fault detection, wiring, connectors, wiring enclosures, inverter bypass switches, source circuit combiner boxes, grounding connections and battery cables and battery containers. In some cases, the connected loads are considered to be part of the BOS, and in other cases the loads are not considered to be a part of the system. The distinction is generally made when the system is installed to operate a specific load, as opposed to being installed to be one contributor to the operation of all loads.

Certain components are regulated by codes or standards. Array mounts, for example, must meet any wind loading requirements of applicable building codes. Battery compartments are covered in the *NEC*. In certain environments, small components may need to be resistant to corrosion from exposure to salt air or may need to be appropriate for other environmental considerations. If a PV system is part of a building-integrated structure, then a number of other codes and standards may become applicable.

The *NEC* [15] specifies the requirements for choosing most of the electrical components. *NEC* requirements will be described and used in detail in the next few chapters. ASCE-7 [17] is essential for guidance in the design of array mounts and structural components of a PV system. Structural design is covered in Chapter 5.

3.9.6.2 Switches, Circuit Breakers, Fuses and Receptacles

All switches, circuit breakers and fuses used in the DC sections of PV systems must be rated for use with DC. Switches, circuit breakers and fuses used in AC circuits must be rated for AC use. Switches have both current and voltage ratings. If a switch is used to control a motor, it must be rated to handle the horsepower of the motor at the operating voltage of the motor.

Circuit breakers must be sized in accordance with *NEC* requirements. For example, although the maximum current ratings for #14, #12 and #10 THHN wire are 25, 30 and 40 A, respectively, the maximum fuse or circuit breaker sizes allowed for use with these wire sizes are 15, 20 and 30 A, respectively. Larger wire sizes may be fused at their rated ampacities (i.e., I_{max}). If a fuse or circuit breaker (overcurrent protection device, OCPD) is not available at the ampacity rating of the wire, for ampacities up to 800 A the next higher ampacity of OCPD may be used. For example, #6 copper wire with 75-degree insulation is rated at 65 A, but, since a 65 A OCPD is not a standard size, a 70 A unit may be used, with certain exceptions.

For motors, it may be necessary to install a fuse or a circuit breaker with a rating that exceeds the circuit ampacity to accommodate the starting current of a motor. When this is the case, the motor must have a form of overload protection that will disconnect the motor if the motor current exceeds approximately 125% of its rated running current, depending on the size and type of motor. Details of wiring for motors and motor controllers are covered in great detail in *NEC* Article 430.

Different voltages require different receptacles. While it is not very likely that 12 VDC will damage a piece of 120 VAC equipment except possibly a motor or transformer, it is almost certainly true that 120 VAC will damage a piece of 12 VDC equipment. It is thus necessary to use different attachment caps and receptacle configurations for different voltages.

Since the *NEC* will be referenced in nearly all the design examples in later chapters, specific applications of *NEC* requirements will become more clearly evident as the examples are developed. Further discussion of overcurrent protection, disconnects and wire sizing in accordance with *NEC* requirements will be covered in detail in these examples.

3.9.6.3 Ground Fault, Arc Fault, Surge and Lightning Protection

Presumably, current will leave the PV array via the positive conductor and the same amount of current will return to the array via the negative conductor. This will be the case, provided that no alternate return paths are present.

The *NEC* requires that metallic frames and other metal parts of PV systems be connected to the ground. The conductors used for this purpose are called *grounding conductors*. In older systems, but almost never in newer systems, the positive or negative conductor (but not both) is also connected to the ground at some point along the system. If so, then the conductor connected to the ground is called the *grounded conductor*. Since the grounded conductor is connected to the ground at only one point, current will flow in the grounded conductor, but will not flow in any of the grounding conductors, since there is no closed circuit in which the current can flow when the system is operating properly.

However, if for some reason the ungrounded conductor was connected to the ground, then there would be an alternate closed path in which current would be able to flow through the grounding conductors. Such a condition is classified as a ground fault. If this is the case, then Kirchhoff's current law requires that the current in the ungrounded conductor will equal the sum of the currents in the grounded and grounding conductors. The net result is that no longer will the currents in the positive and negative conductors be equal. The greatest danger in DC ground fault currents in a PV system is the means by which they are established. If the ground fault results from a loose connection, the connection may begin to arc and become a fire hazard.

NEC 690.5 requires ground fault protection to be provided (GFDI) to disconnect the array in the event of a difference in current between the positive and negative DC array conductors leading to the controller/inverter/loads, unless the system is a small system with all DC output circuitry isolated from any buildings.

NEC 690.11 requires that all DC circuits operating at a maximum system voltage of 80 V or greater be protected by a listed DC AFCI. Arc faults can occur as series faults, such as arcing across a broken wire, or as parallel faults, such as arcing between two conductors. The AFCI device must trip in either situation. The AFCI device provides additional protection over and above the protection provided by a GFDI. While GFDI devices sense a difference in current between positive and negative (supply and return) conductors of approximately 0.5 A for small systems, a difference in supply and return current is not necessary to activate an AFCI device. The AFCI device electronically detects the electronic signature of an arc and responds by opening the circuit and providing a visual indication that the device has detected a fault.

Another important component used to protect the array and inverter is the surge suppressor. Surge suppressors are similar to Zener diodes in that they are made of material that is essentially insulating until a predetermined voltage appears across the material. At this point, avalanche breakdown occurs and the surge suppressor acts as a current shunt. Metal oxide varistor (MOV) types of surge suppressors respond in nanoseconds and can bypass many joules of surge energy. However, a disadvantage of MOV surge suppressors is that they draw a small current at all times and generally the failure mode of a MOV device is a short circuit. SiO (SOV) surge suppressors are an improvement in terms of the MOV disadvantages. While it is not required by the *NEC*, it is good practice to incorporate listed DC surge protection into system design at a location close to the common system ground point to protect system electronic components. In fact, some inverters incorporate ground fault, arc fault and surge protection at the inverter input. The NEC now requires AC surge protection to be installed near the service entrance to prevent any backfed surges from power line disturbances.

Additional protection from lightning strikes can be obtained with lightning rod systems. Contrary to popular belief, lightning rods do not attract lightning. Rather, they present a sharp point (or points) that are connected to the ground. If a potential difference should appear between the rod and the surrounding atmosphere, a very high electric field builds up around the point to the extent that a corona discharge takes place, equalizing the charge between the air and the ground. Corona discharge is also known as glow discharge. Lightning is known as arc discharge. The arc discharge is violent, but glow discharge is essentially harmless since the charge is equalized over a much longer period of time.

Inverter bypass switches are sometimes used in battery backup systems to bypass the inverter in the event that maintenance is needed on the inverter and an alternate connection between grid and standby loads is required. Inverter bypass switches will be included in Chapter 7.

Source circuit combiner boxes are used to combine the outputs of individual source circuits (strings) of a PV array into a single PV output circuit that feeds the inverter DC input. When string inverters are used, often the DC strings are combined at the inverter input. When an inverter has multiple DC string inputs, often it will have more than one MPPT to accommodate strings with different orientations. A typical string inverter may have three MPPT inputs, each of which can accommodate two strings. The combiner boxes or inverter inputs include either fuses or circuit breakers at each MPPT input circuit to protect the strings from backfed current from other strings in the event that a problem should occur in one of the strings. Note that modules connected in DC strings are connected in series, since the currents of all modules are expected to be identical.

Source circuits (strings) are also used to combine the AC power outputs of microinverters. Microinverters are designed with an MPPT input and are installed at the location of the PV module. They deliver the module power to the string as an AC current source operating at a fixed AC voltage determined by the utility line voltage, which will normally be either 240 or 208 V single-phase AC. Note that since microinverters act as current sources, they are connected in parallel, since currents in parallel add. Individual microinverter output currents do not need to be identical to other microinverter output currents in the string. In the case of microinverters, combiner boxes often also incorporate a communications device that keeps track of production data in real time and cumulative and identifies any individual microinverter that may be acting strangely.

All non-current-carrying metallic parts of the system, such as array mounts, PV module frames, equipment enclosures, junction boxes and conduit, must be grounded. Grounded means that these items must be at ground potential. The *NEC* has established the sizes of conductors and the types of connectors that must be used for grounding equipment. As system designs are completed in later chapters, careful attention will be paid to the selection of proper grounding equipment.

Homework Problems

3.1 If $I_\ell = 12$ A, $I_o = 10^{-10}$ A and T = 300 K for a PV cell, determine the maximum power point of the cell by differentiating the expression obtained by multiplying (3.1) by the cell voltage.

3.2 Determine the range of operating voltages for which a 60-cell module will have a power output within 90% of maximum power. You may assume $I_o = 10^{-9}$ A and $V_{OC} = 0.600$ V at an operating temperature of 300 K. Assume all cells in the module are identical.

3.3 A PV module is found to operate at a temperature of 60°C under conditions of $T_A = 30°C$ and $G = 980$ W/m². Determine the NOCT of the module.

3.4 Plot V_m versus T for $-25 < T < +75°C$ for a 72-cell Si module for which each cell has $V_m = 0.54$ V at 25°C. Use a typical value for $\Delta V_m/\Delta T$.

3.5 Plot P_m versus T for the module of Problem 3.4 if $I_m = 9.85$ A at 25°C. Use a typical value for $\Delta I_m/\Delta T$.

3.6 Two 72-cell PV modules are connected in series. One is shaded, and one is fully illuminated such that the I–V characteristics of each module are as shown in Figure p3.1.
   a. If the output of the two series modules is shorted, estimate the power dissipated in the shaded module.
   b. If the two modules are equipped with bypass diodes across each 12-series cells, estimate the power dissipated in the shaded module.

3.7 For the circuit of Figure 3.11, explain the on-off sequencing of the MOSFET switches to produce a symmetrical square wave output. Sketch the result.

3.8 Determine whether the DC source in Figure 3.11 remains on constantly during the switching. If not, explore modifications to the circuitry to ensure that the DC source is delivering constant DC current.

3.9 Show how a DC–DC converter can be used as a part of an inverter designed to have a square wave output of 120 V with a 48-V DC input. The idea is to design the inverter without a transformer. Sketch a block diagram, showing some components to clearly express your design.

3.10 For the five-level H-Bridge of Figure 3.13, determine T_2 if $T_1 = T_0$, $V_{rms} = 120$ V, and
   a. $V_{dc} = 150$ V

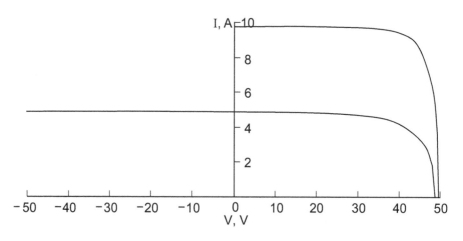

FIGURE P3.1

Introduction to PV Systems

    b. $V_{dc} = 155$ V
    c. $V_{dc} = 160$ V
    d. $V_{dc} = 165$ V
    Note that Excel or MATLAB can be very useful for this problem.

3.11 For the five-level H-Bridge of Figure 3.13, determine a set of values of T_0, T_1 and T_2 that will minimize the THD of the waveform if $V_{dc} = 160$ V and $V_{rms} = 120$ V. A convenient tool is to construct the waveform in SPICE using series pulse voltage sources and then to run a Fast Fourier Transform of the resulting waveform.

3.12 For the five-level H-Bridge of Figure 3.13, determine the maximum value for V_{dc} if the design requires $T_2 > 0.025T$ and $T_0 > 0.025T$.

3.13 Referring to Figure 3.13a, design a nine-level H-Bridge, using four capacitors and an appropriate number of series switches in each string. Show the waveform and the switching needed to obtain each of the nine levels of the output voltage.

3.14 For the nine-level H-Bridge output waveform, derive a formula for the rms value of the output voltage.

3.15 Construct the nine-level H-Bridge output waveform using series pulse voltage sources in SPICE and perform a Fast Fourier Transform on the waveform to explore the harmonic distortion of the waveform. Keep V_{dc} and V_{rms} constant, while varying the duration of the different levels of the waveform to minimize the THD of the waveform.

3.16 Make a list of loads that might confuse an inverter that is in the "sleep" mode. Explain why each load causes a problem.

3.17 If an MPPT operates at 98% efficiency with a PV array that has the characteristics of Figure p3.2, using the load curves as shown in the figure
    a. Determine the additional power available to each load when the array operates at 1000 W/m².
    b. For each load, determine the charge controller input current and output current at 1000 W/m².
    c. For each load, how long would the system need to operate at 1000 W/m² to recover the cost of the MPPT if the MPPT costs $500 and the PV-generated electricity has a value of $0.40/kWh?
    d. Repeat parts a–c if the array operates at 500 W/m².

3.18 The following data are given for a series of gasoline-powered electrical generators:

Rated output	1500 W	2300 W	3000 W	4500 W
Fuel tank size	2.9 gal	2.9 gal	4.5 gal	4.5 gal
Run time/tank	9 hr	9 hr	8.3 hr	5.6 hr

Calculate the kWh/gal for each of these generators under rated load conditions.

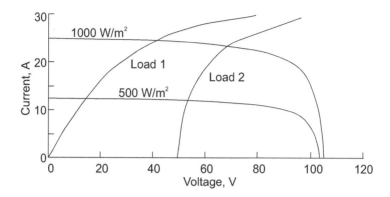

FIGURE P3.2

REFERENCES

[1] Streetman, Ben and Banerjee, Sanjay, *Solid State Electronic Devices*, 7th ed., Pearson, March 2014.
[2] Yang, E. S., *Microelectronic Devices*, McGraw-Hill, New York, 1988.
[3] Kandil M. Kandil, Majida S. Altouq, Asma M. Al-asaad, Latifa M. Alshamari, Ibrahim M. Kadad, Adel A. Ghoneim. *Stand-Alone Photovoltaic Systems: A Handbook of Recommended Design Practices*, Sandia National Laboratories, Albuquerque, NM, 1995.
[4] National Renewable Energy Laboratories, System Advisory Model, 2022.11.21 version.
[5] Information on New Energy Installations in the United States in 2021: www.eia.gov/todayinenergy/detail.php?id=46416
[6] Information on a Wide Selection of Lower Power Inverters: www.powerinverters.com.
[7] Information on Utility-Scale Inverters: www.solectria.com/pv-inverters/utility-scale-inverters/
[8] IEEE Standard 1547–2018 Standard for Interconnection and Interoperability of Distributed Energy Resources with Associated Electric Power Systems Interfaces: https://standards.ieee.org/ieee/1547/5915/
[9] Information on Generation of PWM Waveforms: https://wiki.analog.com/university/courses/electronics/electronics-lab-pulse-width-modulation
[10] Solaredge Inverter Installation Manual (111 pp)cps 125: https://knowledge-center.solaredge.com/sites/kc/files/se-inverter-installation-guide.pdf
[11] CPS 125 kW Installation and Operation Manual: www.chintpowersystems.com/wp-content/uploads/2020/09/CPS-SCH100-125KTL-DO-US-600_Manual_-Sept-2020.pdf
[12] Information on Battery Resistance vs Temperature for Several Battery Types: www.researchgate.net/figure/Dependence-of-internal-resistance-versus-temperature-for-lithium-based-batteries-LiFePO_fig15_316171277
[13] Information on Lead-Acid Battery Terminal Voltage vs SOC and Charging/Discharging Rate: www.scubaengineer.com/documents/lead_acid_battery_charging_graphs.pdf
[14] Information on MPPT Efficiency: www.solar-electric.com/learning-center/mppt-solar-charge-controllers.html/
[15] *NFPA 70 National Electrical Code*, 2020 ed., National Fire Protection Association, Quincy, MA, 2019.
[16] Welcome to Generator Joe Home Page: www.generatorjoe.net/store.asp (a wealth of information on a wide variety of fossil fuel generators).
[17] ASCE Standard 7–16, *Minimum Design Loads for Buildings and Other Structures*, American Society of Civil Engineers, Reston, VA, 2016.

4 Grid-Connected Utility Interactive PV Systems

4.1 INTRODUCTION

As the cost of PV systems continues to decrease, utility-interactive systems have now been at grid parity in most locations since about 2018, meaning that the cost of PV-generated kWh is less than or equal to the cost of kWh purchased from the grid. Furthermore, increases in consumer concern over global climate change have resulted in an ever-increasing interest in reducing CO_2 emissions by replacing fossil fuel-generated electricity with PV-generated electrical energy. This feedback loop, coupled with the increased demand for grid-connected and grid-backup systems, as well as the financial incentives offered by a number of utilities, states and countries, has resulted in a healthy market for PV systems. This increased demand has enabled PV module and balance of system component manufacturers to scale up manufacturing facilities to take advantage of economies of scale to further reduce system costs.

In addition to cost reductions, the increased demand for PV systems has led to significant improvements in the reliability of PV system components, designs and installations. As a result, not only are PV systems decreasing in cost, but they are increasing in reliability, resulting in longer warranties. By 1999, the technical issues associated with connecting PV systems to the utility grid had essentially been solved. In 2000, IEEE adopted Standard 929–2000 [1]. Any PV system meeting the performance criteria of IEEE Standard 929, using power conditioning units/inverters (inverters) listed under UL 1741 and installed in accordance with the current *National Electrical Code* (NEC), automatically met all established technical performance criteria. Since then, IEEE has adopted further refinements to Standard 929–2000, which are reflected in IEEE 1547–2018 and IEEE 1547.1–2020 [2]. IEEE 1547–2018 is a full revision of IEEE 1547–2003. The 2018 revision addresses the rapid deployment of grid-connected PV and other renewable energy systems.

Grid-connected utility-interactive PV systems are capable of supplying power to the utility grid. There are essentially two classes of grid-connected systems—straight grid-connected with no battery backup and grid-connected with battery backup. Grid-connected systems with battery backup can be further separated into DC-coupled systems and AC-coupled systems, and, in some cases, battery backup systems do not sell back to the grid but only purchase power from the grid to charge the system batteries.

This chapter will focus on the design of straight grid-connected PV systems. Straight grid-connected systems are the most straightforward systems to understand and are currently the most commonly installed systems. Battery backup systems will be covered in Chapter 7, and stand-alone PV systems will be covered in Chapter 8. Before beginning the design process, it will be helpful to take a closer look at some of the codes and standards that govern the design and installation of PV systems.

4.2 APPLICABLE CODES AND STANDARDS

A number of codes and standards have been created to ensure the safety of electrical systems. These codes and standards also generally address the efficiency and reliability of systems. Since a PV system is capable of generating sufficiently high voltages to present a potential electrical safety hazard, PV systems are included in these codes. Perhaps the two most common codes and standards that deal with PV systems are the *NEC* [3] and IEEE 1547–2018 [2]. The PV engineer should be familiar with both. But the list does not end here. PV systems have a mechanical component and an electrical component, and, hence, a number of mechanical and structural codes also apply. If the PV system is to be integrated into the construction of a building, the list of codes and standards becomes even longer, now explicitly including fire codes. Table 4.1 [2–10] provides a *partial* listing of the many codes and standards that may be applicable to any particular PV installation. Note that by the time this section is read, some or all of the codes and standards listed may have been revised or updated.

4.2.1 THE *NEC*

4.2.1.1 Introduction

The *NEC* is published by the National Fire Protection Association and is updated approximately every three years. This text will use the 2020 edition of the *NEC* as a reference. The *NEC* consists of a collection of articles that apply to considerations such as wiring methods, grounding, motor circuits and nearly every conceivable topic in which electrical safety and efficient utilization is a consideration, including Articles 690, 691, 705, 706, 710 and 712, which deal specifically with PV systems and their various modes of operation and interconnection.

TABLE 4.1
Partial Listing of Codes and Standards That May Apply to PV Systems or Components [2–10]

Reference #	Title/Contents
IEEE 1547–2018	IEEE Standard for Interconnection and Interoperability of Distributed Energy Resources [2]
NEC 2020	*National Electrical Code*/Wiring methods (comprehensive) [3]
IEEE 1561–2019	IEEE Guide for Optimizing the Performance and Life of Lead-Acid Batteries in Remote Hybrid Power Systems [4]
IEEE 1526–2020	Tests to Determine the Performance of Stand-Alone Photovoltaic Systems [5]
UL 61730-1&2	Standard for PV module safety qualification [6]
IEEE 1562–2021	Guide for Array and Battery Sizing in Stand-Alone PV Systems [7]
IEC TC 82	A Collection of Standards for PV Cells and Wafers [8]
UL 1741	Standard for Inverter Safety Based upon IEEE 1547 [9]
ASCE 7–16	Design Loads for Buildings and Other Structures [10]

Local and State Current Building Codes
Local, State and National Fire Prevention Codes

Grid-Connected Utility Interactive PV Systems

The *NEC* specifies the sizes and types of switches, fuses and wires to be used and specifies where these items must be located in the system. These components are necessary not only to protect the end user but also to protect the maintenance technician. As specific PV system design examples are discussed, compliance with the *NEC* will be incorporated into the design process discussion. In particular, proper wire sizing to limit voltage drop in connecting wires to acceptable limits, proper use of switches, circuit breakers and fuses, types of insulation and conductors, types of electrical conduit, electrical safety devices for detecting and interrupting ground faults and arc faults, and proper grounding will be considered.

Normally, copper wire will be used in PV system wiring. Aluminum wire is allowed by the *NEC*, but generally it is used only over longer distances for carrying higher currents when the cost of using copper would be prohibitive. For example, since copper is significantly heavier than aluminum, the insulators on power poles would need to be considerably stronger if copper wire was used instead of aluminum. Aluminum has a lower conductivity than copper, it oxidizes faster than copper, and, unless terminals are tightened properly, it is more likely to loosen in a connector. Furthermore, aluminum oxide is not readily visually distinguishable from aluminum. Since aluminum oxide is a good insulator, special care must be taken when terminating aluminum to ensure that the exposed aluminum is free of oxidation. If aluminum wire is used in a system, devices with terminations approved for use with aluminum must be used.

The intent of this section is to highlight several *NEC* requirements. The reader who goes beyond the walls of the classroom into the design of PV systems should invest in the latest edition of the *NEC* to be sure to have all the latest requirements on items such as ground fault detection and interruption, source and output circuit maximum voltage ratings, acceptable conductor placement, rapid disconnect methods and storage batteries.

4.2.1.2 Voltage Drop, Wire Sizing and Overcurrent Protection

Conductors for general wiring are typically characterized by the DC resistance per 1000 ft (304.8 m) and rated current-carrying capacity (ampacity) for copper or aluminum conductors with insulation rated at 60°C (140°F), 75°C (167°F) or 90°C (194°F). The ampacities are for no more than three current-carrying conductors in the conduit at a temperature of 30°C (86°F) or less. If more than three current-carrying conductors are in a conduit, or if the conductors are operated in an ambient that exceeds 30°C (86°F), their ampacity must be derated. Ampacities for single conductors in free air are higher than those in conduit, since heat is not trapped inside the conductor by crowding with other conductors or conduit. Information on resistance and ampacity can be found in *NEC* chapter 9, table 8 [3]. Information on derating for conduit fill is in *NEC* table 310.15(C)(1), and information on derating for ambient temperature can be found in *NEC* table 310.15(B)(1) [3].

The *NEC* requires that the total voltage drop in feeder and branch circuits be less than 5%, with the drop in either feeder or branch circuits limited to no more than 3% [3]. A feeder circuit is a circuit that provides power to an electrical distribution panel. In a common residential electrical service, the feeder circuit is the wiring between the electric meter and the circuit breaker panel to which branch circuits are connected. The branch circuits are the circuits that provide power to the individual electrical loads, such as lighting, refrigerators, dishwashers and air conditioners. The PV equivalence of branch circuits can be considered to be the PV source circuits that connect the PV array to the power conditioning equipment, and the PV equivalence of feeder circuits can be considered to be the PV inverter output circuits that connect to the utility. In any case, a good PV system design generally requires a voltage drop in any PV circuits to be less than 2%, with an overall voltage drop of less than 3%.

Many tables exist in various design manuals that list the maximum distance a certain size conductor can be run with a given current and still not produce excessive voltage drop. For the engineer with a calculator, however, all one needs to do is recognize that a circuit consists of wire in both directions so that a load located 50 ft (15.24 m) from a voltage source will need 100 ft (30.48 m) of wire to carry the current to and from the load. If d is the distance from source to load in ft and V_s is the source voltage, Ohm's Law can be used to calculate the percent voltage drop in the wire via

$$\%VD = 100\frac{I}{V_s}\left(\frac{\Omega}{kft}\right)\left(\frac{2d}{1000}\right) = \frac{0.2Id}{V_s}\left(\frac{\Omega}{kft}\right), \tag{4.1}$$

as long as the circuit is a DC circuit or a single-phase, two-wire AC circuit. Three-phase voltage drop calculations will be introduced in Section 4.9.

Example 4.2.1: A 20-watt, 12-VDC LED lamp is located 50 ft (15.24 m) from a 12-V battery. Specify the wire size needed to keep the voltage drop between the battery and the lamp under 2%.

Solution. First, solve (4.1) for (Ω/kft) to obtain

$$\Omega/kft = \frac{(\%VD)V_s}{0.2Id}.$$

Then, determine the load current from P = IV, assuming the load voltage to be essentially equal to the supply voltage. Substituting the known values on the right-hand side yields Ω/kft = (2×12)/(0.2×1.67×50) = 1.437. The *NEC* shows that #12 (3.31 mm²) wire has too much resistance, so it is necessary to use #10 (5.261 mm²). The actual voltage drop with #10 (5.261 mm²) wire can now be found from (4.1) to be 1.68%.

Note that although #10 wire will carry 40 amperes, the current in the circuit is limited to 1.67 amperes due to the voltage drop limitation. It is very important to be aware of the need for larger wires in low-voltage systems. Equation (4.1) will find considerable use in examples to follow in this and later chapters. Note also that if the wire resistivity is given in Ω/km, then Ω/kft = (Ω/km)/3.28.

Because of the problem with voltage drop at low voltages and the correspondingly larger wire sizes necessary, it is generally desirable to operate PV systems that deliver any significant amounts of power over any reasonable distances at voltages higher than 12 volts. If the previous 20-watt load was connected to a 24-volt system, the load current would be halved, and the load voltage would be doubled. This allows the resistance of the wiring to be four times higher, or as much as 5.648 Ω/kft, which means that #16 (1.31 mm²) wire is now adequate. From a total cost standpoint, however, the cost of the additional 12-V battery will exceed the difference in cost between #10 wire and #16 wire. Presumably, the cost of the PV modules will be the same since the amount of power required has not changed.

For concealed wiring, the minimum wire size is #14 (2.08 mm^2). Smaller wire sizes are normally used only for portable cords, for attaching single loads or for low-voltage control and sensor wiring. Once the final wire size has been determined, proper overcurrent protection will then likely be needed. Careful determination of necessary overcurrent protection will be included in the design examples that will follow.

4.2.1.3 Wire and Conduit Types, Applications and Deratings

The *NEC* is an important source of information for the PV design engineer, since it clearly defines acceptable PV system design practice. While articles mentioned previously deal mostly with PV systems, other articles such as Article 240 on overcurrent devices, Article 250 on grounding and Article 310 on conductor ampacities are also important to the PV system designer. Other parts of the *NEC* also apply to specific installations or installation methods. Table 4.2 summarizes the components

TABLE 4.2
Summary of Contents of *NEC* Articles Specifically Related to PV Systems

(Article) Part	Contents
(690) I	**General**: Scope, definitions, general requirements, ground-fault protection, AC modules
(690) II	**Circuit Requirements**: Maximum voltage, circuit sizing and current, overcurrent protection, arc-fault circuit protection, rapid shutdown
(690) III	**Disconnecting Means**: Disconnection of PV equipment, fuses, disconnect types, installation and service
(690) IV	**Wiring Methods**: Methods permitted, component interconnections, connectors, access to boxes, ungrounded PV systems, DC systems
(690) V	**Grounding and Bonding**: Grounding configurations, ground fault protection, location of connections, equipment grounding and bonding, grounding electrode system
(690) VI	**Marking**: Modules and AC modules, DC PV power source, point of common connection, PV systems with energy storage, identification of power sources, rapid shutdown switch
(690) VII	**Connection to Other Sources**: Refers to Articles 705 and 712
(690) VIII	**Energy Storage Systems**: Refers to Article 706, conditions for exception to charge controllers
(691)	**Large-Scale PV Electric Supply Systems**: > 5 MW inverter generating capacity not under utility control
(705) I	**General**: Scope, definitions, identification of power sources, supply-side and load-side connections
(705) II	**Microgrid Systems**: System operation, primary power source connections, microgrid interconnect devices
(706) I	**Energy Storage Systems**: General, definitions, system requirements, listing, maintenance, storage batteries, maximum voltage
(706) II	**Disconnecting Means**
(706) III	**Installation Requirements**
(706) IV	**Circuit Requirements**
(706) V	**Flow Battery Energy Storage Systems**
(706) VI	**Other Energy Storage Technologies**
(712)	**DC Microgrids**: General, definitions, circuit requirements, disconnecting means, wiring methods, marking, protection, systems over 1000 V

of *NEC* Article 690 and lists some of the other *NEC* articles that apply to PV system installations.

While many different types of insulation exist, as shown in *NEC* Article 310, type THWN-2 is a high-temperature (90°C = 194°F), moisture- and oil-resistant, thermoplastic insulation that is commonly in use for wiring protected by conduit. Type THHN insulation is almost always also rated as THWN-2. However, often the termination points for the wiring are only rated for 75°C (167°F). If this is the case, even if insulation may be rated for more than 75°C, the 75°C ampacity for the wire must be used to protect the terminals from overheating.

Electrical conduit is frequently used to protect wiring from physical damage. Numerous types of rigid and flexible conduit are identified in the *NEC*. Of all the types, rigid and flexible PVC and metallic conduit are commonly used for PV installations. When the wire is in conduit, it is protected, but it is also subjected to more heat buildup due to I^2R losses from multiple current-carrying wires next to each other. Thus, when more than three current-carrying wires are in a conduit and/or if the conduit is subjected to elevated temperatures, *NEC* Chapter 310 provides derating factors for each of these conditions. Note that equipment grounding conductors (EGC) are not counted as current-carrying, and in certain special situations where neutral conductors do not carry current comparable to the phase conductors, neutrals may be excluded from the current-carrying classification.

For example, for four to six current-carrying conductors in a conduit, the conductor ampacity must be derated to 80%. For seven to nine current-carrying conductors, the derating factor is 0.7, and for 10–20, the derating factor is 0.5. Thus, a #10 THWN-2 copper wire has a rating of 40 A for three or fewer wires in a conduit. However, with terminations rated at 75°C (167°F), the 75°C-rated ampacity of the wire (35 A) must be used as a starting point. Then, if the conduit contains eight conductors, not counting ground wires, the rating drops to $0.7 \times 35 = 24.5$ A.

For operation at elevated temperatures, an additional derating factor must be applied. Temperature derating factors are listed in *NEC* tables 310.15(B)(1) and (2). The factors depend on both the ambient temperature and the temperature rating of the insulation. For example, at an ambient temperature between 114°F and 122°F (46–50°C), the derating factor for type THWN-2 insulation is 0.82. The design examples that follow in this chapter as well as in Chapters 6 and 7 will introduce specific sections of the *NEC* that apply to the specific designs as the need arises.

4.2.2 IEEE STANDARD 1547–2018

4.2.2.1 Introduction

Prior to the adoption of IEEE Standard 1547–2003, grid-connected PV inverters were required to comply with IEEE Standard 929. IEEE Standard 929 was developed specifically to address concerns of utilities regarding the quality of power delivered to the grid and the need to disconnect the PV system from the utility grid in the event of utility power failure. IEEE Standard 1547 was developed "to provide a uniform standard for interconnection of distributed resources with electric power systems," by providing requirements for performance, operation, testing, safety considerations

Grid-Connected Utility Interactive PV Systems 91

and maintenance of the equipment. Thus, in addition to covering electronic inverters as used in PV systems, IEEE Standard 1547 covers distributed resources that use synchronous machines and induction machines, such as wind and low-head hydropower. If more than one type of distributed resource, up to an aggregate capacity of 10 MVA, is connected at a common point of utility connection, then all must meet the requirements of this standard at the connection point.

IEEE 1547–2018 [2, 11] is a consensus standard that was developed by more than 120 industry experts in a four-year effort, including a 389-member public ballot pool and more than 1500 comments. Approval required a 75% majority vote of the committee, and the actual vote came out as 93% approval. A summary of its achievements includes

- More coordinated operation under normal conditions
- Maintain grid safety
- Grid support under abnormal conditions
- New guidance for interoperability and open communications
- New guidance for intentional islands
- Striking a balance between the need for large and small installations

While the PV system designer only needs to specify that the inverter must be listed to UL 1741, which is based on IEEE 1547, to be assured that it will meet all utility interconnect requirements for power regulation, voltage regulation and current regulation within its controller, it is interesting to explore some of the requirements of IEEE 1547 to appreciate the amount of creative engineering that was involved in creating the standard and the subsequent amount of creative engineering that is necessary to meet the standard with an inverter design. In fact, the engineer who loves to speak in acronyms should probably obtain a copy of IEEE 1547–2018 to add another dozen or so acronyms to her vocabulary.

4.2.2.2 Specific Requirements

IEEE 1547–2018 is essentially a functional standard as opposed to a prescriptive standard. It specifies functions that must be performed and functions that "may" be performed by a distributed energy resource (DER) rather than specifying how the function must be implemented. It is up to the designer to figure out how to satisfy the functional requirements. The standard delineates the general requirements of voltage and power quality from a DER, voltage regulation, integration of grounding, synchronization with the grid, distributed secondary spot networks, disconnect from the grid when the grid is de-energized, monitoring, isolation, electromagnetic interference (EMI), withstanding surges, voltage rating of interconnecting device and response to abnormal grid conditions. Abnormal grid conditions include over voltage, under voltage, over frequency, under frequency and reconnection to the grid after the abnormal condition is cleared. The following discussion will apply to PV and/or energy storage inverters compliance with IEEE 1547–2018.

Previous versions of IEEE 1547 focused primarily on those grid conditions where the DER would be required to disconnect from the grid. IEEE 1547–2003 listed clearing times that decreased as the utility voltage decreased or increased. For example, if the utility voltage dropped to 50% or lower than its nominal value or if it exceeded 120%, the inverter needed to shut down within 0.16 seconds. Between 50% and 88%, two seconds were allowed and between 110% and 120%, one second was allowed. Between 88% and 110%, the inverter was allowed to operate. These percentages are applied to typical "normal" operating voltages of 120, 208, 240, 277, 480 or 600 V. The listed clearing times were maximum limits for inverters rated at less than 30 kW and were default clearing times for larger inverters [2].

As multi-megawatt PV installations became more and more common, it also became more and more evident that if the grid frequency were to decrease as a result of the grid approaching its peak load, the last thing these large systems should do is to shut down and make the situation worse. Thus, voltage and frequency ride-through provisions were introduced to require, or, in some cases, allow, the PV and/or energy storage system or any other DER to remain in operation while communicating with the grid to help stabilize the grid. In fact, it was also observed that thousands of smaller DER systems could also do their part in assisting with grid stabilization under peak loading conditions. As a result, even microinverters are now designed to comply with IEEE 1547–2018.

The opposite effect was also recognized. With large amounts of DER connected to the grid, it is also possible that available generation might exceed system load requirements. Alternatively, what if a big cloud all of a sudden causes a 50% reduction in the output of a multi-megawatt DER system? This could also lead to grid instability unless the grid is smart enough to either quickly curtail loads or quickly find replacement megawatts, either from alternate generation or from energy storage sources. In fact, IEEE 1547–2018 opens up the grid stabilization process for the use of energy storage systems to be controlled by the utility to either store excess grid energy or deliver energy to the grid, depending on the needs of the grid. It is interesting to note that electric vehicle chargers are included as possible grid stabilization sources, either by storing excess grid energy or by delivering available EV energy to the grid if the grid asks for it. IEEE 1547–2018 lists mandatory and optional ride-through requirements for three performance categories of abnormal grid conditions.

- Category I includes essential bulk system needs and is attainable by all state-of-the-art technologies.
- Category II includes full coordination with all bulk system power system stability/reliability needs.
- Category III includes all bulk system needs and distribution system reliability/power quality needs.

Then, there is the part about operating at 100% power factor, which was a historic requirement of IEEE 1547. Again, it soon became evident that PV systems with inverters that can communicate with the grid about the need for power factor correction could provide an additional stabilizing contribution to the grid. Operating the DER system at a leading power factor to compensate for a lagging power factor on the grid could help reduce line losses at the distribution and maybe even the transmission level, thus again improving grid efficiency and stability. Since the DER

Grid-Connected Utility Interactive PV Systems

inverters operate as current sources, synchronized by utility voltage, adjustment of the phase of the current waveform with respect to the voltage waveform is a straightforward electronic exercise.

If the grid drops out completely, for whatever reason, this is NOT a time to require DER sources to remain in operation unless they can be isolated from the grid. This isolation was initially applied to backup power systems in occupancies served by a single utility feed. The isolation requirement has now been expanded to allow continued grid operation in isolated "islands" that might be equivalent to neighborhoods. Once again, this requires precise control and communication links between the customer and the utility.

When a DER does shut down due to the loss of utility power, the standard has been for the DER to continuously monitor the utility voltage and frequency until they have been stable for five minutes. This requirement has now been modified to allow almost immediate return of DER power to the grid under the control of the utility as it sequentially brings power generation or storage sources back online after the reason for the loss of utility has been isolated or eliminated.

IEEE 1547 also defines necessary power quality requirements for PV inverters. The maximum DC component of an inverter output must be less than 0.5% of the rated inverter output current. The inverter must not cause objectionable flicker for other grid users, generally defined as not causing their lights to flicker. The final power quality measure is the accuracy of the sine wave output, defined in terms of the harmonic content of the waveform. Since excessive harmonics can cause damage to certain equipment, total harmonic distortion is limited to a maximum of 5%. In addition, amplitudes of odd harmonics below the 11th harmonic must not exceed 4% of the fundamental frequency, while odd harmonics above the 35th are limited to 0.3%. In other words, the higher the frequency, the more potentially damaging, so keep their amplitudes low. Even harmonics can be even more damaging and are thus limited to less than 25% of the odd harmonics in the same frequency ranges [2].

The final requirement of IEEE 1547 is that if an unintentional island should occur in which a collection of inverters is connected to the grid and grid power is lost, all of the inverters must be able to distinguish the grid from other inverters such that all inverters in the island shut down within two seconds of loss of grid power.

It is relatively straightforward to detect islanding, since PV inverters that are synchronized with the grid voltage act as nearly ideal current sources. Figure 4.1 shows how a typical inverter will generate a current waveform that is slightly off-grid frequency. As long as the grid voltage is present, the inverter adjusts the current waveform frequency at zero crossings of the grid voltage. If the grid is lost, then the inverter current drifts in frequency until it exceeds either the upper or the lower frequency limit. This algorithm, developed at Sandia National Laboratories, is called the Sandia frequency shift (SFS) [12]. A similar algorithm, the Sandia voltage shift (SVS) [13], is used to detect the output voltage of the inverter going out of range and shutting down the inverter until the grid is restored.

IEEE 1547–2018 not only mandates communication among grid interactive systems but specifies three protocols, one of which must be supported by compliant equipment. The protocols are IEEE Std 2030.5, IEEE Std 1815 and SunSpec Modbus. The information exchange among system components includes system

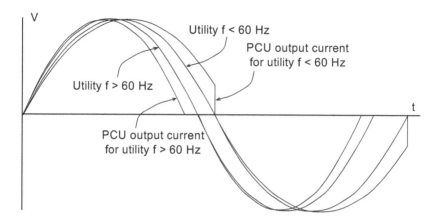

FIGURE 4.1 Comparison of utility voltage and Power Conditioning Unit (PCU) current for under-frequency and over-frequency utility voltage.

nameplate information, present operating conditions, present capacity and ability to perform functions and provisions for updates to functional and mode settings. Response times must be less than 30 seconds.

4.2.2.3 Comparison of PV Inverters to Mechanically Rotating Generators

When considering the connection of a PV source to the grid, it is important to distinguish between the electrical characteristics of an inverter and a conventional rotating generator. First of all, most utility interactive inverters are best modeled as dependent current sources, while rotating generators appear as voltage sources. In the event of a short-circuit fault, a rotating generator can deliver a very large current, limited only by the ability of the prime mover to keep the generator rotating. Any energy stored as rotational energy can be dissipated into a short circuit as electrical energy.

On the other hand, if a short circuit occurs at the output of an inverter, say, at the point of utility connection, a little more current than the full-load value will flow *from* the inverter. On the contrary, if the output circuitry of the inverter becomes a short circuit, then a high current can flow *to* the inverter from the utility. The inverter output circuit breaker is thus required to protect the inverter from the world, as opposed to protecting the world from the inverter. Proper overcurrent protection at the output of an inverter can minimize inverter damage in the event of a fault within the inverter itself.

Because the inverter acts as a current source, it is easier to ensure that the inverter will meet the standards for utility interconnection. The reason is that the utility is close to being an ideal voltage source. Hence, the inverter can sense the utility voltage and frequency and inject current only if the voltage and frequency fall within prescribed limits. This same circuitry can be used to ensure that the current is injected in phase with the utility voltage. This assures a high power factor for the inverter output, unless the utility asks for something different. The sensing circuitry has high-impedance inputs and can remain connected to the utility at all times to monitor the voltage and frequency stability of the utility.

It should be noted that the voltage values apply to the point of utility connection, also known as the point of common coupling (PCC) for the inverter. If the inverter is located some distance from the PCC, there may be a voltage drop on the line between the inverter and the PCC. If so, compensation can be made at the inverter output, since the inverter output voltage in these cases will be higher than the voltage at the PCC for the inverter to deliver power to the PCC.

4.2.2.4 Islanding Analysis

The greatest concern over islanding occurs when more than one inverter is in operation within the island. In this case, it is possible that they will support a feedback situation in which each inverter, sensing the combined output of the other inverters, thinks the output of the other inverters constitutes the grid. This possibility is enhanced under worst-case load conditions. It is thus useful to consider what might constitute worst-case load conditions.

Worst-case load conditions occur if, when the island is created by the utility fault, the island voltage does not change quickly to a steady-state fault value. This will occur under two conditions: resonance at the utility frequency and island motor loads with low damping, such as grinding wheels that continue to rotate even though power is removed from the motor.

If a circuit is at resonance with a relatively high Q, then it will continue to oscillate at its natural resonant frequency until the oscillation is ultimately damped out. As long as a motor continues to rotate, it will act as a generator and return its back emf to the grid. Except for those used in wind turbines, induction motors are normally inefficient generators, but synchronous motors are very efficient. The generation frequency, of course, depends on the rotation speed, so as the motor slows down, the generated frequency changes.

Circuit analysis textbooks generally prove that any parallel Resistance-Inductance-Capacitance (RLC) circuit will be underdamped if it has a Q > 0.5. Equation (4.2) represents the response of an underdamped parallel RLC circuit.

$$v(t) = V_m e^{-\alpha t} \cos(\omega_d t + \phi), \qquad (4.2)$$

where V_m is the initial amplitude, $\alpha = \frac{1}{2RC}$, $\omega_o = \frac{1}{\sqrt{LC}}$, $\omega_d = \sqrt{\omega_o^2 - \alpha^2}$ and ϕ is a phase angle that depends on initial energy storage in the inductor and capacitor. The Q of the circuit can be found from

$$Q = \frac{\omega_o}{2\alpha} = \omega_o RC \qquad (4.3)$$

It is interesting to use (4.2) and (4.3) to calculate how long it will take for the amplitude of the voltage to decrease to the inverter trip limit as a function of Q.

The two important parameters of the equation are the exponentially decaying amplitude part, $V_m e^{-\alpha t}$, and the natural frequency, ω_d. Assuming the energized grid has a natural resonant frequency outside the trip limits, the inverter will sense the frequency departure and disconnect within the prescribed time. If the natural resonant frequency is within the trip limits, then the inverter will need to disconnect when the voltage falls outside the trip limits. For the case of the motor/flywheel load, again, as soon as the motor speed drops enough to move the motor output frequency outside the trip limits, the inverter will trip. On the other hand, if the inverter keeps the motor running, it may never trip. Given that neither the inverter nor the motor will normally have any provisions for stabilizing the frequency or rotation, this occurrence is extremely unlikely.

It is useful to determine the number of cycles required for the amplitude of the island load voltage to decrease to 50% of its initial value. The results are obtained by setting $V_m e^{-\alpha t} = 0.5 V_m$. Solving for t yields the result

$$t = \frac{\ln 2}{\alpha}. \quad (4.4)$$

But it can be shown (see Problem 4.3) that $\alpha = \pi f_o/Q$. Thus, the time for the voltage amplitude to fall to half its starting value is given by

$$t = \frac{Q \ln 2}{\pi f_o}. \quad (4.5)$$

It also can be shown that (Problem 4.4) the relationship between resonant frequency, ω_o, and natural resonant frequency, ω_d, is

$$f_o = \frac{\omega_o}{2\pi} = \frac{\omega_d}{2\pi\sqrt{1-\frac{1}{4Q^2}}} = \frac{f_d}{\sqrt{1-\frac{1}{4Q^2}}}. \quad (4.6)$$

So, finally, solving for t in terms of the period of the natural frequency,

$$t = \frac{Q \ln 2}{\pi f_d}\sqrt{1-\frac{1}{4Q^2}} = \left[0.2206\sqrt{Q^2-\frac{1}{4}}\right] T_d. \quad (4.7)$$

Since T_d represents one cycle at the natural resonant frequency, the coefficient of T_d represents the number of cycles, N, that it takes for the voltage amplitude to decay to half its initial value. Table 4.3 tabulates the number of cycles needed for a signal to decay to half its original amplitude as a function of the Q of the circuit.

The previous analysis was based on the assumption that the utility island load had only initial stored energy. If energy continues to be added to the load in a synchronous manner, the load will continue to oscillate in a manner not very different

TABLE 4.3
The Number of Cycles for the Voltage to Reach Half the Original Amplitude in an Underdamped, Decaying, Parallel RLC Circuit as a Function of the Q of the Circuit

Q	1	2	3	4	5	6	7	8	9	10
N	0.19	0.43	0.65	0.88	1.10	1.32	1.54	1.76	1.98	2.20

from an electronic class C amplifier, depending on the conduction angle of the current source. The time to decay is thus prolonged, perhaps indefinitely, depending on the match between the load and the inverter output. Thus, this is the worst-case condition that must be overcome, and this is why the SVS and SFS, as previously discussed, were developed.

4.2.2.5 Grid Support

It is more common for the grid voltage and frequency to drop due to peak loading effects than for the grid to be lost completely. Under peak loading conditions, it makes little sense to drop out PV generation when the grid frequency or voltage drops. Disconnecting a large system, or even a large number of small systems, merely makes the situation worse. Modern systems, especially those with energy storage, are now being depended on to provide additional stability to the grid, either in the form of power factor correction or delivery of peaking power. As mentioned previously, IEEE 1547–2018 now includes provisions for utilities to control power factor and power output in agreement with the system owners. These provisions also include electric vehicles as sources of grid support.

4.2.3 OTHER ISSUES

4.2.3.1 Aesthetics

Although not a part of Article 690, *NEC* 110.12 requires that "electrical equipment shall be installed in a neat and workmanlike manner." This addresses the importance of a professional appearance for an installation and helps to remove any potential objections to the aesthetics of the installation. The aesthetics of a PV installation are of particular concern to architects and building owners who want to make the PV system look like it is an intended part of a structure as opposed to looking like an add-on. The PV community has responded by building integrated PV products such as shingles, tiles, windows and laminates, as well as framed modules with "more aesthetic appeal" than other framed modules, to address the wishes of the design community.

4.2.3.2 Electromagnetic Interference

The U.S. Federal Communications Commission requires that any electronic device with an electronic clock that runs at a frequency greater than 9 kHz must meet its standards for EMI [14]. EMI occurs in two forms—conducted and radiated. Conducted EMI is coupled directly through connections to the power line. Radiated EMI consists of radio frequency signals transmitted by the device directly to the surroundings. Both types of EMI, if sufficiently strong, can interfere with other electronic devices.

Since all modern utility interactive inverters are microprocessor controlled, all have internal clocks that can be expected to run at frequencies in the MHz range. Hence, all are subject to the radiative and conductive emission standards as set forth in FCC Part 15 of the Code of Federal Regulations. Each unit must have a label that indicates compliance. Conducted emission will normally not be a problem because of the low pass filters needed at the output of an inverter to average out the synthesized sinusoidal signal. High-frequency currents that might otherwise be conducted to the utility interconnection are thus suppressed by this filter.

Radiated emission is caused by clocks and other high-frequency currents circulating in sufficiently long conductors on the printed circuit boards of the inverter. The long, sometimes looped, conductors act as antennas and radiate the clock pulse signal. These same conductors act as receiving antennas for signals from other devices. Since the clock is not sinusoidal, it contains many harmonics that spread across the spectrum. An important part of the suppression of radiated emission is the inverter cover. Hence, it is important to avoid operating the inverter without its cover for the prevention of electrical shock but also for FCC compliance.

4.2.3.3 Structural Considerations

The primary consideration in the structural design of utility interactive photovoltaic systems is safety. This includes the safety of individuals as well as the protection of structures and other property that could be damaged as a result of mechanical failure. Building codes are adopted and enforced by local jurisdictions to help ensure a safe environment in and around buildings. A major difference exists between the installation of utility-interactive versus stand-alone photovoltaic systems in that codes are much more likely to be enforced for the former. This is primarily because of the need to electrically connect the interactive system to the grid, which can result in multiple inspections by (a) the electric utility service provider and (b) local code officials, which may include both electrical and building inspections. Stand-alone PV systems need only be inspected as required by local building officials.

As will be discussed in Chapter 5, large-area arrays may present a significant wind load on a structure. They are also subject to corrosion and degradation by ultraviolet sunlight components. Furthermore, they may be subjected to other conditions such as rain, snow, ice and earthquakes. While most arrays are tested for mechanical performance and while most array mounts have been pre-engineered to withstand the worst of loading conditions, it is still important for the engineer to determine that the specific array mounting is adequate with respect to existing structural parameters and local weather conditions. The American Society of Civil Engineers (ASCE) standard procedures and formulas for computing the various types of mechanical forces should be followed. In addition to the *NEC*, the PV engineer should keep abreast of the periodic updates to *Minimum Design Loads for Buildings and Other Structures (ASCE 7–16)* [10].

Of the various types of forces mentioned earlier, aerodynamic wind loading generally presents the most concern. At most locations around the world, the force effects due to wind loading are much higher than the other forces acting on the structure. For example, both dead- and live-weight loads added by a photovoltaic array to a building are usually less than 5 pounds per square foot (psf) (239 pa). In contrast, the computed wind loads are typically between 24 and 55 psf (1149 and 2633 pa) but sometimes greater than 80 psf (3830 pa), depending on location and the specific array mounting configuration. Earthquake zones also require special attention as discussed in *ASCE 7* [10].

Grid-Connected Utility Interactive PV Systems

In addition to the previous considerations, the design engineer must select and configure the array of mounting materials such that corrosion and ultraviolet degradation are minimized. Also, if the array is mounted on a building, the structural integrity of the building must not be degraded, and building penetrations must be properly sealed such that the building remains watertight over the life of the photovoltaic array. Structural considerations and calculations are covered in detail in Chapter 5.

4.2.3.4 Fire Protection and Other Codes and Rules

Two other important design considerations include fire protection rules and flood or high-water zone requirements for equipment locations. When plans are submitted to building departments for approval, in addition to electrical and structural review, they are generally sent for fire review and, in the case of flood zones, review of proposed locations of equipment. If the job will be near a corrosive environment, such as an ocean, special corrosion-proof enclosures will be required.

Some states have laws that prohibit local ordinances from restricting solar access, and, in other cases, it is necessary to prove the solar system is not in view from street level. In any case, the local building department, regardless of location, is generally the best resource for all things that need to be considered before an installation can be permitted.

4.3 DESIGN CONSIDERATIONS FOR STRAIGHT GRID-CONNECTED PV SYSTEMS

4.3.1 DETERMINING SYSTEM ENERGY OUTPUT

Depending on the technology, PV modules in common use may have sunlight-to-electrical conversion efficiencies ranging from 6% to over 20%. Assuming 20% overall module efficiency, this would mean that, for irradiance of 1 kW/m^2, each square meter of PV area would generate 200 W if the modules are operating under standard test conditions. Hence, a 10 kW array would require 50 m^2 (538 ft^2) of unobstructed roof space, assuming the array to be roof mounted. For most dwellings, this area comes close to being the maximum area that may be south facing, especially if room is provided for access to the modules for maintenance, to comply with fire code(s) and to minimize wind loading. Hence, 10 kW is a reasonably practical average size for residential PV systems, and systems under 10 kW are generally considered to be small systems.

The next question is how many kWh can the system be expected to generate? The answer depends on the location of the system and on the overall conversion efficiency from DC array output to AC inverter output. While some areas can generate upward of 1500 kWh/kW/yr, other areas or orientations may be in the range of 1000 kWh/kW/yr. A 10-kW PV array can thus be expected to generate somewhere between 10,000 and 15,000 kWh/yr, depending on location and orientation. Figure 4.2 shows loss mechanisms that occur between the generation of electrical power by the individual modules and the final delivery of power to the utility grid.

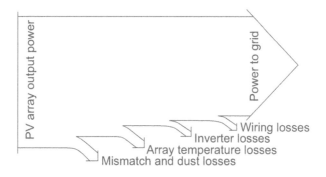

FIGURE 4.2 Loss mechanisms between module output and inverter output.

Historically, it was difficult for manufacturers to produce identical PV modules, even of the same model number. It was not uncommon for modules of the same model number to have a tolerance in rated output power of ±10% or more. Currently, however, improvements in manufacturing quality control, combined with I–V testing of each individual module, enable manufacturers to market modules that are guaranteed to produce at least their rated power and possibly up to 5 W in excess of rated power. Any module testing at 5 W more than rated power gets sorted out and marketed with a 5 W higher rated power. As a result, at this point in history, minimal array degradation results from module mismatch, whereas, up to about 2008, it was prudent to include a 10% mismatch factor when estimating array performance.

Module output does degrade as modules age, but manufacturers generally warrant annual power decreases to be less than 0.5%. Thus, a 25-year-old module should still produce at least 87.5% of its original rating. Other than mismatch and aging, the only other significant degradation factor for most modules is dust or dirt on the modules, which can be minimized by cleaning the modules as necessary.

A number of computer programs are available for calculating hourly, daily, monthly and annual PV system performance. Some are more detailed than others, such as including more worldwide weather and sunlight data, and have a price tag. Others are free (i.e., the government already paid for them). Perhaps the most popular free program for PV system performance analysis is the System Advisor Model (SAM), which can be downloaded from the U.S. National Renewable Energy Laboratory (NREL) [15]. By inputting location and array orientation information, along with specific system component information (module models and inverter models), estimated monthly and annual system performance and savings can be obtained. A simplified version of this simulation program is based on the PVWatts model that does not incorporate the performance specifications of modules and inverters.

Since the average residential electrical consumption is in the 1000 kWh/mo range, depending on the location once again, it appears that installation of a 10-kW PV system on every rooftop can produce sufficient kWh to exceed the demands of the dwellings upon which the PV systems are installed. Of course, the PV systems will be generating electrical energy only while the sun is shining, so if all electricity used is to be PV generated, obviously massive storage capability will be required. One way to reduce the storage needs is to reduce nighttime electrical consumption. This, of course, is contrary to most current utility marketing schemes, since it is economically beneficial to them to operate existing large nuclear and some of the remaining fossil generation at nearly constant capacity all the time. This requires nighttime electrical use to distribute their load more evenly.

One major advantage of distributed electrical generation on rooftops is the elimination of power plant siting concerns. Most consumers tend to be happy to install PV systems on their roofs, whether they own the system themselves or whether the system is owned by someone else. Customer surveys consistently have shown customer support for cleaner generation, even if it has a premium price. Now that, in most instances, PV-generated electricity carries a price tag below utility rates, customer interest is no longer limited to those who are willing to pay extra for their systems. Furthermore, if the choice is between a PV system on their roof versus a coal or nuclear plant nearby, the PV system becomes an even more attractive choice for most consumers. For those who perceive PV systems as aesthetically unappealing, it must be remembered that beauty is in the eye of the beholder.

4.3.2 Array Selection and Installation

The preferred orientation of a PV array will normally be south facing (in the northern hemisphere), mounted in a fixed position and tilted at 90% of latitude. This positioning will usually result in maximum annual kWh generation by the array. However, generally little is lost if the array is within 20° of south with a tilt within ± 15° of latitude. In fact, since about 2020, in many areas, parts of an array facing in any other direction than south will still generate enough annual energy to justify the installation. A check with NREL SAM will verify this statement for any specific location. For a rooftop array on a sloped roof, it is now unusual to use any form of mounting other than stand-off mounting that simply attaches the array parallel to the roof plane with about 5 to 6 inches of space below the array to allow for better cooling and thus better performance of the array. Specific modules used in the array will likely be chosen from a wide selection, but certain features will need to be matched to the particular installation.

Availability—While a specific module may be very popular, it might also be difficult to obtain.
Compliance with Standards—International standards for modules include IEC 61215 and IEC 61646.
Cost—While cost is important, it should not be the most important consideration.
Dimensions—Roofs are generally everything but perfect rectangles. The dimensions of a module, everything else being equal, may determine how many might fit on the preferred roof planes.
Efficiency—The higher the efficiency, the more power that can be installed in the same space. Efficiency may depend upon the cell materials used in the module fabrication.

Location of Manufacture—Depending on where and how manufactured, some modules may have larger carbon footprints or other environmental or social issues associated with their manufacture. These issues are generally not listed on a module data sheet.

Warranty—Most warranties are 25 years, but some are better than others.

Wind—The uplift and downward forces that a module can tolerate are regionally specific. International standards provide minimum requirements. Regional requirements may be more strict, and, in some cases, module uplift ratings are based on the use of specific module racking systems and module orientations.

4.3.3 Inverter Selection

When selecting an inverter, it is important to remember to specify a utility-interactive unit, since there are a wide variety of units available. Some are straight utility interactive, some can be used as utility interactive as well as stand-alone and some are intended for stand-alone use only. While a wide range of performance specifications may be available for any unit, it is important to identify those parameters that are most significant. Features to consider may include any or all of the following:

IEEE 1547 and UL 1741—If the inverter has a UL 1741 listing [9], then it has been tested to IEEE Standard 1547 and consequently complies with *NEC* requirements. For small systems, all voltage and frequency trip points are factory preset and are not field adjustable.

Power rating—Maximum inverter output power is normally specified at an ambient temperature of 25°C (77°F). Surge power rating is less important for a utility interactive inverter, since any surges encountered in the load can generally be met by the utility. For a stand-alone inverter, surge capability is more important. Utility interactive units with battery backup act as stand-alone units if the utility disconnects, so, in this case, the surge capacity becomes relevant.

Peak efficiency—It is important to note the input power over which the stated peak efficiency is obtained. Since the utility-interactive inverter always is loaded by the utility connection, if it is designed with maximum power point tracking, it will deliver maximum power to the grid over a wide range of PV input power and will operate close to peak efficiency over most of the maximum power point tracking range. Modern straight grid-connected inverters should have peak efficiencies in excess of 98% over most of the operating range.

Maximum power point tracking (MPPT) range—Look for the widest possible range of input voltages over which the unit is capable of tracking maximum power for the greatest flexibility in PV array design. Some inverters have multiple MPPT inputs.

No-load power consumption—No-load power consumption should be very low, although normally a straight grid-connected inverter will operate at maximum power output during the day since the grid always provides a load for all available PV power, but, during nighttime hours, a straight grid-connected inverter should consume only a small amount of power to keep its control circuitry energized.

Nighttime losses—If the unit operates on power supplied by the PV system, then there will be no nighttime losses. If the unit has a stand-alone mode of operation, then one can normally expect it to have a "sleep" mode that consumes very little power if no loads are connected. Straight grid-connected units with power ratings below 10 kW will generally have power losses of a few watts.

Transformerless—Nearly all modern inverters now employ transformerless circuitry. This decreases the price, decreases the weight and the size, increases the efficiency and generally increases overall performance, including the elimination of noise that might be present in a unit with a transformer.

Warranty—Many quality inverters carry a ten-year warranty, sometimes extendable. Check what the warranty covers. In fact, many microinverters now carry a 25-year warranty.

Code compliance—The UL 1741 listing indicates compliance with IEEE 1547. It is possible to purchase inverters that incorporate additional *NEC* compliance components, such as ground fault protection, arc-fault protection, input and output disconnects and fusible combiners for multiple string inputs, so strings will not need to be fused in a remote or difficult-to-access location and weatherproof packaging.

Data monitoring—Most grid-connected inverters provide optional data monitoring capability by digitizing system parameters and making them available on an output bus. Some monitoring is available at the inverter, and some units provide for remote data logging via telephone dial-up, internet or other communication means. Data monitoring will be at the inverter level for some systems and at the module level for other systems.

Rapid shutdown—Since 2014, the *NEC* has required that PV systems mounted on buildings have a means of rapid shutdown, which may be incorporated into inverter design, especially the requirement for quick discharge of capacitors. The rapid shutdown system must shut down every system component that is located more than 1 ft from the edge of the array in all directions.

Although the preceding properties tend to be thought of in terms of relatively high-power string inverters, perhaps the most commonly employed inverters for residential installations are microinverters, for reasons to be enumerated in several of the design exercises later in this and following chapters. Individual microinverters generally have maximum output power ratings in the range of 240–500 watts but incorporate all the features of larger string inverters. DC input voltage and current ranges are generally designed to accommodate as many different modules as possible.

4.3.4 Other Installation Considerations

Prior to commencing the installation, since the system will very likely be inspected by both the utility and the local municipality, it is important to first complete all the necessary permitting paperwork. This may offer an excellent opportunity for the engineer to perform an educational public service by answering any questions either authority may have about the installation.

Persons not familiar with IEEE 1547, UL 1741 and Articles 690 and 705 of the *NEC* may be reluctant to accept the straightforward installation allowed by these codes and standards. It is possible that the most time-consuming part of the installation may be convincing the inspectors that the installation will be safe, competent and code compliant. Fortunately, electrical inspectors tend to be well-versed in the *NEC* and are generally interested in new technology.

It is also conceivable that a zoning restriction on the installation of rooftop solar systems on the street side of a building may exist. The permitting process may then involve a trip or two to zoning boards or town council meetings, perhaps with an attorney or a delegation of solar enthusiasts.

Certain locations may require special equipment enclosures, such as when enclosures are exposed to salt air. Another location-specific concern is flooding and the minimum height above ground level necessary for equipment mounting. For systems with energy storage, each system may have different fire restrictions, such as not mounting it on the opposite side of a bedroom wall. Spacing between equipment enclosures and height above ground level may also be restricted. Reading the installation manuals carefully is very important.

4.4 UTILITY INTERCONNECTION OPTIONS

4.4.1 Supply-Side (Line Side) Connections

The fundamental requirement of the connection of a PV system to the utility is to avoid overloading any wiring, busbars or overcurrent protection devices. This is where *NEC* 705.11 and 705.12 enter the picture.

The *NEC* allows for a PV system to be connected on either the line side or the load side of the system's main circuit breaker. Note that the line side does NOT mean on the line side of the revenue meter. Note also that, in the 2020 *NEC*, the line side connection (line side tap, supply-side tap) appears under section 705.11. The line side tap generally follows the tap rules of *NEC* 240.21(B). The basic requirement is that the amount of PV current available to the tap should not exceed the ampacity of the feeder conductor. It is important to recognize that the PV tap is a *source* tap that adds electric power to the system, while a traditional tap is a load tap that uses power from the system.

Connecting on the line side of the main circuit breaker means the main circuit breaker limits the load side current, regardless of whether the origin of the current is the PV system, the utility or a combination of the two. A line side tap is a convenient way to interconnect nearly any size PV system to the utility, provided that the line side of the main circuit breaker is accessible and the available PV current does not exceed the size of the utility feeder. This is NOT the case in a number of different combination load centers where the line side of the main circuit breaker is not accessible, so other means of achieving the interconnect must be explored. One interesting electrical service configuration involves a meter enclosure that has no main circuit breaker and insufficient space to add a supply-side tap within the meter enclosure. At least two possibilities exist for overcoming this roadblock.

The first step is to determine whether a service upgrade is needed. Sometimes the meter is fed from an overhead utility feed with undersized feeders and undersized riser conduit. In this case, the meter enclosure and riser conduit may need to be replaced along with the feeder conductors. When this is done, it will also likely be desirable to incorporate a main disconnect as a part of the outdoor service equipment.

In this case, it is often easier to replace the meter enclosure with a meter center that includes a main circuit breaker and a busbar with spaces for branch circuit breakers and feedthrough lugs to feed an existing indoor service panel. If a line side tap is the objective for the PV interconnect, then it is important to be sure that the equipment choice includes feeders between the meter and the main circuit breaker that are suitable for tap connections. One way to ensure this is to use separate enclosures for the meter and the main disconnect.

If it is not necessary to replace the meter enclosure, another option is to add a junction box between the meter enclosure and the existing Main Distribution Panel (MDP). This will generally allow the existing feeders to reach the junction box, but a new set of feeders will be needed from the existing meter enclosure to the junction box, where the existing feeder, PV and new feeder are connected.

Two important requirements for a line side tap are that the wiring between the PV system output and the tap must be in conduit and there must be a fused, load break, service-rated disconnect switch, preferably within 10 ft (3.05 m) of the tap connection. The purpose of this disconnect is to provide overcurrent protection for that section of the feeder between the tap and the disconnect, and, possibly, as an AC disconnect for the PV system output power. In some cases, this disconnect will also serve as a rapid shutdown switch for the PV array.

In addition to the line side tap, if there is no main circuit breaker, but, instead, a busbar connected directly to the utility meter with up to six positions for circuit breakers, it may be possible to use one of these positions for a *line side connection*. All this means is that the PV system is connected to the main busbar via a backfed circuit breaker, sized at 125% (or the next larger size) of the rated maximum PV system AC output current. In effect, this does nothing to change the load on the system. It simply adds one more source of power to assist the utility in meeting the needs of the system loads.

The largest PV source that can be connected is determined by the PV source not being able to overload the busbar or the utility feeder if the total output of the PV system is fed back to the utility. No AC disconnect is needed for a line side connection, since the backfed circuit breaker serves as the disconnect. In some cases, however, the local Authority Having Jurisdiction (AHJ) or utility may require an unfused lockable disconnect on the load side of the line side connection, especially if the PV system is in utility Class II (AC power rating per utility formula is > 10 kW).

4.4.2 Load Side Connections

NEC 705.12 describes a number of methods of connecting on the load side of the main circuit breaker without overloading any wires, busbars or overcurrent protection devices on the load side of the main circuit breaker.

Since essentially all PV inverters electronically limit their output currents to maximum rated values, the rated output current of an inverter is all that one can expect to get from that inverter under any operating conditions, since the inverter is designed with internal current limiting. This is consistent with the fact that the input to the inverter from PV modules is also currently limited by the amount of sunlight that shines on the modules. In fact, due to sunlight and temperature effects, a module rarely produces its rated maximum output current. What this means is that if the output of an inverter sees either an open circuit or a short circuit, it will not trip the overcurrent device at its output. Instead, it will think the utility is gone and it will shut down instead and wait five minutes for the utility to return.

Thus, for wire sizing and overcurrent protection, 125% of the rated inverter output current is used. The next highest overcurrent protection device is then used to protect the wiring from any fault current that may flow from the utility side of the connection.

Various load side connections include

1. Feeder tap connections per 705.12(B)(1–2)
2. Busbar connections per 705.12(B)(3)(1–6)

Thus, all one needs to do is to figure out how to use each of these provisions and when one may be preferred over another when more than one may be possible.

Two of the most popular load side connections are NEC 705.12(B)(3)(1), the 100% rule, and NEC 705.12(B)(3)(2), the 120% rule. Both use backfed circuit breakers on the busbar of a distribution panel. For use of the 100% rule, the sum of 125% of the PV current plus the size of the main circuit breaker for the busbar must be less than the rating of the busbar. For example, if the maximum PV output current is 24 A, then, 125% of 24 A is 30 A. If the busbar is rated at 200 A and it is protected by a 150 A circuit breaker, then 150 A + 30 A = 180 A, which is less than the 200 A rating of the busbar. This connection thus qualifies as a legitimate connection of PV to the utility under the 100% rule. Since the maximum current on the busbar is limited to 180 A at any point on the busbar, the PV connection can be made at any available circuit breaker space on the busbar.

The 120% rule states that *if the backfed PV source circuit breaker is located on the busbar at the end opposite the utility connection*, the sum of the rating of the main circuit breaker plus 125% of the maximum PV output current must not exceed 120% of the busbar rating. Thus, for the previous example, suppose the main circuit breaker size is 200 A instead of 150 A with everything else the same. In this case, 125% of the PV output current (30 A) plus the size of the main circuit breaker (200 A) is 230 A. This is less than 120% of the busbar rating (240 A) and is also a code-compliant installation.

But doesn't this last interconnection constitute a potential 15% overload of the main busbar? Not really. The reason is that the current from the utility flows downward on the busbar, diverting to loads as they may require it, such that the total current on the busbar from the utility decreases as it flows down the busbar toward the bottom. On the other hand, the current from the PV source flows up the busbar,

Grid-Connected Utility Interactive PV Systems

diverting to loads as they may require it, such that two possible situations may occur while the PV system is producing power.

The first scenario involves all of the PV power going to loads on the busbar and current from the utility supplying only any additional current that may be needed by the loads. In this case, the PV current offsets the need for the same amount of utility current, thus resulting in less current entering the top of the busbar from the utility than might be needed at night under the same load conditions without help from the PV.

The second scenario involves the PV generation to exceed the load requirements of the occupancy. In this case, since some of the PV current is not needed by loads connected to the busbar, this remaining PV current has nowhere to go, except back to the utility for someone else to use. In this case, the busbar current will be 20% or smaller than the busbar rating. This should make electrical engineers very happy, since this is a classic example of the application of Kirchhoff's current law.

The 100% and 120% rules along with other load side rules as well as line side connection rules will be applied to the design examples that follow in this and the following chapters. The various rules of interconnection are well worth the time spent by PV engineers in learning them, since these rules account for a measurable percentage of plans rejected by permitting authorities, and it is then up to the design engineer to defend the interconnection choice. In doing so, it is particularly important that the engineer is correct.

4.5 DESIGN OF A SYSTEM BASED ON DESIRED ANNUAL SYSTEM PERFORMANCE USING MICROINVERTERS

4.5.1 INTRODUCTION

PV system design may be targeted at the achievement of a variety of objectives. In this example, it will be assumed that the objective is to provide a specified percentage of the electrical consumption of an occupancy. In this case, the building will be a residence in Oklahoma City, OK, USA, that has an average monthly electrical consumption of 1000 kWh. The design goal is to provide 99% of the annual electrical consumption with a grid-connected PV system, which amounts to an average of 990 kWh/mo. Since it is desired to provide 99% of the annual consumption, this amounts to essentially 12,000 kWh/yr.

The dwelling has 770 ft^2 (71.5 m^2) of unobstructed, unshaded, south-facing roof that has a 5:12 slope. In other words, for every 12 horizontal units, the roof rises five vertical units. This means the roof slope is $\tan^{-1}(5/12) = 22.6°$.

While a completely formal analysis might involve estimating all the factors that affect DC–AC conversion efficiency, along with looking up the peak sun hours for each month for Oklahoma City, for this example, NREL SAM [15] will be used instead. By opening up SAM, choosing the PVWatts with no financials option and selecting Oklahoma City as the location, with a south-facing roof with a 22.6° slope, it is possible to iterate array sizes to determine the approximate array size needed to produce the annual 12,000 kWh.

4.5.2 Array Sizing

One easy way to proceed is to start with an array size of 1 kW, using a DC:AC ratio of 1.4 and assuming overall system losses of 12%. If this is done, the annual AC kWh production is found to be 1578 kWh. Thus, to generate 12,000 kWh annually, it would appear the array size will need to be 12,000 ÷ 1578 = 7.604563 kW. Just to confirm this, entering 7.604 kW as the array size yields 11,998 kWh/yr.

Before moving on to select an inverter and modules, the reader should experiment with the entries in the SAM dialog boxes to explore the variations in monthly and annual system output for various values of array tilt and array azimuth. The first thing to notice is that up to 20° either side of the south does not make much difference in annual system performance. If the array is west-facing, then it can be expected to produce approximately 10,342 kWh/yr, which means the array size would need to be increased to (12,000 ÷ 10,342)×7.604 kW = 8.823 kW to produce 12,000 kWh/yr.

It is also interesting to note that any tilt within about ±15° of latitude (35.4°) also does not make much difference in annual results. At a tilt of 35.4°, annual performance is estimated to be 12,040 kWh, a very slight increase. Thus, mounting parallel to the roof is clearly the best option from structural, cost and aesthetic perspectives. Since the system is grid-connected, normally there will not be concern over monthly performance, as long as the desired annual performance is achieved.

Another very interesting simulation to perform with this array is to look at the hourly data over the year. This is easily done by selecting *time series* to obtain simulations of array output each of the 8760 hours of the year. When *time series* is selected, numerous parameters can be displayed, one or two at a time, as indicated on the right-hand side of the screen. For this example, choose AC *inverter power* at the bottom of the list and then zoom in on the highest power peak shown over the year. Note that peak power at any hour never exceeds 5.5 kW. On a few occasions, the array power levels off at 5.5 kW for a short time. This indicates the inverters reaching their maximum output power. Projection of the power beyond the leveling point shows only a very small loss of energy during this short period.

One way to do this more precisely for the year is to simulate again by simply changing the DC:AC ratio to 1.0 and noting the difference in annual kWh production. This number is important when selecting an inverter for the system. In fact, returning to the SAM System Design page, under System Parameters, the program shows a rated inverter size of 5.43 kW, which is consistent with the simulation results and the assumed DC:AC ratio of 1.4.

It should be remembered that the data used in SAM simulations is based on a typical meteorological year (TMY), which means that the actual system output on any single day will probably not be as predicted by the model, since the model is based on months of past years. But on a monthly basis, the prediction for monthly output is pretty good, and the prediction for yearly output is even better, since it essentially averages all the daily/hourly data over an entire year.

4.5.3 Module Selection

The challenge at this point is to find a module such that enough of the selected modules will fit on the roof to achieve the design. This process takes note of the selection

Grid-Connected Utility Interactive PV Systems

criteria previously listed. Starting with 7604 W as the array size, one method is to divide 7604 by various module sizes to find a reasonable module power. Since 400 W modules are relatively common, 7604 ÷ 400 = 19.01 modules. This is a pretty good start. Just to be sure about the sizing, using 405 W modules will result in a 7695 W array for modules of the same physical size. The module dimensions are 67.8 × 44.65 × 1.18 in (1722 × 1134 × 30 mm).

Considering that the spacing between modules to allow for clamps will be about 0.5 in (12.7 mm), the width of a row of 19 modules mounted in a portrait position (the long side up and down) will be very close to 19 × (44.65 + 0.5) = 857.85 in = 71.49 ft (21.79 m). The total array area will be 71.49 ft × (67.8 in) ÷ 12 = 404 ft^2 (37.53 m^2), which is less than 770 ft^2. However, since the hip roof is trapezoidal, before making a final decision on the modules, it is probably a good idea to check the fit. Figure 4.3 shows one possible array layout that will most likely be suitable for most roof surfaces. Note that this is only slightly more than 20% of the plan view area of the entire roof (60 ft) × (33 ft) = 1980 ft^2 (183.9 m^2) needed to produce a year's supply of electricity.

4.5.4 Inverter Selection and Connection

Since a DC:AC ratio of 1.4 has been used in the simulation, then to achieve the design performance, an inverter rated at 405/1.4 = 289 W will suffice. Interestingly enough, one particular microinverter is rated at 290 VA, 240 V and 1.21 A maximum output. This microinverter complies with all the requirements of UL 1741, including all of the recent IEEE 1547 additions. Since it mounts beneath the module, when the grid is lost or when the system is disconnected from the grid, there is no longer a synchronizing signal for the unit and it shuts down. Thus, the entire system is shut down within the array boundaries, satisfying the system's rapid

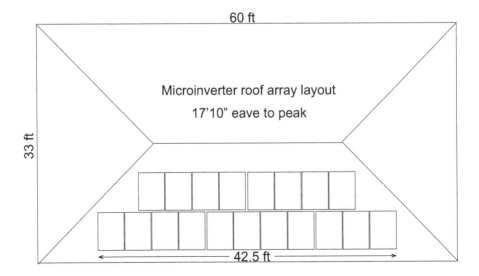

FIGURE 4.3 Module layout on the south-facing roof for the microinverter system.

shutdown requirement. It has a 25-year warranty and can easily be incorporated into a system with energy storage at a later time. In addition, each unit has its own MPPT, GFDI and AFCI circuitry. As long as it is available, it appears to be a good choice.

The DC input of the microinverter is limited to modules that have I_{SC} less than 15 A and V_{OC} less than 60 V. The selected module has V_{OC} = 37.3 V and I_{SC} = 13.77 A, thus satisfying the needs of the microinverter. DC connections are simple snap-together connectors that meet the *NEC* requirements for DC disconnects.

Wiring among the modules is done with the listed cabling available from the microinverter manufacturer. The microinverter output is connected in parallel with the outputs of other microinverters in a string and the string is terminated at a backfed 2P20A circuit breaker at a combiner box that also houses system instrumentation. Individual microinverter connections to the string are also by means of listed snap-together connectors, which serve as AC disconnects for the individual microinverters.

Since the maximum continuous current fed to a 20 A circuit breaker is limited to 16 A (80%), this means that up to 13 microinverters of this model can be connected in a single string. Since this system has 19 microinverters, it will need to be broken down into at least two strings that will be fed in conduit from a rooftop junction box to a combiner box near the point of the utility connection. A #10 copper EGC is connected to the module frames and racking to ensure no live parts are exposed on the roof. The wiring from the rooftop junction box to the combiner box is generally in conduit and is often increased from #12 to #10 as a result of deratings for more than three current-carrying conductors and perhaps also an ambient temperature derate. Generally, the length of this conduit is short enough to minimize voltage drop in this wire run.

4.5.5 Utility Interconnection

A few calculations are helpful to determine the best method of interconnection with the utility. Step one is to determine the maximum inverter output current. In this case, it will be 19×1.21 A = 22.99 A. Then calculate 125% of this current to calculate wire sizes, overcurrent protection sizes and, finally, which interconnect method will be the best. Since 125% of 22.99 A is 28.74 A, #10 copper will be the minimum size wire and the interconnect overcurrent protection will need to be at least 30 A. From earlier examples, it is known that 28.74 A is acceptable for a 100% rule interconnect as long as the sum of 28.74 A plus the size of the main circuit breaker is less than the rating of the distribution panel busbar.

Otherwise, if the PV output is connected at the opposite end of the distribution panel busbar, then the sum of 28.74 A plus the size of the main circuit breaker needs to be less than 120% of the busbar rating. This connection would be acceptable for a 200 A service with a 200 A busbar. Either of these situations could be likely for the occupancy used in this example.

On the other hand, if there are no spaces for circuit breakers left in the distribution panel, then a line side connection may be necessary. For simplicity, since this is the first serious design example, it will be assumed that the occupancy electrical service consists of a 200 A main circuit breaker and a 200 A main distribution panel. The 120% rule will be followed by installing a 2P30A circuit breaker at the end of the distribution panel opposite the main circuit breaker. In some cases, it will be necessary to move an existing circuit breaker or two to provide the needed space at the bottom of the busbar.

Wiring from the combiner box to the backfed circuit breaker in the main distribution panel can often simply use the circuit breaker as a combination AC Disconnect and Rapid Shutdown device. If this circuit breaker is outside, it is compliant with NEC requirements for accessibility. In some cases, however, an additional unfused, lockable disconnect will be required by either the local permitting authority or the local utility. In this case, if required, it would be a 2P30A load break unit. Load break means it can open the circuit while current is flowing through it.

This completes the design of this system. Figure 4.4 shows a single-line schematic diagram of the overall system. Note that single-line drawings (SLD) are commonly used to simplify the layout of a system. The numbered ellipses represent conduits, the details of which can then be listed in a table. Table 4.4 shows the details of the wiring in the conduits. Note that neutral and EGCs are listed as conduit contents wherever it is appropriate. In the case of conduit #4, the neutral and EGC are bonded together in the MDP. This is the only connection between the neutral and the EGC in the system, and, since this is a single point ground, the neutral and EGC are not effectively in parallel, and, as a result, current will flow on the (grounded) neutral but not on the EGC unless a ground fault occurs.

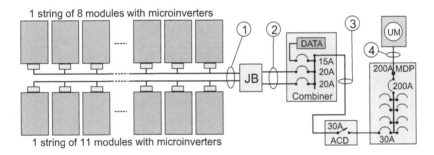

FIGURE 4.4 Single line drawing of a microinverter system.

TABLE 4.4
Summary of Wire and Conduit Sizes for the 7695-W Microinverter System

Conduit #	Description	Conduit Size
1	Manufacturer listed #12 open wiring + #10 EGC	Free air
2	(4) #10 THWN-2 (L1, L2) + #10 EGC	¾"
3	(3) #10 THWN-2 (L1, L2, N) + #10 EGC	¾"
4	(3) 2/0 THWN-2 (L1, L2, N) (No EGC)	1½"

4.6 DESIGN OF AN OPTIMIZER SYSTEM BASED UPON AVAILABLE ROOF SPACE

4.6.1 Introduction

A common challenge in PV system design is to fit as much power onto the roof as possible. With a roof that is "just the right size and just the right shape," that is, rectangular, this may not seem to be much of a challenge. But, in many areas of the country, finding a south-facing, rectangular roof, is like finding a needle in a haystack. Odd roof shapes, shading and roof protrusions all influence the placement of PV modules. In this example, the roof of Figure 4.5 will be used.

Roof diagrams may come in two forms. One form is a diagram showing the actual measurements of the roof. The other form is a plan view, where the projection of the roof onto the horizontal plane is shown. The plan view was used in the previous microinverter system example. Plan views, believe it or not, are what are found on sets of plans for construction. The plan view is also what is seen in satellite photos, such as on Google Earth. Plan sets (but not Google Earth) also generally indicate the slope of the roof in terms of rise:run ratios, that is, the ratio of vertical distance to horizontal distance. The slope of the roof is then simply

$$\theta = \tan^{-1}(y/x), \qquad (4.8)$$

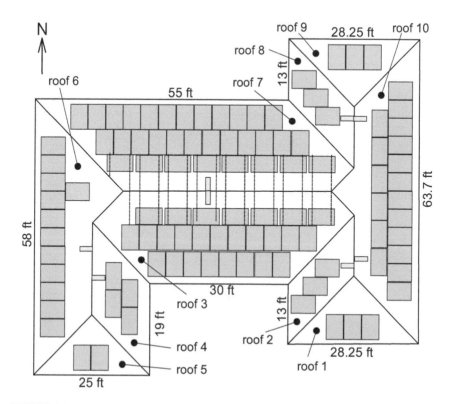

FIGURE 4.5 Array layout for optimizer design example.

Grid-Connected Utility Interactive PV Systems

where y is the vertical rise and x is the horizontal run of a right triangle. The horizontal dimension, x, is the projection of the actual roof dimension, r, in the direction of the slope. The actual dimension in the direction of the slope is the hypotenuse of the right triangle formed by x, y and r. Thus, if x is known from the plan view and the ratio y:x is known, then r can be found from

$$r = x/\cos \theta. \qquad (4.9)$$

While (4.8) and (4.9) may not be exactly rocket science, if the slope is not taken into account in the plan view by the designer, then the installer may end up with some surprises when laying out the array on the actual roof. The roof dimension perpendicular to the slope direction on the plan view is the same as the actual dimension. All that the designer needs to do is to first recognize which view is being used. If the plan view is being used, then it is convenient to use a "plan view" of the module. A module plan view simply adjusts the length of the side of the module that runs in the direction from eave to peak of the roof using (4.9) with x as the eave-to-peak dimension of the module. Chapter 5 will introduce the information needed for designing the racking system for the array. At this point, the challenge will be to locate the modules. The racking design will be left for Chapter 5.

4.6.2 ARRAY SELECTION

Figure 4.5 shows the actual dimensions of the roof perimeter. The array layout was done by converting the modules to plan view for both portrait and landscape orientations using (4.8) and (4.9) before locating them on the roof.

The proposed modules were then fitted onto the roof in what looks like a logical configuration, within any constraints that may be imposed by shading, protrusions, wind loading or other loads such as snow loads. In fact, there may even be a constraint imposed by the location of the roof trusses. If the designer is willing to try more than one module type, each with different dimensions, then scale model drawings can be made of each module and scaled drawings can be made for each possible module so it can be fitted to the roof. The key to maximizing PV power production in a limited space is to find the most efficient module in a particular family. The array shown in Figure 4.5 uses a 420 W module that is the most efficient in its "family."

The array was assembled with modules in both portrait and landscape orientation, taking into account assumed requirements for a 36-inch (91.4 cm) wide access zone on each side of the ridge of each roof section. The small rectangles were used to indicate the border of the access area. In anticipation of the need to anchor roof attachments into the trusses, note that the landscape modules are shown balanced between two trusses, spaced 48 inches apart. The roof section slopes are all 22.6° as in the last example. It is assumed the trusses run from eave to peak for each of the ten roof sections and that it may be necessary to move some of the landscape modules on the west roof section slightly unless the module is rated for mounting with the rails parallel to the long side of the module. Locations of trusses are shown for roof sections 3 and 7 to show where the mounting rails for the landscape modules can be located for truss attachment.

TABLE 4.5
Summary of Simulated Performance of 44.52 Kw Array Used in Optimizer Example

Roof	# Modules	kW	Azimuth	kWh/yr	kWh/kW/yr
1	3	1.26	180	1758	1395
2	3	1.26	270	1488	1181
3	26	10.92	180	15,188	1395
4	4	1.68	90	2019	1202
5	2	0.84	180	1172	1395
6	12	5.04	270	5951	1181
7	32	13.44	0	12,659	942
8	3	1.26	270	1488	1181
9	3	1.26	0	1187	942
10	18	7.56	90	9086	1202
Totals	106	44.52		51,996	1168 (avg)

Table 4.5 summarizes the simulated annual array performance for each roof section. The system is assumed to be located in Nashville, TN. Note that the average monthly kWh production is 4333, which, considering the conditioned space area of the building is approximately 3600 sq ft, will likely put a nice dent in their electric bill.

A few of the interesting takeaways from the table include comparing 1395 kWh/kW/yr for the south-facing roof sections with 1202 for east-facing, 1181 for west-facing and 942 for north-facing. One might expect identical east and west results, but since the simulation includes temperature and cloud cover, it appears that it may be a bit cooler and less cloudy on average in the morning. Note also that the annual average kWh/kW production has been calculated as the ratio of annual kWh production to total system kW.

It is interesting to note that Nashville receives a large fraction of its electrical power from the Tennessee Valley Authority hydroelectric project, meaning (2023) residential electrical rates are approximately $0.11/kWh. Chapter 9 introduces the concept of levelized cost of electricity (LCOE). The LCOE enables comparison of various sources of electricity.

4.6.3 INVERTER SELECTION AND MATCHING STRINGS TO THE INVERTER

Inverter selection and then determining the appropriate string size are done together. Since the roof has ten different sections with four different orientations, this is a nearly perfect application of a system that uses an optimizer for each module. This way, if strings include modules facing different directions, maximum power point tracking at the module level can still be done. It is convenient to begin with the array power and an idea of what inverter size might be suitable. Since the site is single-family residence, it can be expected the residence will have a 120/240 V single-phase service.

Grid-Connected Utility Interactive PV Systems

Since the 106-module array has a DC power rating of 44,520 W, it is reasonable to begin with an assumed DC:AC ratio of 1.3. This would result in a total inverter output power of 34,246 W. One possible candidate is an 11.4 kW unit. Three of these would have an overall rating of 34.2 kW. On the DC side, the nominal input voltage is 350 V, with a maximum allowable input voltage of 500 V. Each unit can handle a maximum input power of 15,350 W and has a maximum input current of 34.5 A, with up to three inputs. The inverters are transformerless and require ungrounded arrays. On the AC side of the inverter, the maximum continuous output current is 47.5 A. Again, note that for the very short time intervals when the inverter inputs may exceed the rated inverter output power, the inverter simply limits its input power by communicating with the optimizers to move away from maximum power points.

One of the functions of the yet-to-be-selected optimizers is to equalize string voltages, since the module outputs are connected directly to the optimizers and the optimizers are connected in series in the strings. Note that in this case, voltages are being added in the series-connected strings, whereas microinverter currents are added in parallel-connected microinverter strings. Microinverters operate at the same output voltage and optimizers operate at the same output current for each string. This does not necessarily require equal string lengths, but, unlike the microinverter systems, there is both a maximum and minimum string length for optimizer systems.

Now that the design rules are established, the design can be completed with the following steps:

1. Since the inverters each can handle 3 strings, divide the 106 modules into 9 strings. The solution may not necessarily be unique. Since 106 is not evenly divisible by 9, the strings cannot have equal numbers of modules. However, 8 strings of 12 and one string of 10 will work. With this combination, two inverters will have 3 strings of 12 and one inverter will have 2 strings of 12 and 1 string of 10.
2. With these numbers, it is helpful to make one more check by calculating the string powers to be sure they do not exceed the inverter limit of 15,350 W or an average of 5117 W per string. Assuming the unlikely possibility that all modules are operating at rated power, the total maximum input power of an inverter with 3 input strings of 12 modules will be 15,120 W. Since this is less than 15,350 W, 3 strings of 12 modules are acceptable for inverter inputs.
3. The next step is to select optimizers. Each optimizer must have DC input ratings that match the module-rated V_{OC} and I_{SC} values. In this case, V_{OC} = 37.17 V and I_{SC} = 14.03 A. Again, the optimizer choice is not unique. Often it will simply depend upon availability. For the present module, one optimizer that meets the requirements has maximum module V_{OC} = 60 V, I_{SC} = 14.5 A and maximum module rated power = 440 W. DC output limits are 15 A and 60 V but not simultaneously, since this would be 900 W.

If a string of 12 modules and optimizers operates with all modules producing the same power, and if the nominal inverter input voltage is 350 V, then each optimizer can be expected to have a DC output voltage of 350 ÷ 12 = 29.17 V. If each module is producing 70% of rated power, string power will be 3528 W. This means the string current must be I = P ÷ V = 3528 ÷ 350 = 10.08 A.

If modules are producing different amounts of power due to orientation or shading, then the optimizer voltages adjust accordingly, using a current that enables the string voltage to operate at the same voltage as the other strings. In essence, the inverter is solving a system of simultaneous equations to balance all the powers, currents and voltages.

Suppose that, of the 12 modules in the string, 3 are operating at 270 W, 3 are operating at 300 W, 3 are operating at 330 W and 3 are operating at 360 W. This brings the string's total power production up to 3780 W. If the string is operating at 350 V, then the string current will be 10.8 A. Individual optimizer voltages can now be determined by dividing the module output power by the string current, with the result as shown in Table 4.6. Note that if this table is done in Excel, any arbitrary power may be assigned to each module, including zero. At some point, at least one optimizer output voltage will reach 60 V, and this is where the inverter will need to decide whether to reduce its input voltage.

Since each string has a power limitation, the maximum length of a string will depend upon the rated power of the modules used, plus the fact that the inverter will also limit its input current to 34.5 A. This current does not necessarily divide equally among strings, as already noted. In any case, since the maximum voltage per optimizer is 60 V, if the inverter wants to operate with a 350 V input, then the string should have at least 350 ÷ 60 modules, which, in this case, rounds up from 5.8 to 6 modules.

TABLE 4.6
Optimizer Output Voltages When Modules Have Different Power Production

Module	P	Opt V	Module	P	Opt V	Module	P	Opt V
1	270	25	13	300	42	25	300	46.88
2	270	25	14	200	28	26	350	54.69
3	270	25	15	200	28	27	200	31.25
4	300	27.78	16	200	28	28	150	23.44
5	300	27.78	17	200	28	29	190	29.69
6	300	27.78	18	200	28	30	100	15.63
7	330	30.56	19	200	28	31	100	15.63
8	330	30.56	20	200	28	32	100	15.63
9	330	30.56	21	200	28	33	250	39.06
10	360	33.33	22	200	28	34	220	34.38
11	360	33.33	23	200	28	35	160	25
12	360	33.33	24	200	28	36	120	18.75
Totals	3780	350		2500	350		2240	350
String V	350			350			350	
String I	10.8			7.143			6.4	

Grid-Connected Utility Interactive PV Systems

Next, since the optimizer maximum output current is 15 A and the nominal inverter input voltage is 350 V, the PV power connected to the string should not exceed 15 A × 350 V = 5250 W. With 420 W modules, this would appear to mean a maximum of 12 modules per string. However, the inverter's maximum input voltage is 500 V, beyond which the inverter input could be damaged. Thus, the string length could theoretically be increased to 17 modules, but this would be a rather risky choice. However, this analysis shows that 12 is a comfortable string length for this particular combination of module, optimizer and inverter. Now that the inverters, modules, optimizers and string lengths have been determined, the rest of the system design can be completed.

4.6.4 INVERTER OUTPUT TO UTILITY INTERCONNECTION

Since there are three inverters in this system, the typical design procedure is to first combine the inverter outputs and then feed the combined output to the utility connection point. This, of course, also means that the interconnect method must be determined. Included in this part of the design will likely be an outside AC disconnect that will shut down all the inverters if a rapid shutdown or inverter maintenance is needed.

Since the rated AC output current of each inverter is 47.5 A, the design current for wire sizing and overcurrent protection will be 125% of this value, or 59.375 A, which might be a clue to the inverter power rating. The result is that #6 Cu wiring and a 60 A circuit breaker will be needed. Combining three of these will require a 200 A combiner panel, which, is simply a 200 A distribution panel that houses three backfed 60 A circuit breakers. The 120% rule for this system would require 178.125 A plus the main circuit breaker value to be less than 120% of the busbar rating. With a 400 A busbar, 125% of the busbar rating would be 500 A. Thus, if the main circuit breaker is less than 321.75 A, then the 120% rule can be applied. Given the size of this residence, it is possible that it might have a 400 A main service panel with a 300 A main circuit breaker. If this were the case, then a backfed circuit breaker opposite the utility could be used. However, generally in a residential system, the main service is divided into several small sub-service panels that are fed out of the main service panel when the service exceeds 225 A.

Alternatively, if the main service consists of a utility meter connected via feeder cables to a 200 A or larger distribution panel, a line side tap can be made to the feeders just ahead of the main circuit breaker in the main distribution panel. This offers a perfect opportunity to show the line side tap as an acceptable utility interconnect.

To complete the line side tap, a fused disconnect will be needed that is rated as service entrance equipment. This means a 200 A disconnect with 200 A fuses will be an appropriate size. This disconnect will require wiring rated to carry the 178.25 A, which means 3/0 (85.01 mm^2) Cu rated at 75°C will be needed between the combiner panel and the AC Disconnect and from the AC Disconnect to the line side tap.

Figure 4.6 shows the complete one-line schematic of the system, except for EGCs neutrals and conduit sizes. Since modules in strings do not need to be on the same roof plane with the same orientation, it is left as an exercise for the reader to explore various options for connecting the modules in strings in a manner that might make

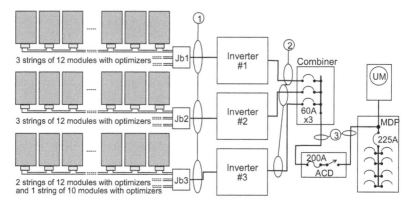

FIGURE 4.6 SLD for optimizer example.

an installer happy. Note that all rooftop wiring among the modules consists of open wiring with listed #12 (3.31 mm²) wires provided by the manufacturer with the appropriate terminations for connecting to the optimizers. The wiring is run along with a #10 (5.261 mm²) ground wire beneath the modules supported by special clips up to rooftop junction boxes where it is junctioned with #10 THWN-2 wiring for the conduit from the rooftop to the inverter.

Note that the inverters are equipped for monitoring individual module performance as well as overall system performance on an instantaneous and cumulative basis. Since all the communication is on the DC side of the inverter, the inverters isolate themselves from any interference among individual communication systems. Note also that since the power from the array to the inverter is DC, a metallic conduit is required for this connection by the *NEC*. Conduit sizes are determined from tables in the NEC.

Specifically, conduit 1 is a ¾″ (19 mm) metallic conduit, since it encloses DC wiring. It contains 6 #10 (5.261 mm²) Cu current-carrying wires, three of which are positive and three of which are negative plus one #10 EGC that grounds the module frames and the racking from the 3 DC input strings. The #10 conductors are based on assumed string currents of 15 A, which is the optimizer maximum output current. The ampacity involves derating the 15 A for conduit fill (80% for six current-carrying conductors) and for ambient temperature [91% for 100 °F (37.8°C)]. Thus, to determine the necessary string ampacity, 15 must be successively divided by 0.8 and then 0.91. The result is 20.6 A. Note that #12 (3.31 mm²) THWN-2 has adequate ampacity, but it is generally traditional to use #10 wire for PV strings, so why should we be different? The #10 EGC is adequate for up to 60 A overcurrent protection at the inverter input. Maximum inverter input is 34.5 A.

Conduit #2 (one for each inverter) contains two #6 (13.3 mm²) THWN-2 sized at 125% of the inverter output current for L_1 and L_2 plus a #10 equipment ground. To fit these three wires into the conduit, which may be PVC since the current is now AC, the minimum size of conduit #2 is ¾ inch.

Conduit #3 needs to carry up to 178.25 A, as mentioned earlier, using 2 3/0 (85.01 mm²) THWN-2 Cu and a #6 Cu equipment ground between the combiner panel and the AC disconnect. The conduit size is 1 ½″ (38.1 mm).

Grid-Connected Utility Interactive PV Systems   119

Existing wiring from the utility meter to the main distribution panel includes three 3/0 THWN-2 Cu (L1, L2 and N). The neutral is connected to the equipment ground in the main distribution panel, the single grounding point for the neutral at the occupant side of the electrical service.

4.7 DESIGN OF A STRING INVERTER-BASED SYSTEM

4.7.1 Introduction

String inverter systems, once the mainstay of utility interconnected systems, are now less common but still have some important applications. Modern straight grid-connected string inverters generally are equipped internally with a number of important features, including MPPT, GFCI, AFCI, RS and monitoring and communications capability. Some have multiple MPPT inputs to accommodate strings on different roof planes. The important thing to remember is that MPPT is achieved at the string level instead of at the module level. This means that all modules in any given string must have the same exposure. If one of the modules in a string is shaded, it affects the performance of the entire string.

The advantage of using a string inverter is primarily cost. Although it is advantageous to use microinverter and optimizer systems when orientation and shading can be a problem, these individual components add to the cost of a system. As a result, most large, ground-mounted systems will generally use strings that are directly connected to the inverter. Further cost advantages of ground-mounted systems are that they do not require the installation of rapid shutdown components and they are often operated at DC voltages up to 1500 V and 600 V on the AC side. A larger system will be designed later in this chapter. The purpose of this smaller system is to show the additional details that need to be considered for a system with the strings directly connected to the inverter.

Since the modules do not have microinverters or optimizers to match their output characteristics, including thermal effects on V_{OC} and V_{mp}, it is necessary to make all these checks during the design process to be sure that the string will produce voltage and current at levels consistent with the normal operating range of the inverter. The total rated DC array power connected to the inverter also needs to fall within the inverter maximum array input power limits.

Otherwise, from the inverter output to the utility interconnect, the design process is essentially the same as with the previous straight grid-connected systems in this chapter.

4.7.2 System Design

This system will be located in Duluth, MN, where it gets a bit colder than Nashville. The ASHRAE design temperatures in Duluth are −14°F (−25.6°C) and 77.9°F (25.5°C). Figure 4.7 shows a relatively simple array on a south-facing roof that meets the requirements for using a string inverter. Modules are identical 440 W units and all face south at a tilt of 22.6°. Important module performance parameters are V_{OC} = 41.11 V, I_{SC} = 13.71 A, V_{mp} = 33.89 V and I_{mp} = 13.01 A. Temperature coefficients are −0.280%/°C for V_{OC}, −0.40%/°C for V_{mp} and the Nominal Operating Cell Temperature (NOCT) of the modules is 45°C. Since the total array-rated power is 12,320 W, using a DC:AC ratio of 1.3 suggests looking for an inverter rating of about 9476.923 W.

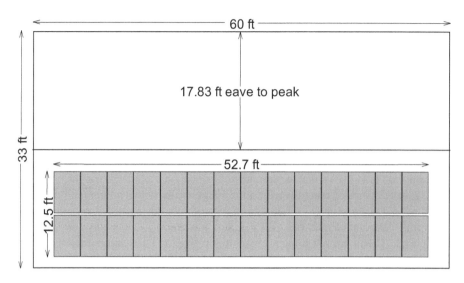

FIGURE 4.7 12.32-kW string inverter example array layout.

At the time of this writing, string inverters less than 10 kW and greater than 100 kW popped up in significant numbers, but 10-kW inverters that will handle a DC array of 12,320 W were a bit sparse. However, an 11,400 W inverter was found that is designed for use with arrays ranging from 9.1 to 13.7 kW. This inverter has a DC input voltage range of 80–600 V, with a maximum input voltage of 600 V. For design purposes, the nominal input voltage is 420 V and the maximum useable input current is 33 A for one MPPT input and 18 A for the other. The minimum string V_{mp} is 240 V. The maximum AC output current at 240 V is 47.5 A.

4.7.2.1 Array to Inverter

The first step in the design is to determine the maximum and minimum string length. The maximum string length is determined by the string voltage at its coldest temperature and can be calculated using

$$V_{OC}(\text{max}) = V_{OC}(25°C)[1 + (T_{min} - 25)(\Delta V_{OC}/\Delta T)]. \quad (4.10)$$

In this case, as soon as sunlight is seen by the array, it begins to get warmer. As a result, T_{min} for a module occurs when there is no incident sunlight on the array, which is typically at about sunrise on that really cold day. Thus, $V_{OC}(\text{max}) = 41.11[1 + (-25.6 - 25)(-0.0028)] = 46.93$ V. Since the maximum string voltage must be limited to 600 V, this means the maximum number of modules in the string will be 600 ÷ 46.93 = 12.78, which must be rounded down to 12.

Next, the minimum string length is found by calculating the shortest string length that will meet the inverter minimum MPPT input voltage. This is a two-step process. First, the maximum cell temperature of the module needs to be determined from

$$T_C = T_A + (\text{NOCT}-20)G/0.8 \quad (4.11)$$

Grid-Connected Utility Interactive PV Systems

In this case, G must be set to 1.0 to establish full sunlight irradiation on the module. $T_C = T_{max}$ is then found to be 56.75°C when $T_A = 25.5$°C. The next step is to calculate the minimum value for the module V_{mp} from

$$V_{mp}(\min) = V_{mp}(25°C)[1 + (T_{max} - 25)(\Delta V_{mp}/\Delta T)]. \quad (4.12)$$

The result is $V_{mp}(\min) = 29.6$ V. Thus, the minimum string length will be $240 \div 29.6 = 8.11$, which must be rounded up to 9. Thus string lengths must be between 9 and 12 modules. For an array of 28 modules, one possible set of string lengths can be 9, 9 and 10. To validate this possibility, the inverter inputs need to be examined.

The inverter has one MPPT input that can handle 33 A and another that can handle 18 A. Each has its own MPPT. Since the modules in the strings are in series, for design purposes the value of current to use is the module I_{SC}. This suggests that the two nine-module strings (27.42 A) can connect to the 33 A input and the 10-module string (13.71 A) can connect to the 18 A input. Is this lucky, or what?

Once again, wire sizes are based upon $13.71 \times 1.25 \div 0.8 \div 0.91 = 23.54$ A and again, #10 (5.261 mm²) THWN-2 will be used between the rooftop junction box and the inverter. The metal conduit size needs to be a minimum of ¾" (19 mm). Since the array wiring is a single conductor in free air, it needs to be rated at $13.71 \times 1.25 \div 0.91 = 18.83$ A. In free air, #12 wire with 75°C insulation rating is rated at 35 A. As long as voltage drop is not a problem and as long as the insulation is rated for free air use (usually type PV), this should be adequate for running the individual strings to the junction box.

The final step in array design is to incorporate rapid shutdown devices in the strings. For this particular inverter, it is possible to replace the rooftop junction box with a rapid shutdown initiator device. The device must be installed no more than 1 ft (0.3 m) from the array. The device controls all three strings. Rapid shutdown is initiated by the inverter upon loss of utility voltage. This completes the system design from the array to the inverter.

4.7.2.2 Inverter to Utility Interconnection

The rated maximum inverter output current is 47.5 A at 240 V. Again, the wire is sized on the basis of 125% of the rated inverter output current, which is 59.375 A. This is just the perfect value for using #6 (13.3 mm²) THWN-2 wiring for the two-phase wires and a #10 for the EGC. Note that this inverter does not use a neutral. This means that if the load is not balanced evenly, any neutral current will need to be supplied by the utility. Whatever method of utility connection is chosen will need 60 A of overcurrent protection, either as fuses or as a circuit breaker.

The residence has a 200 A electrical service, but the electrical load calculation per *NEC 220* shows the load to be 155 A, even though the 200 A main distribution panel is protected by a 200 A circuit breaker. If the 120% rule is used with 125% of the maximum PV output, the sum of the main circuit breaker plus 125% of the PV load is 259.375 A, which exceeds 120% of the busbar rating.

In this case, since the calculated load on the occupancy is 155 A, the main circuit breaker could be replaced with a 175 A unit. Then the sum of the main circuit breaker plus the PV will be 234.375 A, which is less than 240 A.

Thus, it is possible to use the 120% rule as long as the main circuit breaker is replaced with a 175 A unit.

Once again, the NEC requires a PV disconnect in a "readily accessible location." To some local authorities and some utilities, this means outdoors. And it also sometimes means lockable. If this is the case, then an unfused disconnect will be needed between the inverter and the main distribution panel's backfed circuit breaker. If the outdoor and/or lockable requirements are not enforced, then it is possible to feed directly from the inverter to the backfed circuit breaker as long as this circuit breaker is readily accessible, which normally means in a garage. The backfed circuit breaker then serves as the PV system disconnect and also effectively serves as a rapid shutdown device since it removes utility voltage from the inverter. The final system design is shown in Figure 4.8. Table 4.7 summarizes all wiring and conduit sizes.

4.7.3 OTHER OPTIONS

Most modern inverters provide various options for monitoring system output and sometimes for utility control of various inverter functions, such as power factor. This inverter is no exception. It provides both wireless and wired options for monitoring system performance and communicating with the outside world. It is simply a matter of settling upon whatever options make both the system owner and the utility happy.

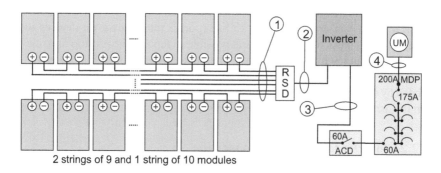

2 strings of 9 and 1 string of 10 modules

FIGURE 4.8 String inverter system SLD.

TABLE 4.7
Summary of Wire and Conduit Sizes for the 12.32 kW-String Inverter System

Conduit #	Description	Conduit Size
1	Manufacturer listed #12 open wiring + #10 EGC	Free air
2	(6) #10 THWN-2 (+, −) + #10 EGC	¾"
3	(3) #6 THWN-2 (L1, L2, N) + #10 EGC	¾"
4	(3) 2/0 THWN-2 (L1, L2, N) (no EGC)	1½"

4.8 DESIGN OF A NOMINAL 100-KW COMMERCIAL ROOFTOP SYSTEM THAT FEEDS A THREE-PHASE DISTRIBUTION PANEL

4.8.1 INTRODUCTION

Thus far, the systems that have been designed have been systems that connect into single-phase distribution systems. For the reader who has been wondering when an example of a larger, three-phase system would be presented, the wait is over. Furthermore, this example illustrates that a design does not necessarily need to begin with the modules. Since this system will be on the roof of an otherwise relatively unshaded commercial building except for a few rooftop fans and air conditioning units, an optimizer system will be used to ensure maximum system output when a few modules are being shaded by various rooftop objects. Another procedure that will be needed in this example will be to confirm that the percent voltage drop in any of the longer wire runs will not be excessive.

In general, any PV system between 10 and 500 kW is considered to be a medium-sized system. While systems at the lower end of this range may connect to a 208, 240 or 277 V (U.S. versions) single-phase AC supply, as system sizes increase, it becomes more likely that they will be connected to 208, 240 or 480 V three-phase power distribution systems. These connections can be made with multiple inverters having single-phase outputs or with single, larger, inverters with three-phase outputs or with multiple inverters, each with three-phase outputs.

In this example, the system will be designed around two 40-kW inverters that feed a 277/480 V three-phase power distribution system. The system will be located near Dallas, TX. The latitude of Dallas is 32.8°. The summer high design temperature is 96.2°F (35.7°C), and the winter low design temperature is 25°F (−3.89°C).

4.8.2 INVERTER

The inverter choice was easy. The wife of the brother of the owner of the building has a cousin who has just struck a fantastic deal on the price of a 40-kW inverter with a friend of his. Not only that but the inverter specifications indicate that the inverter has a maximum DC input power of 70,000 W, an input operating voltage between 840 and 1000 V, 48.25 A maximum input current, an output voltage of 277/480 V that can be connected as either three-wire delta or four-wire Wye and a host of attractive additional features such as built-in DC input safety switch, arc-fault protection, rapid shutdown, DC and AC surge protection, smart energy management and three different communications options. The inverter communicates with individual module optimizers. Optimizers need to be matched to the selected modules. The rated inverter output current is 48.25 A. All this in a 78.2 lb (35.47 kg) package.

4.8.3 MODULES AND DC SYSTEM WIRING

To stay within budget, the owner has decided that, if his roof has room, he will try to fit about 100 kW on it to produce a monthly average of about 13,000 kWh if the array is tilted at 15 degrees and facing south. Literally dozens of modules could be selected for this project. Again, price, availability, suitability, and so on enter the selection process. One convenient commercial size module is a 500-W unit that is characterized by V_{OC} = 51.7 V, I_{SC} = 12.28 A, V_{mp} = 42.8 V and I_{mp} = 11.69 A. Its dimensions are 2187 mm × 1102 mm × 35 mm (86.1 in × 43.39 in × 1.38 in). The temperature coefficient of V_{OC} is −0.25%/°C, and the temperature coefficient of V_{mp} is −0.40%/°C. Thus, it will require 200 modules to complete the 100-kW array.

String sizes will depend upon the inverter operating DC voltage, the voltage and current range of the optimizer input and output and the total string power. One optimizer that can be used has a 500 W maximum input power, 60 V maximum input voltage and maximum module I_{SC} of 15 A. Optimizer output limits are 60 V and 15 A. As a double check on the module maximum output voltage, using (4.10) shows the maximum module V_{OC} for this location is 55.43 V.

Since the optimizer maximum output voltage is 60 V and the nominal DC input operating voltage of the inverter is 840 V, the string length must be at least 840÷60 = 14 modules. Each inverter has four DC inputs with a maximum input current of 48.25 A overall. Since the optimizer maximum output current is 15 A, this suggests three strings of 15 A each or four strings of 12 A each. Thus, if each string operates at 840 V, then the string power will be 15×840 = 12600 W for each of the three strings. Using 500 W modules, this amounts to 12,600÷500 = 25.2, which rounds down to 25. Thus, string lengths can range from 14 to 25 modules.

If each inverter is connected to half the modules, this suggests using four strings of 25 modules each. Although this may seem like too much input power, it is important to remember that the inverter allows for 70,000 W for the array, in recognition of the fact that most of the time, the array will be operating at less than 80% of its rated capacity. Thus, connecting 50,000 W to the input of a 40,000 W inverter is well within the inverter limits. If, for whatever reason, the array power should exceed 40,000 W, the inverter limits its output current to the rated value.

Figure 4.9 shows the roof with 6 rooftop air conditioning units that measure 6′ × 6′ × 4′ (1.83 × 1.83 × 1.22 m) high. The dotted areas show the shading patterns around the units on the first day of winter. These patterns are easily determined by using (2.9) and (2.10) for the system location and calculating the shadow direction and length for a 4-ft high object at 9 a.m. and 3 p.m. sun time. The roof diagram also shows a 3 ft high parapet around the edge of the roof with a minimum 4-ft access walkway around the perimeter as well as allowing room for maintenance on the units. Since space was available on the east and west sides of the roof, a larger spacing between the array and parapet was used to take advantage of longer hours of sun in the morning and afternoon when the sun is lower in the eastern or western sky.

Figure 4.9 also shows the proposed array layout. The modules are rack mounted at a 15° tilt, and, thus, the plan view dimension shown is the dimension of the module short side ×cos15°. The beginnings and endings of the proposed string layout are designated by x.y.z, where x is the inverter number, y is the string number and z is the module number. Note that all string wiring up to the junction box will be open wiring in free air supported by cable hangers behind the modules. Wiring in conduit between the junction boxes and inverters must be in metallic conduit since it carries DC. Locations of DC, such as junction boxes and conduits, must be appropriately labeled.

Grid-Connected Utility Interactive PV Systems

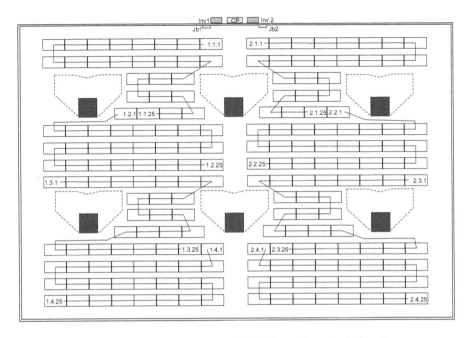

FIGURE 4.9 Array and String Layout for 100 kW Three-Phase Optimizer System

To check voltage drop in the string wiring, the one-way length from module 1 to module 25 and then to the rooftop junction box is approximately 160 ft. Thus, (4.1) shows that if #12 (3.31 mm) wire is used for the array open wiring (as supplied by the manufacturer), using I = 15 A and V = 840 V shows a %VD of 1.13%. An additional 0.11%VD is added if #10 (5.261 mm) Cu is run from the junction box to the inverter.

As a check on the use of #10 wiring, note that the conduit between the rooftop junction box and inverter will contain eight current-carrying wires, thus requiring a derating of 0.7. In addition, the high summer ambient temperature of 96.2°F requires an additional derating factor of 0.88. Thus, the minimum 30°C ampacity of the wire must be adjusted to 15 ÷ 0.7 ÷ 0.88 = 24.4 A. In fact, since the ampacity of #12 Cu with 75°C terminations is 25 A, and given that the actual current will rarely reach 15 A in every conductor in the conduit and since the temperature will rarely exceed 96.2°F and if the ambient temperature really does exceed 96.2°F, the array output will be significantly reduced below rated value, choosing #12 would not be an unacceptable choice unless it resulted in too much %VD. However, the most compelling reason for using #10 is that most installers are accustomed to using it.

One final word about derating when either microinverters or optimizers are used. It's the *either/or* rule. Simply put, when sizing wire, first evaluate 125% of the maximum current. Then evaluate the maximum current divided by each of the derating factors. The larger of the two results is the one to use. Once upon a time, both were required to be used consecutively.

It is important to note that, since the strings are wired in series, it could be very tempting to save wire by using a shortcut for the return wire from module 25 to module 1. However, this temptation is to be avoided like the plague, since doing so results in a loop of wire that acts as an antenna for attracting a huge voltage surge from a nearby lightning strike. The reader certainly remembers well that the voltage induced in a loop is equal to the loop area times the time rate of change of the magnetic field through the loop. Even though the inverter is equipped with surge protection at the input, there is no reason to challenge it every time lightning is nearby.

As a final word on lightning, note that all metal parts of the system are grounded. Some might think this would attract lightning, but, once again, one needs to go back to remember that the electric field surrounding a charge is inversely proportional to the square of the distance from the charge. For a sharp point, the electric field can build up to a high value and when it reaches the dielectric breakdown voltage of the surrounding air, it dissipates gradually, thus effectively discharging any difference in charge between atmosphere and ground. In fact, why else would lightning rods be installed on buildings? Certainly not to attract lightning. Finally, remember how sharp the corners are on the modules. This makes them very effective at discharging excess charge buildup.

4.8.4 AC System Wiring

Figure 4.9 also shows the locations of the inverters and a combiner panel/load center for the outputs of the two inverters. The inverters are mounted about 5 feet above ground level to enable easy access. The rated output current of 48.25 A will require #6 (13.3 mm) wiring with a #10 (5.261 mm) EGC. Since the four-wire option is available, a neutral wire will be included in the wiring. The 4 #6 THWN-2 and the #10 Cu ground will require a 1" (25.4 mm) conduit, which can now be PVC if the contractor desires, since this is the AC wiring side.

The combiner panel can be a standard 125 A three-phase load center rated for 277/480 V. Outputs from each inverter will terminate at a 3P60A circuit breaker. The output of the load center is then connected to a 200 A, three-phase unfused disconnect, which will serve as both a system disconnect and a rapid shutdown switch. When the inverters lose utility power, they communicate with the optimizers to shut down the array.

The building's main service panel is a 600 A, 277/480 V three-phase load center. Luck would have it that it has space left for a 125 A circuit breaker at the bottom of its busbar. Since 120% of 600 A is 720 A, this means that up to 120 A of PV can be connected. And, surprisingly enough, 125% of the overall PV output is 120.625 A. However, due to the two 60-A circuit breakers in the combiner panel, the output current is limited to 120 A. The 125 A circuit breaker is simply the next largest size circuit breaker that will meet the necessary ampacity.

Admittedly, this is a close call, and it may be required by local authorities to use a different connection method. This could be either a line side tap or any number of other options to be explored. If the line side tap is possible as the next most desirable connection, the unfused disconnect would be changed to a fused disconnect, fused at 125 A.

The wiring between the combiner panel and the point of utility connection will need to be 4 #1 (42.41 mm^2)Cu THWN-2 and a #6 (13.3 mm^2) Cu equipment ground in 1 ½″ (38.1 mm) conduit. If a line side tap is used, there will likely not be a neutral wire in the conduit between the fused disconnect switch and the line side tap.

As a final note, since the AC wiring distances are relatively short, no attempt has been made to evaluate any AC voltage drop for this system. The next system to be considered, however, will involve relatively long distances between the inverter output and the next part of the system and will include three-phase voltage drop calculations.

Figure 4.10 shows the final system wiring diagram.

4.8.5 Final System Design and Operation

As the final word on this system, it is interesting to look at the additional features of the inverters. They support RS485, Ethernet and optional cellular communication. Thus, three ways for the inverters to communicate with the installer, the owner and the utility. Because of the optimizers, each module performance can be individually monitored, along with overall system performance, instantaneously and cumulatively. The system can be commissioned with a cell phone. For the system to identify all of its components accurately, it requires an accurate setup. Another convenience of this is remote diagnostics of the system.

Of the additional features previously mentioned earlier in this example, the Smart Energy Management feature deserves additional discussion. Deloitte Insights [16] observes in a 2021 article that "converging trends will likely accelerate industrial companies' adoption of energy management solutions and potentially boost their interaction with electric utilities and the grid." The bottom line is that, by communicating with other loads in a building as well as communicating with the utility, the inverters can be a part of a much more sophisticated energy management system

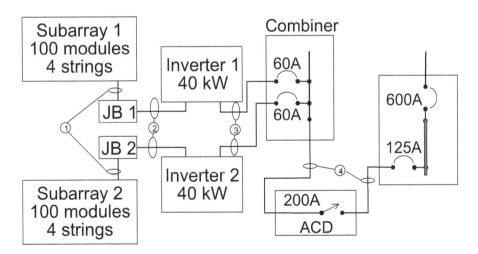

FIGURE 4.10 Final SLD for the 100-kW optimizer, three-phase rooftop system.

that works to level the grid energy output to make the grid more efficient. The fact that Deloitte is in the investment business is a solid testimony to the fact that solar is becoming an important economic driver in the economy and is becoming more and more of interest to commercial and industrial sites.

4.9 DESIGN OF A NOMINAL 5-MW AGRIVOLTAIC SYSTEM

4.9.1 Introduction

The next design example will be a 5-MW ground-mounted system on agricultural land near Springfield, IL. Springfield has a summer high temperature of 88.2°F (31.2°C), a winter low of 3.6°F (−15.8°C), and a latitude of 39.7°.

Large systems usually comprise multiples of small systems, such as the system of the last example. Smaller Megawatt systems often use inverters in the 125 kW range with either 1000 or 1500 V DC inputs. Systems in the 100 MW range will likely use inverters in the 20 MW category. Combining the outputs of the chosen inverters in the most efficient manner then becomes the challenge. Other considerations not normally included in medium-sized systems are weather stations, grid stability and connecting into distribution level or transmission level voltage levels.

With larger systems, sometimes the output will be fed directly to the utility distribution system rather than to a building system. In these cases, it will be necessary to transform the PV system output to distribution voltage levels, which are typically 7.6/13.2 kV.

The advantage of using multiple smaller inverters is the ability to locate the inverters close to the subarrays that supply each inverter. Another advantage of using multiple small inverters is the lower cost of AC components, such as transformers, switches, circuit breakers and fuses. Furthermore, with inverter level performance monitoring, if one inverter out of 80 in a 10 MW system should fail, it will only reduce system output by 1.25% but still will be located by the monitoring system, provided that the monitoring system is working properly.

If the inverters can be conveniently located among the array, DC wiring can still be kept reasonable, but it is still important to recognize that the inverter DC input voltage will generally be higher than the inverter output voltage, meaning input wires can be longer than output wires for a given %VD. The following system will be designed around a 125 kW inverter that incorporates 1500 V strings and a ground-mounted array.

One common feature of large ground-mounted systems is the concern about how the land might otherwise have been used. One solution to this concern is adding agriculture to the PV system such that the two applications can support each other in a win–win operation. The concept is based upon the idea that there is some farmland out there that is now getting minimal use but might be used for some sort of specialized agriculture. Then there's the PV, with an anticipated lifetime of 40 or so years, after which it can be either removed relatively easily or replaced with the next generation of energy production or storage. The reader might be interested in reading about a similar system in Iowa.

Iowa State University has been built to incorporate experimenting with crop optimization as well as beekeeping for pollination and honey production [20].

Grid-Connected Utility Interactive PV Systems

4.9.2 Grid Stability Analysis

Systems of this size are generally connected to the utility at distribution level voltage and power levels. Each utility substation includes a number of distribution branches and the power flow in each of these branches is monitored closely. In the past, the biggest concern has been the amount of power flowing out of the substation on the feeders to be sure it does not exceed the feeder capacity. With renewable generation added to the grid, the concern now extends to the amount of power flowing back toward the substation and what will happen if the source is suddenly interrupted. For that matter, is the substation or the transmission line feeding the substation sufficiently stable to withstand the addition of a relatively large quantity of generation that is dependent upon whether a cloud passes over the sun?

Most engineers who are reading this will have had courses involving stability analysis. It might be electrical, mechanical, hydraulic, or otherwise. In response to a rapid change, will the system be overdamped, underdamped or critically damped? The system will generally include two energy storage mechanisms, such as mass and spring or capacitance and inductance. If the storage overrides the dissipation, the system can become unstable in a manner similar to the islanding discussion in Section 4.2.2.4. Thus, transient analysis of the distribution system is important when large sources are added, especially if their output can change rapidly. This applies to any source, since on a rare occasion, even a fossil source drops out quickly.

The other part of the analysis deals with how the PV power level compares with the load on the distribution line. If the load is too small, is it possible that the PV will meet the needs of the load and continue to flow back to the substation. If this sounds familiar, it is essentially the same as when a smaller system, as previously discussed, generates more power than is needed by the occupancy and the remaining power is sold back to the utility. Thus, a good grid interaction study will check the hourly and seasonal power flows as well as the transient stability of the system. In effect, the study is a load study of the system, which is then compared with the anticipated generation study of the PV system. These studies are generally done by consultants to the PV specialists. It is important to note that these studies can also show the positive effects of large PV generation on the grid, provided that it is properly controlled, including possible curtailment and variable power factor adjustment.

4.9.3 Inverters

Assuming the grid stability study passes the test, then the system design can begin. This time, since it is desired to design a system in the 5 MW range, it is possible to begin with the inverter. The system will use a set of 125-kW inverters. Again, this does not mean that a 5-MW system must be designed with 125-kW inverters. It just happens that a 125-kW inverter with very attractive features is available. The inverter will be used to introduce the reader to the corresponding design procedure to incorporate high DC voltages into the design. Since the availability of very high currents, if subjected to fault conditions, can result in extremely dangerous arcing conditions, arc flash safety and engineering computations will be introduced. The array will be mounted on racks at ground level. Rack mounts will be further discussed in the next chapter.

The choice of inverter may depend on price, availability, weight, physical dimensions, reputation or any number of other considerations. The 125 kW, transformerless, UL 1741 listed inverter to be used in this example is characterized by $V_{OC}(max) = 1500$ V, $870 < V_{mp} < 1300$ V, $I_{IN}(max) = 275$ A, $V_{OUT} = 600$ V and $I_{OUT}(max) = 127$ A. The unit can be configured with up to 20 ungrounded strings with 20 A fused inputs. The units have a California Energy Commission-rated efficiency of 98.5%. They have built-in load break DC and AC disconnect switches and have a utility adjustable power factor between ±0.85 and internal GFDI. In this case, since the entire system will be located outside, no arc-fault protection is required. The weight of the unit is 80 kg (176 lb). All that is left at this point is to decide how many inverters will be needed.

4.9.4 Modules, Inverters and Array

4.9.4.1 Determining String Length and the Number of Inverters

One easy way to end up with 5-MW DC input is to use 10,000 500 W modules. If the same modules from the last example are used in this example, then the performance data of the modules is already known. One additional important item is that the maximum operating voltage is 1500 V. Thus, these modules are compatible with the selected inverter. It's just a matter of deciding upon string length and the number of inverters.

Once again, since no microinverters or optimizers are involved, string length is decided by the value of V_{OC} at the lowest expected operating temperature. This time, minimum string length is not as important, since it is desired to operate the system at the highest possible voltage. Using (4.10) with a winter low temperature of $-15.8°C$ and NOCT = $43°C$ shows that the maximum string length is 26, for which the string V_{OC} will reach 1480 V. Adding another module will add another 56.93 V, bringing the total to 1537 V, which exceeds the 1500 V limit.

With 25 modules in a string, the total string-rated power is $25 \times 500 = 12,500$ W. Thus, the number of strings needed to reach 5 MW will be $5 \times 10^6 \div 12,500 = 400$, which is a nice, round number, compared to using strings of 26.

So far, so good. Next, the number of inverters can be determined. As a ballpark figure, an interesting approach is to start with an acceptable value for the System DC:AC ratio. Usually, 1.3 is a good value that results in minimal losses over the year. Using this criterion, the AC rating of the overall system would be $5 \times 10^6 \div 1.3 = 3.846$ MW. Since the inverter-rated power is 125 kW and since 3.846 MW = 3846 kW, this suggests that a total of $3846 \div 125 = 30.76$ inverters would be a good starting point. If 30 inverters are used, the DC:AC ratio becomes 1.333 and if 31 inverters are used, the DC:AC ratio will be 1.29, making it a toss-up between 30 and 31 inverters.

Given that the number of strings is 400, it is interesting to see how the strings might be divided up among 30 or 31 inverters to see what grouping might be preferred. With 30 inverters, the average number of strings per inverter is 13.333. With 31, the number drops to 12.901. Thus, with either 30 or 31 inverters, some will have different numbers of strings than others.

Grid-Connected Utility Interactive PV Systems

For example, by juggling a bit, it can be determined that with 30 inverters, if 20 of them have 13 strings and 10 of them have 14 strings, the total number of strings in the system will be 400. With 31 inverters, 3 inverters with 12 strings and 28 inverters with 13 strings also add up to 400 strings. Thus, two viable solutions have been found. Before committing to either, it makes sense to explore how the array might be laid out. The array layout can have an influence on the number of inverters selected if using the additional inverter results in some savings in construction costs.

4.9.4.2 Configuring the Array

As anticipated at the beginning of this example, there are a number of constraints on the array, including shading, orientation and access to the space beneath and between as may be needed for the agricultural project. Ideally, for exposure for a fixed array, the array should point south. This means the array rows will run from east to west.

Since this will be a ground-mounted system, it will be assumed that the area will be essentially flat, with no surprises beneath the ground where array footings will be located. Presumably, a geotechnical engineer will have submitted a report on soil information so the array mount can be appropriately designed. Wind load, of course, will be an additional consideration. The design criteria will be to produce close to the maximum kWh/acre, which means the tilt and spacing of the array will be adjusted to maximize array production while keeping a satisfactory row spacing for agricultural use.

Another decision that needs to be made is whether the array should be one-up or two-up and whether the lowest point on the array must allow easy passage of animals beneath the array. One-up simply means a single row of modules on the racking, while two-up means a double row of modules on the racking, as shown in Figure 4.11. During this exercise, the designer can be reassured that there are numerous manufacturers and vendors of fixed-racking solutions. Thus, at this point, it is acceptable to design the desired racking layout and then look around for the right manufacturer.

Where to begin, then, becomes an interesting question. In fact, there are a number of possibilities. As in many design exercises, the process is to try something, see if

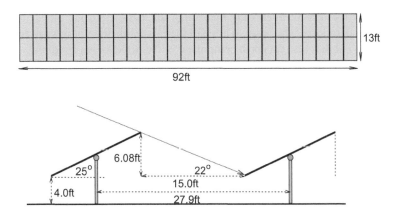

FIGURE 4.11 Determining the layout of strings and row spacing for the 5 MW system.

you like it, and then redesign it if necessary. To save some space in the book, however, this design will show just one option to illustrate the design process.

To begin, the first assumption will be that the lowest part of the array must be at least 4 ft (1.219 m) off the ground. Next, the array will consist of rows of 25 modules each in a two-up configuration as shown in Figure 4.11. This configuration results in twice the difference in height between the highest and lowest point on the array, which then requires twice the spacing between rows, thus allowing more space between rows for agricultural use. Since the module length is 86.1" (2.187 m), two connected at the short sides will have a length, L, of approximately 173" or 14'5" (4.394 m). If the modules are tilted at latitude −15°, the tilt will be, to the nearest degree, 25°. This means the horizontal projection of the modules will have a length of Lcos 25° = 157" = 13'1" (3.988 m). It also means the vertical distance between the top and bottom of the modules will be Lsin 25° = 73" = 6'1" (1.854 m). Thus, if the lowest distance to the ground is 4 ft (1.219 m), the top of the array will be 10'1" (3.073 m) above ground. The plan view dimensions of each row of 50 modules will thus be 91'8" × 13'1" (27.94m × 3.99 m). This dimension can be the building block for the remaining system, assuming space will allow it. Since each rack includes 50 modules, this means it holds two strings. Since there are 400 strings, this means 200 of these building blocks will be needed.

The next exercise is to determine the spacing between rows, as measured between the centers of each row, since this will locate the pilings that support the rows. This will require some shading analysis as well as agreement with the agricultural row spacing requirements. Figure 4.11 also shows an end view of two rows, with an array tilt of 25°, an array spacing of 15 ft (4.572 m) and a spacing between the array supports of 27.9 ft (8.504 m). This shows that with the 15 ft (4.572m) spacing and the 6.08 ft (1.853 m) height difference, the sun altitude below which the rows will begin shading each other will be 22°. Checking the altitude and azimuth formulas, (2.9, 2.10) the result is that on December 21, the sun will be above this altitude between approximately 9 a.m. and 3 p.m. (sun time) or six hours of unshaded operation. On June 21, however, the sun altitude will be above this angle between approximately 5:30 a.m. and 6:30 p.m., giving 13 hours of shade-free energy production.

However, there is an interesting and important item regarding the 13 hours of shade-free operation of the array. Although there is no shading of rows to the north at 5:30 a.m. There is also no north component of the direct beam irradiance on June 21 until about 8:30 a.m. Before that time, the sun is in the northeast and is shining on the backs of the modules. In fact, this is where bifacial modules can improve array performance, since the early morning and late afternoon direct beam sunlight impinges upon the back of the modules. Thus, direct beam sunlight only impinges on the modules when it has a north component, which means between when it moves from northeast to southeast in the morning until it moves from southwest to northwest in the afternoon.

This, of course, does not mean there is no power generation at these times of day. It simply means that any generation is due to the diffuse component unless the modules are bifacial. This could still be an option if the owner wanted to squeeze every last possible drop of energy out of the available sunlight. In Chapter 5, other ground mount configurations, including backtracking arrays, will be discussed as options for capturing more sunlight from the northeast and northwest.

At this point, it is possible to begin placing the building blocks into the space allocated for the system. Since there will be so many possible variations, this discussion will be limited to a simple rectangular space for the system to explore some of the options for choosing inverter locations. Figure 4.12(a) shows one option for laying out the 25-kW building blocks that give reasonable distances between strings and inverters and reasonable spacing between inverters. The rows each show 4 strings of 25 modules, for a total of 100 modules per row. Thus, 25 rows represent 2500 modules, which is 25% of the system. Note that the inverters are located on the north side of the array in easily accessible locations that have vehicular access. In the figure, strings are identified as x.y-z, where x represents the inverter number and y-z represents string numbers, such that 2.8–11 means inverter 2, strings 8–11. Thus, Figure 4.12(a) shows 13 strings each for inverters 1–5, 14 strings for inverters 6 and 7 and 7 strings for inverter 8. The remaining seven strings for inverter 8 will come from the next section of the array.

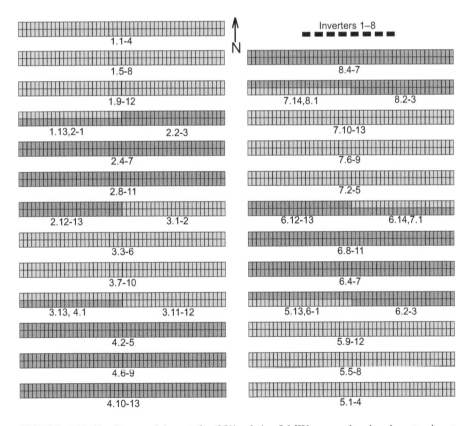

FIGURE 4.12(A) Proposed layout for 25% of the 5-MW array showing inverter input strings.

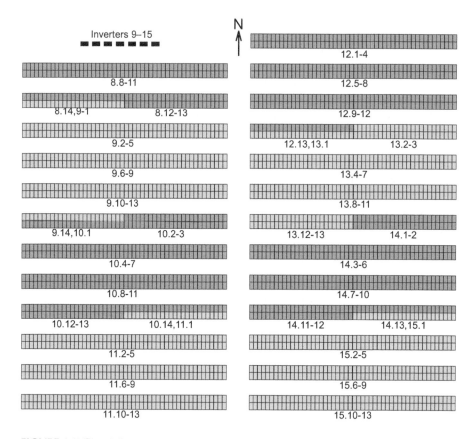

FIGURE 4.12(B) Adjacent group of 2500 modules to complete half the system with 15 inverters.

Figure 4.12(b) shows the adjacent 25% section of the array and includes the remaining half of the strings for inverter #8 as well as all the strings for inverters 9–15. The combination of these two figures constitutes half the total system. It will be assumed that the remaining half will be identical, which means that 30 inverters will be used in the total system.

4.9.4.3 Wire Sizing and Voltage Drop Considerations

With this layout, the longest distance between the end of a string and its corresponding inverter is about 550 ft (168 m). For voltage drop calculations, it is also helpful to include the %VD within each string by noting the equivalent one-way length of the string wiring, which is about 75 ft (22.9 m). This distance takes into account that the modules are half-cell modules that have their positive and negative contacts opposite each other in the center of the back of the module. This eliminates nearly 2 ft (0.61 m) of wire behind each module, since the module is carrying the current.

Assuming an operating voltage of $25V_{mp}$ (1070 V) and an operating current of I_{mp} (11.69 A), the worst case %VD for any string, if #10 (5.261 mm^2) Type PV open wiring is used along with #12 (3.31 mm^2) cables on the modules, will be 1.49 + 0.32 =

1.81%, which is less than the maximum of 2.0% that is generally the design goal for these systems. Thus, given that both #12 open wiring and #10 wire in conduit have adequate ampacity for 125% of the system operating current, an adequate solution for string and inverter location appears to be at hand as long as the method used for this quadrant of the system is continued for the other three. Wiring is thus now complete from array to inverters. The next step is to decide what happens with the wiring from the inverter outputs.

4.9.5 AC Wire Sizing, Voltage Drop, Disconnects and Overcurrent Protection

4.9.5.1 Introduction

The inverter manufacturer specifies 175 A overcurrent protection for the inverter output. Recall the inverter outputs are three-phase 600 V L-L with an optional neutral for instrumentation use. The maximum output line current is 127 A if the inverter is operating at either 0.85 leading or lagging power factor. Unity power factor maximum output current is 120.3 A. The next step in the design is to determine the AC wiring and additional components of the system.

Figure 4.13 shows the proposed layout of the remaining system components. Note that this is ONE way the system can be implemented, but there are many more possible options. The overall idea is to combine the outputs of the system inverters and then connect this combined output to a transformer. Whether to use a single transformer or to use several smaller units will depend on site-specific conditions. In this example, two medium voltage transformers (600V to 12 kV) will be used. The sizes of the transformers will be determined on the basis of the system's AC power output. Another determining factor is the exact location of the utility feeder. In this case, it will be assumed that the utility feeder runs overhead, about 500 ft (152.4 m) north of the array, parallel to the array.

4.9.5.2 Calculating Three-Phase %Voltage Drop

Before calculating the wire sizes, it will be helpful at this point to determine the correct formula for percent voltage drop in three-phase systems. Figure 4.14 shows a balanced three-phase, Y-connected system. The voltage drop in the wiring to and from the load is simply IR, where IR has two parts: $I_L R_L + I_N R_N$, where I_L is the line current, I_N is the neutral current, R_L is the line resistance and R_N is the resistance of

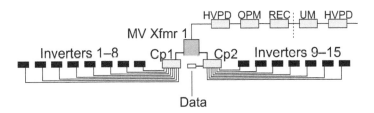

FIGURE 4.13 Remaining system components between inverter outputs and the utility interconnect.

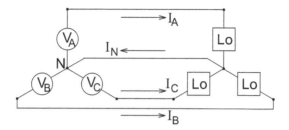

FIGURE 4.14 Three-phase Y-connected source and Y-connected load.

the neutral. In this balanced Y-connected system, where all the loads (Lo) are equal, the voltage drop in the wiring between the A-phase source and the A-phase load is $I_A R_A + I_N R_N$. Note since the load is balanced, the line current in each phase is equal and the neutral current is the sum of the line currents, which, for a balanced load, is zero. Thus, $I_N R_N = 0$.

The resistance per 1000 ft (or resistance per km) of the phase wires can be found in tables that can be found in numerous locations, including the *NEC*. For convenience, as in the single-phase case, it can be represented as Ω/kft or Ω/km. Thus, if the length, d, of the wire is known, $R = (\Omega/\text{kft})\text{d} \div 1000$ can be used to determine its resistance. The %VD is then given by the ratio of the voltage drop in the wire to the source voltage, divided by 100 to convert a fractional result into the percent result. The result is, for *balanced* three-phase loads only, %VD = $100 I_L (\Omega/\text{kft}) \div 1000 \div V_{LN}$ or, also noting that $V_{LL} = \sqrt{3} V_{LN}$,

$$\%VD = \frac{0.1 I_L d\left(\frac{\Omega}{\text{kft}}\right)}{V_{LN}} = \frac{0.1732 I_L d\left(\frac{\Omega}{\text{kft}}\right)}{V_{LL}}. \quad (4.13)$$

4.9.5.3 Wire Sizing

The system layout as shown in Figures 4.12 and 4.13 is intended to be friendly to everyone: to the installer because all wiring is based upon integral numbers of source circuits to each rack along with reasonable access space for installing all system components, to the owner because it is designed for minimal cost and maximum performance, and to the designer because of the symmetry that enables repetitive calculations, to the inspector because of the neatness and to the teams responsible for commissioning and maintenance of the system. All wire sizing will begin with the calculation of NEC-required wire ampacities and then voltage drop calculations will be made to be sure that no total %VD from source circuit to inverter input will exceed 2%. Another consideration with large systems is the extent to which exposed wiring will be used vs the use of wiring in conduits. Given the dual use of this property, wiring beneath arrays up to inverters will be open wiring, protected by cable trays under the arrays, but from inverter to combiner to transformer will be via trench and conduit.

Grid-Connected Utility Interactive PV Systems

Since wiring between the inverters and the combiner box is based upon 125% of the inverter output current, it needs to be rated at a minimum of 1.25×127 A = 158.75 A. If 90°C (194°F) wire and terminations are used, then 1/0 (53.49 mm^2) Cu can be used if it meets %VD requirements. With equipment rated at 75°C (167°F), a more expensive 2/0 (67.43 mm^2) wire would need to be used.

If the inverters are located as shown in Figure 4.13, then the distance from the inverter furthest from the combiner panel to the combiner panel is approximately 130 ft (39.6 m). Using $V_{LL} = 600$ V, $I_L = 127$ A and $\Omega/\text{kft} = 0.122$, (4.13) gives %VD = 0.58. As a result, all of the wiring between Inverters and the combiner can be 1/0 Cu. The wiring from each of the inverters to combiner would thus be 3 1/0 THWN-2 Cu and an #6 (13.3 mm^2) equipment ground in 1½" (38.1 mm) Schedule 80 PVC conduit. If a neutral is required between inverter and combiner for instrumentation reference, it will need to be at least the size of the EGC. This should be coordinated with the instrumentation design.

4.9.5.4 Combiners and Overcurrent Protection

For each group of 15 inverters, one combiner will combine the outputs of 7 inverters and the other will combine the outputs of 8 inverters as shown in Figure 4.13. The combiners might also be known as load centers or as switchgear. Effectively, they are large electrical distribution panels that are being connected backward. In other words, the power comes in where it would usually go out and it goes out where it would usually go in. The important part is that they are large enough and have a main circuit breaker in each along with circuit breakers for each inverter connected into the combiner. Perhaps a more subtle issue is that the circuit breakers must allow current to flow in either direction. This is the case for most AC circuit breakers, but some DC circuit breakers are labeled Line and Load, meaning they are only designed for power flow from line to load.

The size and rating of the combiner is determined by the total current to the combiner from the inverters. Thus, for 7 inverters, the maximum total current would be $7 \times 127 = 889$ A and for 8 inverters, the total current would be 1016 A. Since the overcurrent protection is generally sized at 125% of the expected current, this would be 1111 A and 1270 A. Standard sizes for panelboards are 1200 A and 1600 A. Thus, unless some sort of output regulation is employed for the 1270 A, it would require the use of a 1600 A panelboard. Standard sizes for circuit breakers suitable for use in this design for the panelboard output protection are also 1200 A and 1600 A.

For currents higher than 800 A, the *NEC* requires that conductors protected by these circuit breakers must be rated at least as high as the size of the overcurrent protection. For conductors rated below 800 A, the overcurrent protection used can be the next larger standard size. This is important with high currents, since a 1200 A circuit breaker is not large enough for 1270 A, and a 1600 A circuit breaker would require considerably larger wire than would be required by a 1270 A circuit breaker. Fortunately, many of the circuit breakers 800 A and larger are adjustable in 10% increments. Thus 90% of a 1200 A circuit breaker is 1080 A, which, unfortunately, is less than 1111 A. This means that for the smaller combiner, a 1200 A main circuit breaker will be needed and the wiring from combiner to transformer must have an ampacity of at least 1200 A. The larger panelboard will need a 1600 A circuit breaker. If this circuit breaker is adjusted to 80%, its rating drops to 1280 A, which, fortunately, is slightly higher than the required 1270 A. As a result, this is an acceptable solution, which now enables determining the sizes of the wires from the panelboards to the transformer.

Up to this point, all wiring of previous examples has consisted of single conductors per line. However, when ampacities of 400 A or more are required of conductors, often the conductors are wired in parallel, since it is much easier to work with smaller diameter conductors and also since the "skin depth" of larger diameter AC conductors tends to limit the current in larger conductors to less than might be expected. Skin depth accounts for the fact that when a conductor is carrying 60 Hz current, the current tends to migrate to the outer ¼" (0.64 cm) of the wire cross section. In fact, this is the major reason for DC transmission lines over long distances. Keeping this in mind, it is now time to determine suitable wire sizes from combiner to transformer. Before doing so, it is first helpful to have some information on the transformer, which now can be specified.

4.9.5.5 Transformers to the Utility

The total maximum output power of each group of 15 inverters is 1.875 MW. Since maximum inverter output power is rarely achieved, a reasonable choice for the transformer will be an outdoor, 2 MW, 600 V to 12 kV pad mounted unit with 90°C terminals. Transformer input will need to match inverter output. In this case, inverter outputs can be connected as either Y or delta. The utility will most likely want the transformer output on the high voltage (primary) side to be Y-connected with a grounded neutral connection and the customer will likely want it to be filled with environmentally friendly oil. But first, the size of the wiring between combiners and transformer needs to be determined.

For the 1111 A connection, since the conductors need to be sized for 1200 A, of the several options, either three sets of 500 kCM (253 mm^2) or four sets of 300 kCM (152 mm^2) will work. The respective overall ampacities of these wire sets (90°C wire and terminals) are 1290 A and 1280 A. Note that when paralleling conductors, all must be the same size and all must be the same length. Otherwise, current will not be uniformly distributed among them. The word "sets" is also important. It means that, for example, with the 300 kCM conductors, 4 separate conduits will be needed, each of which will include L_1, L_2, L_3 (optional N) and G. The reason for this is cancellation of magnetic fields. When a conduit has the same current flowing in one direction as in the opposite direction, the currents essentially cancel out each other's magnetic fields, except for some leakage at close distances. This holds for either single phase or three-phase wiring.

The *NEC* requires that all sets of wires include EGCs that are sized according to the size of the overall circuit overcurrent protection. This means that for a 1200 A circuit breaker, the EGC size must be at least 3/0 (85.01 mm^2) Cu. For 1600 A, the EGC size increases to 4/0 (107.2mm^2) Cu. Note that since the EGC normally does not carry any current, there is no temperature requirement on its insulation. Also, if it is decided to include a neutral, since the neutral carries no current, it may be the same size as the EGC, but not smaller. Thus, if a conduit contains three 300kCM and a 3/0 EGC, the size will need to be at least 3″ (76.2 mm) for SCH80 PVC or 2½″ (63.5 mm) for conduit with thinner walls.

For the group of 8 inverters, assuming a 1600 A circuit breaker adjusted to 1280 A, 4 sets of 300 kCM exactly fit the need, as discovered with the previous calculation. Thus, even though the two combiners are two different sizes, each is wired to the transformer with the same sets of conductors with one exception that has also already been noted. The EGC for the 1280 A system needs to be 4/0 Cu instead of 3/0 Cu and one must be included in each of the 4 conduits.

Finally, note that to achieve this wiring configuration, each combiner output circuit breaker will need to be able to accommodate 4 connections per pole. Thus, it is important to be sure the proposed circuit breaker can accommodate this many wires. Most, if not all, of them can. This also means at the input of the transformer, there must be busbars that are capable of terminating 8 wires per phase, 4 from each combiner. Yes, that's 24 terminations. In addition, terminations must be available for the EGCs and, if neutrals are used, for the neutrals.

Also note that since the voltages are high and the components are close there is no need to do a voltage drop check on the wiring between the combiners and transformer or beyond the transformer to the utility.

Assuming nearly 100% efficiency for the transformer, the line current on the primary (high voltage) side will be $600 \div 12,000 = 0.05$ times the 3810 A per phase low voltage side current, which is the sum of the output currents of the two combiners. Thus, the design line current on the high-voltage side of the transformer will be 190.05 A.

The transformer output current will often be routed up a utility power pole to a high voltage disconnect (HVPD in Figure 4.13) on the owner side of the utility interconnect. From there it goes to another utility pole via a single conductor in free air wiring to a utility-grade power output monitor on the owner's side. Then to a recloser on the owner side, then to a production meter on the utility side and to another disconnect on the utility side. The purpose of the recloser is to detect and interrupt transient faults to protect the system. After this final point, it may end up continuing along distribution level overhead or underground lines, per the option of the utility.

The wiring on the high voltage side of the transformer that is in conduit will need to be a minimum size of 1/0 (53.49 mm^2) for 90°C rated Cu or 3/0 (85.01 mm^2) for 90°C Al. When the wiring becomes a single conductor in free air between utility poles, it can be 1/0 (53.49 mm^2) Al. Note that aluminum overhead wiring will often have a steel core to add strength to the wire. The steel core carries minimal current due to the higher resistance of steel, but particularly due to "skin effect" that causes the current in an AC wire to crowd toward the outside of the wire cross section.

At this point, everything has been wired up to the utility. The only part that is left is to connect a small single-phase, 600:120/240 V, transformer in the seven-inverter combiner to produce a small amount of single-phase, 120 V power for a set of instruments. Monitoring systems can monitor a wide range of variables, such as source circuit current and voltage, inverter input power, inverter output power, inverter self-diagnostics, weather data, such as temperature, irradiance and irradiation, power purchased from the utility and power sold to the utility. A wide range of communication protocols are available, including hard-wired data communication, radio frequency wireless data communication and powerline carrier data communication. Most inverters monitor input and output voltage, current and power to accomplish functional tasks of the inverter, such as maximum power point tracking and self-diagnostics. Thus, the quantities of interest are digitized and made available to display devices, either on the inverter or remote to the inverter [17].

Web-based monitoring systems are becoming very popular. In general, PV system data and sometimes weather data are collected, digitized and then transferred to a controller/display unit and then to a broadband router as shown in Figure 4.13. Some manufacturers provide their own communications equipment, others make data available at a port to be used with third-party hardware and software and some do both.

As the number of performance-based systems increases, the need for utility-grade monitoring also increases. In some cases, third-party monitoring is required by the performance-based contract. As a result, equipment manufacturers are rapidly developing data monitoring capabilities for their own products as well as providing for third-party monitoring of performance as well as fault diagnostics. If the system does not meet the performance objectives, the system installer must pay the owner the difference. With this type of contract, the installer will want to be the first to know if there is a problem with a system.

Before turning the system on, it is a good idea to explore what might happen if there is a problem. In particular, at any part of the system subject to high voltage and current, a short circuit could cause an extremely dangerous condition resulting from intense heat from an electrical arc. Thus, it is relevant at this point to pay some attention to arc-fault calculations.

4.9.6 Arc-Fault Calculations

At voltages exceeding 120 V, it is possible to strike an arc between two conductors of different potentials, as any electrician who has lost part of the head of a screwdriver in a live switch enclosure can attest. For that matter, any time a switch is switched off, interrupting current, at least a small arc is created just at the instant the contacts break. If the source of the arc is capable of delivering high currents at high voltages, the arc can be substantial and possibly can be sustained and sufficient to be life and limb threatening. The arc itself can reach temperatures as high as three times the surface temperature of the sun [18]. If the arc is sustained and if the distance between arcing components is large enough, substantial power can be contained in the arc that will produce intense heat that will radiate away from the arc. If a person is too close, significant and, possibly, fatal, burns may result. Arc flash can occur as a result of human error when working on live components, or it can also result from equipment failure, such as switching off a switch that has corroded contacts as a result of not being used over a long period. It is thus important to take precautions for possible equipment failure when switching a large switch either off or on.

The most common arc flash incidents involve high-energy AC circuits and/or components, primarily because there are a lot more of them. Imagine standing in front of an oven with a 60,000 W heating element. It certainly would not take long to notice that the oven is turned on. Now imagine this 60,000 W being contained in a relatively small space at a temperature of nearly 20,000°C (36,000°F). Planck's radiation law, as discussed in Chapter 2, shows that the radiated power density is proportional to T^4 (T in K). As a result, one need not even come in contact with the arc to sustain significant injury. The question is how far away one must be to avoid second-degree burns.

IEEE Standard 1584–2018, Guide for Performing Arc Flash Hazard Calculations, provides computational methods, based upon 1800 empirical studies, to determine the level of energy and the distance required to avoid second-degree burns for AC arc flash [19]. To determine the key quantities of interest, one must first estimate the arc flash current, the incident energy and the flash boundary. To make these estimates, one needs estimates of the maximum fault current, clearing time of the protective device and the expected distance between a worker and the energized object. It is also important to distinguish AC voltage sources, such as utilities or inverters from DC current sources, such as PV source circuits or PV output circuits. Since the calculations are based upon empirically derived data, they are more than trivial. It is recommended that anyone who might be designing equipment capable of producing life-threatening arc flash consider obtaining suitable software to assist in the calculation. Since the purchase price of this standard is listed at $1230.00, it is likely that most required arc flash calculations will now be performed by experts and that at some point building departments and utilities will have established current and voltage levels above which arc flash calculations will be required. Stay tuned.

4.9.7 SYSTEM COMMISSIONING

Up to this point, an important part of PV engineering has been taken for granted. Presumably, all components used in a PV system have been factory tested and can be assumed to operate as specified. Thus, once the system has been designed and installed, it has been assumed that it will work as predicted.

While this is mostly true for small systems with small numbers of components, when systems have thousands of modules, hundreds of source circuits and a lot more wire and enclosures, it is much easier to miss something during system installation. Thus, for larger systems, it is good practice to have a third party conduct a thorough commissioning procedure on the system. As a minimum, system commissioning should include

- A visual check of all components and connections
- Verification of proper polarity of all source circuits
- Measurement of V_{OC}, V_{mp}, I_{SC} and I_{mp} for all source circuits and conversion to STC values using correction for cell temperature and irradiance at the time of measurement
- A high-voltage insulation check on all wiring to be sure no connections or insulation is faulty
- A check to verify that all module frames and mounting frames are adequately grounded.
- Inverter performance testing
- Measurement of I–V curves of each source circuit (optional)

The visual check should reveal any loose clamps or hardware, any open module connections, any missing junction box covers, whether expansion joints have been incorporated into long conduit runs and whether array wiring is neatly and securely fastened to the modules or array mount or cable tray(s). It should also be a check on missing fuses and whether all switches are in the open position.

Verification of proper polarity is very important and easily accomplished by voltage measurements within the combiner boxes with the fuses open. It is important to take proper safety precautions to minimize any chance of injury from arc flash when making any of these or the following measurements on live equipment.

Measurement of source circuit voltage and current parameters will indicate whether all modules are functioning properly and whether there are any loose or high-impedance connections between any modules. All source circuit electrical parameters, when normalized to STC, should be within a few percent of each other. In making these measurements, typically a specialized instrument can be used that will automatically cycle through the desired measurements for each source circuit and store the results in memory in a CSV format that can be imported into an Excel file. The important point to remember when performing these measurements is that when modules are shaded by clouds, their current drops instantaneously, but their cell temperature drops gradually. When the sun appears again, the current increases instantaneously, but the temperature rises gradually. It is thus important, to achieve reliable STC values, to wait for the irradiance and cell temperature to stabilize before recording voltage measurements. These measurements should normally be performed between 10 a.m. and 2 p.m. sun time or, at least, when irradiance levels are higher than 800 W/m². To minimize the risk of arc flash injury when making these measurements, it is helpful to have only one source circuit fuse connected at a time when measuring I_{mp} or V_{mp}. When V_{OC} and I_{SC} are measured, all fuses should be open.

The high-voltage insulation check will reveal any potential short circuits or even high-impedance connections to the ground. This check is done with a megger, which is an ohmmeter with a high-voltage internal DC source. At least one commercially available instrument incorporates an insulation test in addition to the voltage and current tests.

When performing this test, usually on an individual source circuit basis from the combiner box, it will likely be necessary to disconnect the grounded conductor connection between inverter and combiner if there is one, since the grounded conductor is grounded via the GFDI fuse in the inverter. If this conductor is not disconnected, the megger test will show a short circuit between the source circuit conductors and the ground. An interesting nuance of performing this test during early morning hours is that sometimes moisture will condense on the internal connections of the combiner box, causing resistance to ground on the order of 10 kΩ, when normally this resistance should exceed 1 MΩ. Repeating the test later in the day or after drying out the enclosure will reveal whether moisture is the problem. Since this test is not irradiance-dependent, except for waiting for moisture to clear, it can be performed when irradiance is less than 800 W/m².

The equipment grounding check can also be performed outside the 800 W/m² window. This check involves connecting a long wire to the grounding terminal of

Grid-Connected Utility Interactive PV Systems

the combiner box and then running this wire to each of the modules of the array, where an ohmmeter is used to check the resistance between the module frame and the mounting frame and the combiner box ground. Since the wire between the combiner box and the module may be several hundred feet long, its resistance must be subtracted from the measured resistance value. The result should be less than 5 Ω for each measurement taken. If it is higher, then an immediate check can be made for any loose mechanical or electrical connections at the module or frame.

The inverter performance check is made last. To perform this check, all source circuit fuses need to be inserted, all combiner box circuit breakers need to be on and all switches to the utility must be closed to enable the inverters to function. At this point, the inverter can be started in accordance with the manufacturer's instructions, which usually involves turning on the DC and AC disconnects and then waiting the five minutes required by UL 1741 for something to happen. Generally, the inverter will have a display that shows input and output voltage and input and output current. If the inverter incorporates the possibility of power factor adjustment, either manually or automatically, this also needs to be noted. Ideally, this check will be an efficiency check of the inverter, which is determined by the ratio of output power to input power. In fact, this measurement is more difficult than one might imagine, simply because of catastrophic subtraction errors. For example, if the inverter efficiency is supposed to be 98%, and there is 0.5% uncertainty in measuring each input power and output power, then it is possible that the *computed* efficiency, based on these measurements, will be lower than 98% or higher than 100%. Thus, an error analysis should be made when the uncertainties in these quantities are known. It is good practice to measure the input power and output power at least five times and then take the average. This results in a lower uncertainty of inverter performance. If any of the inverter checks fail, it is important to have the instruction book and the manufacturer's phone number handy.

If the owner or designer specifies I–V checks as part of the commissioning, instrumentation is available for these checks. Strange I–V checks, having more than one maximum power point or some other unusual or unexpected behavior, point to a source circuit that needs additional attention. Generally, the I–V tester manual will have suggestions for causes of unusual I–V curves.

Homework Problems

4.1 Refer to Table 4.2 or to the *NEC* to:
 a. Determine the wire size needed to limit voltage drop to 3% for a 100-watt, 24-volt load at a distance of 75 feet from the voltage source.
 b. Using the wire size determined in part a, determine the actual voltage drop for the wiring.

4.2 An inverter has a maximum output current of 32 A at 240 V.
 a. Determine the wire size needed to carry this current, assuming it is classified as continuous.
 b. Determine the distance at which the % voltage drop reaches 2%.

4.3 Show that $\alpha = \pi f_o / Q$ for a parallel RLC circuit.

4.4 Show that $f_o = \dfrac{\omega_d}{2\pi\sqrt{1-\dfrac{1}{4Q^2}}}$ for a parallel RLC circuit.

4.5 Design a resonant (RLC) load with Q = 5 that will dissipate 1000 watts at 120 V (rms) at 60 Hz. Then use SPICE or your favorite network analysis program to simulate the SFS and SVS algorithms by connecting a sinusoidal current source to the load and varying the frequency above and below 60 Hz and observing the total response (transient plus forced) as the excitation frequency moves farther away from 60 Hz.

4.6 Using NREL SAM, determine the annual kWh output of the PV array of Section 4.5 if the slope of the roof is 7:12. Use a DC:AC ratio of 1.4. How does this compare with the annual output of the 5:12 slope roof?

4.7 For the 5:12 roof slope of Section 4.5.1, if the 7.6 kW array is facing SE, what would be the expected annual array kWh output? Use a DC:AC ratio of 1.4 and then repeat with a DC:AC ratio of 1.1.

4.8 Using the time series display option of NREL SAM, plot the 7.6 kW south-facing at 22.6° array output on May 15 and use the numeric hourly data to estimate the daily kWh output. Then repeat the exercise for June 15. Comment on the accuracy of these numbers.

4.9 Look up a few more modules that might be used to achieve a 7.6 kW array within a few percent. List their power ratings and their values of V_{OC}, I_{SC}, V_{mp} and I_{mp} under standard test conditions. Then verify, using the specified module dimensions, that the new array will still fit on the roof section of Figure 4.3.

4.10 Another microinverter that might have been specified in Section 4.5.3 and 1.45 is rated at 349 VA A maximum output. Using the modules selected in Section 4.5.2, determine a set of string lengths that will ensure that no string current will exceed 16 A. Then using the DC:AC ratio that results for this microinverter and the 405 W module, simulate the system performance for the original design conditions and compare the predicted annual system kWh results for the two different microinverters. The new microinverter has the same input voltage and current limits as the original choice.

4.11 Check to see whether any changes will be needed in the utility interconnect if the 349 VA microinverter is used instead of the 290 VA unit.

4.12 Create an Excel worksheet with the row and column entries that are shown in Table 3.10. Then enter formulas in the current and voltage cells to calculate the string currents when the string voltage and optimizer power outputs are known. Then adjust module powers to determine the set of module powers for which an optimizer input voltage will reach 60 V. Experiment a bit more by adjusting string lengths to explore how this affects the optimizer input voltages of the string. Then experiment with adjusting optimizer output powers until the maximum string current of 15 A is reached.

4.13 Assume the service of the home with the optimizer system consists of a 400 A meter assembly that feeds two different 200 A distribution panels, each of which has a 150 A main circuit breaker. Each panel has a separate set of 3, 1/0 Cu (L1, L2, N) between the meter enclosure and the main circuit breaker of the distribution panel.

Grid-Connected Utility Interactive PV Systems

Explore possible utility interconnect options. Note that any current fed into a feeder by the PV system cannot exceed the 150 A rating of the feeder conductors. List any assumptions.

4.14 Show two options for connecting the strings in the optimizer example, limiting string number to 9 (3 for each inverter) and string length to 12 modules. This means it will be necessary to create an array layout by either making a copy of Figure 4.5 or sketching out the layout to indicate which modules are connected to each other to form the strings. Note that the optimizers are mounted beneath the modules, so all that is needed is a line that joins all modules in each string. Then number the strings. A look at Figure 4.9 might also be useful.

4.15 Using the module of Section 4.7.2, if an inverter has a maximum input DC voltage of 1000 V and an MPPT range from 350–840 V, determine the maximum and minimum string lengths for Duluth, MN, assuming the inverter is capable of handling the maximum module current.

4.16 Verify the claim in Section 4.8.3 that the maximum module VOC in the Dallas area will be 56.4 V.

4.17 For the inverter of Section 4.8, determine the string maximum and minimum length limits if 400 W modules are used for the array instead of 500 W modules.

4.18 a. For the array of Section 4.8, determine the longest possible distance between the rooftop junction box and the inverter that will keep the %VD in the wiring between the junction box and inverter less than 0.87%.

b. Determine the longest distance between the rooftop junction box and the inverter that will ensure a %VD less than 1.87%. Note that these two distances represent 2% overall %VD and 3% overall %VD.

4.19 An inverter has a balanced three-phase, 277/480 V output and is installed a distance, d, ft from the point of utility connection. If the maximum inverter output current is 14.4 A, and the inverter is located 200 ft from the PUC, determine a wire size that will have adequate ampacity if the wire has 90°C insulation and will have a voltage drop of less than 2% between inverter and PUC.

4.20 Do some research on Agrivoltaics and then list some possible agricultural uses for the PV system of Section 4.9 that would be compatible with the PV system.

4.21 a. Compare the lowest array operating voltage for 26-module strings with the lowest operating voltage of 25-module strings for the system of Section 4.9.

b. What is the minimum string length that will keep the string operating voltage higher than the minimum inverter MPPT input voltage?

4.22 Look up monitoring systems on the internet to get a feeling for what is available.

4.23 Specify all the components for a nominal 6000-watt ground-mounted utility-interactive PV system. Assume the code in effect is the most recent version of the NEC. Use a line side connection for the utility interconnect.

4.24 Specify all the components and show the design for a nominal 5000-watt residential rooftop-mounted, utility-interactive PV system, based on the most recent version of the NEC. The system will be located in Atlanta, GA, and will be standoff mounted on a roof with a 30° tilt. Connect the system on the load side of the main circuit breaker, specifying the allowable sizes of the main circuit breaker and distribution panel busbars if the PUC is to be in the distribution panel. Use NREL SAM to estimate the monthly and annual performance of the system.

4.25 For the roof shown in Figure p4.1, assume there are no obstructions in the area shown and make scale drawings that show the following:
   a. How a set of modules could be mounted within roof zone 1 (in this case, more than 3 ft from the edge) to maximize the array power. Select whatever module you feel will best accomplish the design goal. Be sure the modules can be configured to meet the input voltage requirements of an inverter that has $V_{OC}(max) = 600$ V and $250 < V_m < 520$ V as the MPPT range. Specify the approximate power rating needed for the inverter.
   b. How a set of modules could be mounted on the entire available area of the roof, without extending beyond the roof. Select whatever module you feel will best accomplish the design goal. Be sure the modules can be configured to meet the input voltage requirements of an inverter that has $V_{OC}(max) = 600$ V and $250 < V_m < 520$ V as the MPPT range. Specify the approximate power rating needed for the inverter.

4.26 a. The south-facing portion of the roof shown in plan view in Figure p4.2 has a 5:12 slope. Select a module that has a rating of at least 380 W and layout to scale an array configuration on the south roof. Select a suitable inverter for use with the system.
   b. On the east-facing and west-facing portions of the roof, each of which also has 5:12 slopes, configure modules that use microinverters to fill as much roof as possible.

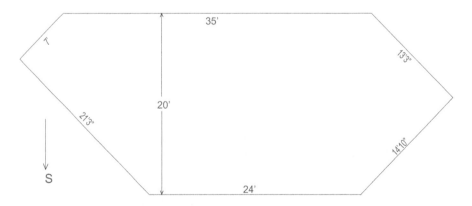

FIGURE P4.1

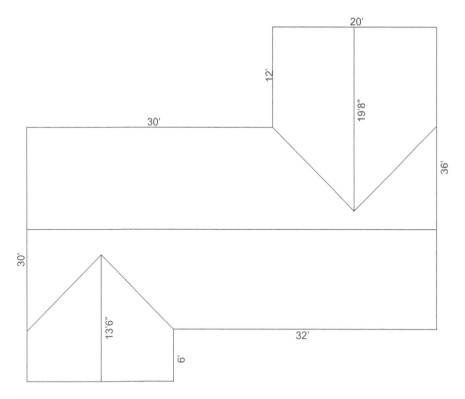

FIGURE P4.2

      c. Assume your configurations of parts (a) and (b) are in Sacramento, CA. Use NREL SAM to estimate total PV production on a monthly and annual basis of the combined set of arrays.

4.27 A 120 kW utility-interactive system is to be installed as a parking lot canopy. It is to feed the grid with a three-phase, 277/480 V balanced output. Design a system, showing all necessary electrical components, including modules, inverter(s), wire sizes and fuse/circuit breaker sizes and locations. Recommend a horizontal layout of modules that will provide shading for cars, and estimate how many parking spots will be covered. Comment on your choice of whether to use a single inverter or multiple inverters.

REFERENCES

[1] IEEE 929–2000, "IEEE Recommended Practice for Utility Interface of Residential and Intermediate Photovoltaic (PV) Systems," *IEEE Standards Coordinating Committee 21, Photovoltaics,* 2000.
[2] IEEE 1547–2018, "IEEE Standard for Interconnection and Interoperability of Distributed Energy Resources," 2018. https://standards.ieee.org
[3] *NFPA 70 National Electrical Code,* 2020 ed., National Fire Protection Association, Quincy, MA, 2019.

[4] IEEE Standard 1561–2019 Optimizing Lead-Acid Battery Performance: https://standards.ieee.org/ieee/1561/7167/
[5] Standard for Testing Stand-Alone PV Systems: https://standards.ieee.org/ieee/1526/7761/
[6] Standards for PV Module Safety Qualification: www.ul.com/services/pv-module-certification
[7] IEEE Recommended Practice for Sizing Stand-Alone PV: https://standards.ieee.org/ieee/1562/10272/
[8] International Technical Commission Technical Committee 82 on Photovoltaics: www.energy.gov/eere/solar/articles/pv-standards-what-iec-tc82-doing-you
[9] UL Standard 1741: www.ul.com/news/ul-launches-advanced-inverter-testing-and-certification-program
[10] ASCE Standard 7-16, *Minimum Design Loads for Buildings and Other Structures*, American Society of Civil Engineers, Reston, VA, 2017.
[11] Narang, David, "Principal Engineer, NREL Power Systems Engineering Center," *Highlights of IEEE Standard 1547–2018*, October 28, 2019.
[12] Information on the Sandia Frequency Shift Anti-Islanding Method: www.sciencedirect.com/science/article/pii/S1110016816303520
[13] Information on the Sandia Voltage Shift Anti-Islanding Method: https://ietresearch.onlinelibrary.wiley.com/doi/full/10.1049/iet-pel.2020.0735
[14] CFR 47, "Federal Communications Commission, Rules and Regulations," *Radio Frequency Devices*, Part 15, 1994.
[15] System Advisor 2022 (SAM 2022.3.14), *National Renewable Energy Laboratory*, Golden, CO, 2022.
[16] "Information on the Evolution of Smart Energy Management and the Grid: Smart Energy Management for Industrials," *Deloitte Insights*. https://www2.deloitte.com>insights>power-and-utilities
[17] Information on Monitoring Platforms for PV Systems: www.energy.gov/femp/monitoring-platforms-solar-photovoltaic-systems
[18] El-Sharkawi, Mohamed, *Electric Safety Practice and Standards*, CRC Press, Taylor and Francis Group, Boca Raton, London, and New York, 2014.
[19] IEEE Standard 1584–2018, *IEEE Guide for Performing Arc Flash Hazard Calculations*, IEEE Standards Association, November 30, 2018.
[20] Kryzanowski, Tony, "Iowa Agrivoltaics: Growing Broccoli, Beekeeping and Solar Panels," *Q1*, 2024: www.altenerg.com

SUGGESTED READING

Brooks, Bill, "Rapid Shutdown for PV Systems: Understanding NEC 690.12," *Solar Pro Magazine*, No. 8.1, January–February 2015.
Brooks, Bill and White, Sean, *PV and the NEC*, 2nd ed., Routledge, New York, 2021.
NFPA 70E: Standard for Electrical Safety in the Workplace®, 2015 ed., National Fire Protection Association, Quincy, MA, 2014.
Shapiro, Finley R. and Radibratovic, Brian, "Calculating DC Arc-Flash Hazards in PV Systems," *Solar Pro Magazine*, No. 7.2, February/March 2014: http://solarprofessional.com/articles/design-installation/calculating-dc-arc-flash-hazards-in-pv-systems
Smith, David, *Arc Flash Hazards on Photovoltaic Arrays*, Colorado State University, Department of Electrical and Computer Engineering, Fort Collins, CO, 2013.

5 Structural Considerations

5.1 INTRODUCTION

Because the primary function of a photovoltaic (PV) system is to convert sunlight to electricity, often the role and importance of the structural aspects of the system are overlooked. Most PV modules are designed to last 25 years or longer. It is important that the other components in the system, including structural components, have lifetimes equivalent to those of the PV modules. It is also important that the structural design requirements of the system be consistent with the performance requirements as well as with the operational requirements of the system. The structural design of PV systems cuts across a variety of disciplines, most notably civil and structural engineering and, to a lesser extent, materials science, aeronautical engineering and architecture. More specifically, structural design involves:

- Determining the mechanical forces acting on the system
- Selecting, sizing and configuring structural members to support these forces with an adequate margin of safety
- Selecting and configuring materials that will not degrade or deteriorate unacceptably over the life of the system
- Locating, orienting and mounting the PV array so that it has adequate access to the sun's radiation, produces the required electrical output and operates over acceptable PV cell temperature ranges
- Designing an array support structure that is aesthetically appropriate for the site and application and provides for ease of installation and maintenance

Each of these elements of the structural system will be discussed in more detail throughout this chapter.

5.2 IMPORTANT PROPERTIES OF MATERIALS

5.2.1 COLUMN BUCKLING

Often structural members used to support a PV array have one dimension, their length, which is much larger than the other dimensions. Such members or columns, if long enough and slender enough, can fail due to buckling under compressive loading. These compressive forces may be considerably smaller than those that would cause failure due to crushing. Although there is no firm rule, generally a long, slender structural member is considered to be a column if its length is more than 10 times as large as its least lateral dimension. For a column made of a material with a modulus of elasticity, E, a minimum moment of inertia (of its cross-sectional area), I, and length, L, there exists a critical compressive force, P, that will cause the column to buckle. This critical load is calculated using the formula:

$$P = \frac{EI\pi^2}{L^2} \tag{5.1}$$

This equation is known as *Euler's formula* [1].

5.2.2 Thermal Expansion and Contraction

Changes in temperature cause materials to expand or contract. For example, the linear deformation of a structural member, including an electrical conduit, of length L can be represented by:

$$\delta_t = \alpha L \Delta T \tag{5.2}$$

where α is the coefficient of linear expansion and ΔT is the change in temperature with units consistent with the units of α.

5.2.3 Chemical Corrosion and Ultraviolet Degradation

A PV system structure that can initially withstand all anticipated mechanical and thermal stresses with an adequate margin of safety might not be a good engineering design. The engineer must also consider the environment in which the system operates and how its component materials interact with the environment and with each other. Failure of metal components by corrosion is as common as failure due to mechanical stresses.

To design for corrosion resistance, knowledge of the various forms of corrosion is necessary.

Uniform attack is the most common form of corrosion. Chemical or electrochemical reactions proceed uniformly over the entire surface area. The metal becomes thinner and thinner and eventually fails. On a tonnage basis, uniform attack is the greatest cause of metal destruction, especially steel. However, it is not particularly difficult to control through proper material selection and appropriate use of protective coatings.

Galvanic corrosion occurs because of the potential difference that exists between two dissimilar metals in contact with a corrosive or conducting solution. Of the two dissimilar metals, the less resistant, or anodic, metal is corroded relative to the more resistant, or cathodic, metal. Table 5.1 [2] presents an abbreviated version of the galvanic series for commercially available metals and alloys. The relative position in the galvanic series depends on the electrolytic environment and on the metal's surface chemistry. To minimize galvanic corrosion, the design engineer should use similar metals or metals close to each other in the galvanic series.

The use of a small anodic metal in contact with a cathodic metal of a larger surface area should be avoided. If two metals far apart in the galvanic series must be used in near contact with each other, they should be electrically insulated from each other. Coating the anodic material may not protect it, because coatings are susceptible to pinholes, causing the coated surface to corrode rapidly in contact with

Structural Considerations

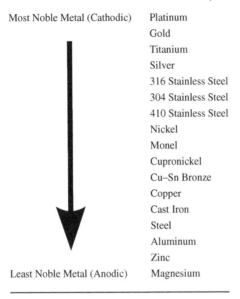

TABLE 5.1
Galvanic Series for Metals and Alloys

Most Noble Metal (Cathodic)	Platinum
	Gold
	Titanium
	Silver
	316 Stainless Steel
	304 Stainless Steel
	410 Stainless Steel
	Nickel
	Monel
	Cupronickel
	Cu–Sn Bronze
	Copper
	Cast Iron
	Steel
	Aluminum
	Zinc
Least Noble Metal (Anodic)	Magnesium

the large cathodic area. If a galvanic couple is unavoidable, a third metal that is sacrificial to both of the other metals may be used. Note that zinc is sacrificial to both aluminum and steel and, consequently, is often used to protect these two most common structural materials.

Crevice corrosion is an intense form that frequently occurs at design details such as holes, gasket surfaces, lap joints and crevices under bolts and rivet heads. Small quantities of stagnant liquid often form in these areas and allow this very destructive electrochemical process to occur. Unfortunately, stainless steel is especially susceptible to crevice corrosion.

Pitting is an extremely localized attack that produces holes in the metal. It is serious because it may lead to very premature failure of structural members. Pitting may require a relatively long initiation period but, once the process is started, it accelerates rapidly.

Intergranular corrosion is a localized attack along the grain boundaries of metals but not (or only slightly) over the grain faces. It is common in steel that has been heat treated at high temperatures or heat sensitized during welding. Intergranular corrosion due to welding is known as *weld decay*.

Selective leaching is the preferential removal of one or more of the alloying elements in a metal by the action of an electrolyte. The most common example of this phenomenon is the selective leaching of zinc from brass, leaving a spongy, weak matrix of copper—a process known as *dezincification*.

Aluminum, iron, cobalt and chromium are also susceptible to leaching. Where and whenever selective leaching occurs, the process leaves the alloy in a weakened, porous condition.

Erosion corrosion occurs when corrosive fluid flows over a metal surface and causes the gradual wearing away of the surface. Usually the velocity of the fluid is high and mechanical wear and abrasion may be involved.

Stress corrosion cracking is caused by the combination of tensile stress and corrosion and leads to the cracking or embrittlement of a metal. The stress may result from applied forces or may be residual. Only specific combinations of alloys and the chemical environment produce stress corrosion cracking. Combinations include aluminum alloys and saltwater, some steel alloys and saltwater as well as mild steel and caustic soda. Preventing stress corrosion cracking involves selecting alloys that are not susceptible to cracking under expected operating conditions. If this is not possible, the structural members should be sized for low stress levels.

Dry corrosion involves the reaction of the material directly with air. Virtually every metal and alloy reacts with the oxygen in the air to form an oxide. This occurs in the absence of any liquid electrolyte and is an important form of corrosion in high-temperature applications.

In addition to corrosion, ultraviolet radiation can cause degradation and deterioration of some of the exposed materials used in PV systems. These include weather-sealing materials, conduits for wiring, wire insulation, some coatings and caulking and other miscellaneous materials. Ultraviolet radiation is that part of the electromagnetic spectrum having wavelengths between about 100 and 400 nm. The ultraviolet portion of the spectrum can be further subdivided into three regions: UV-A (315–400 nm), UV-B (280–315 nm) and UV-C (100–280 nm). UV-B and UV-C cause sunburn (erythema) and pigmentation (tanning). UV-B produces vitamin D_3. Long-term exposure to UV radiation results in loss of skin elasticity. Also, a link has been established between exposure to UV wavelengths below 320 nm and skin cancer. Because the human body is seriously affected by UV exposure, it should come as no surprise that other materials are also affected. In fact, ultraviolet radiation degrades both the optical properties and the physical properties of many materials. Consequently, the design engineer should ensure that all materials in the PV system that are exposed to sunlight, such as wire, conduit and connectors, are resistant to UV degradation.

5.2.4 Properties of Steel

Iron is the base element of all steel. Commercially pure iron contains only about 0.01% carbon and is relatively soft and weak. The addition of carbon significantly strengthens iron. For example, the addition of 0.80% carbon can raise the tensile strength from about 40,000 psi (276 Mpa) to 110,000 psi (758 Mpa) [3]. Steel is an alloy of iron and carbon and usually contains small amounts of manganese and other elements. There are more than 3500 grades of steel, of which nearly 75% have been developed over the last 50 years. The many different grades of steel have many different properties: physical, chemical and environmental. *Carbon steel* is steel that owes its distinctive properties to the carbon it contains. *Alloy steel* is steel that owes its distinctive properties to one or more elements other than carbon, or to the combination of these elements with carbon. Low-carbon steel is often used in the support structures for PV systems. The relatively low cost of these steel products gives them a major advantage over more expensive aluminum, provided they can be protected adequately from corrosion.

Many different methods are used to protect steel from corrosion. In general, these protection or reduction methods include proper selection of materials, design, coatings, inhibitors, anodic protection and cathodic protection [4]. One method of protecting steel structural members is hot-dip galvanizing. Hot-dip galvanizing is a process in which thoroughly cleaned steel or iron is immersed in molten zinc and withdrawn to provide a smooth, even coating that typically has a crystalline appearance. It is by far the most widely used method for protecting steel against corrosion [3].

Zinc is a less noble metal than steel (see Table 5.1) and is sacrificed to protect the steel as part of the corrosion process. How long the zinc protects the steel depends on the thickness of the zinc coating and the environment to which the structural member is exposed. The average minimum weight of zinc coating is 1.5 oz/ft^2 (0.305 kg/m^2) of surface area when hot-dip galvanized according to the specifications of ASTM A123. For highly acidic industrial atmospheres, a 1.5 oz/ft^2 coating may last less than 15 years. For a rural, clean atmosphere, the same coating may last 40 years.

Lead is sometimes added to zinc to produce a surface pattern on steel called *spangle*, which is often used on unpainted surfaces. The addition of aluminum to zinc improves its corrosion protection ability. *Galfan*, which contains 5% aluminum and 95% zinc, and *Galvalume*, which contains 55% aluminum and 45% zinc, are examples of zinc-aluminum coatings that effectively protect steel from corrosion [4].

Paints, lacquers, coal-tar or asphalt enamels, waxes and varnishes are all used as organic coatings for corrosion protection. However, all paints are permeable to water and oxygen to some degree. They are also subject to mechanical damage and eventually break down.

5.2.5 PROPERTIES OF ALUMINUM

Aluminum is the second most commonly used structural material next to steel. Some of the properties of aluminum that make it attractive for use in PV systems follow:

- It is very lightweight, with a density of about one-third that of steel.
- It can be fabricated into many different shapes using a variety of different methods.
- It has a wide range of properties and is available at many different tensile strengths, depending on how it is alloyed.
- It has a high strength-to-weight ratio, thus making it attractive for many applications.
- It has good corrosion resistance and can be used in a wide variety of climates and weather conditions.
- It is available in a wide variety of finishes making it attractive as an architectural metal.

Aluminum is one of the most fascinating of all the elements. It is very chemically active, with a strong affinity for oxygen. In powdered form, it is used as a fuel for solid rocket motors. How can it be that such a chemically active material is so commonly used in an unprotected form?

What makes aluminum unique is its special relationship with oxygen, which destroys other reactive metals such as sodium, and seriously attacks less reactive metals such as iron. In the laboratory, most chemistry students have witnessed the spectacular reaction between sodium and oxygen in water, in which the sodium is left unrecognizable and even more unusable. And everybody frequently encounters rusting steel. Neither of these phenomena occurs with aluminum. Rather, as soon as aluminum is produced, its surface reacts with the oxygen in the air and forms a very thin oxide layer over the entire surface. This oxide layer is hard and tenacious and not only protects the aluminum from attack by other chemicals but also protects it from further oxidation. Because the oxide layer is transparent, it does not detract from the metal's appearance. Aluminum oxide is an insulator and therefore must be removed from aluminum wire before connections are made.

Pure aluminum is relatively weak and only about one-third as stiff as steel. However, aluminum can be strengthened significantly by alloying. The most common alloying additions to aluminum are copper, manganese, silicon, magnesium and zinc. Aluminum and its alloys are divided into two major classes: wrought and cast. The wrought class is broad because aluminum can be formed by virtually every known fabrication process, including sheet and plate, foil, extrusions, bar and rod, wire, forgings and impacts, drawn or extruded tubing and others. Cast alloys are poured molten into sand (sand casting) or into high-strength steel molds (i.e., permanent molds or die casting) and allowed to solidify into the desired shape.

For PV systems, wrought aluminum is more commonly used than cast aluminum. The Aluminum Association has a designation system for wrought aluminum alloys

TABLE 5.2
Designation System for Wrought Aluminum Alloys [5]

Alloy Se	Description or Major Alloying Element
1xxx	99.00% minimum aluminum
8xxx	Other element
2xxx	Copper
3xxx	Manganese
4xxx	Silicon
5xxx	Magnesium
6xxx	Magnesium and silicon
7xxx	Zinc
9xxx	Unused series

(From Kutz, M., *Mechanical Engineers' Handbook*, 2nd Ed, John Wiley & Sons, 1998. Adapted by permission of John Wiley & Sons, Inc.)

that categorizes them by major alloying additions. Table 5.2 [5] shows the various designations. The first digit in the series classifies the alloy-by-alloy series or principal alloying element. The second digit, if different from 0, denotes modification of the basic alloy. The third and fourth digits together identify the specific alloy within the series.

Alloys in the 6xxx alloy series, which use magnesium and silicon as alloying elements, possess a combination of properties including corrosion resistance, that make them well suited for structures and architectural applications. They are also commonly used for marine applications, truck frames and bodies, bridge decks, automotive structures and furniture. From the 6xxx series, 6061 and 6063 aluminum alloys are often used as structural members for PV arrays. 6061 aluminum contains 1.0% magnesium and 0.6% silicon. For the same tempering during fabrication, 6061 aluminum alloy has higher tensile strength than 6063 aluminum, which contains 0.7% magnesium and 0.4% silicon [5]. However, 6063 aluminum has better resistance to corrosion.

5.3 DESIGN AND INSTALLATION GUIDELINES

5.3.1 STANDARDS AND CODES

Prior to the establishment of modern standards, structural members, nuts, bolts, screws and so on, were custom designed and manufactured. For example, one manufacturer would produce ½-inch bolts with 9 threads per inch, while another would use 12 threads per inch. Some fasteners had left-handed threads and had different thread profiles. In the early days of the automobile, mechanics would lay out fasteners in a row as they were disassembled to avoid mixing them during reassembly. A lack of standards leads to a lack of uniformity and precludes efficient interchangeability of parts. It is inefficient and costly.

A *standard* is a set of specifications for parts, materials or processes intended to achieve uniformity, efficiency and a specified quality. Another important purpose of standards is to limit the number of items in the specification and thereby provide a reasonable inventory of tooling, sizes and varieties so custom parts will not be required.

A *code* is a set of specifications for the analysis, design, manufacture, construction or installation of something. The purpose of a code is to achieve a specified degree of safety, efficiency and performance or quality. It should be noted that safety codes do not imply absolute safety, which is impossible to attain. Designing a structure to withstand 120-mph (193 km/h) winds does not mean the designer thinks 140-mph (225 km/h) winds are impossible. It just means the designer thinks they are highly improbable.

The following organizations are involved in developing standards and codes relevant to the structural design of PV systems:

- Aluminum Association (AA)
- American Institute of Steel Construction (AISC)
- American Iron and Steel Institute (AISI)
- American Society of Civil Engineers (ASCE)

- American Society of Metals (ASM)
- American Society of Mechanical Engineers (ASME)
- American Society of Testing and Materials (ASTM)
- Society of Automotive Engineers (SAE)

Building codes are design, construction and installation guidelines. They are adopted and enforced by local jurisdictions to help ensure the safety of individuals and the protection of structures and other property in and around buildings. Building and electrical codes have value only if they are followed and enforced. The process begins prior to the installation of a PV system with an application for a building permit. After the installation, the building official inspects the system to determine compliance with the relevant codes. Unfortunately, many solar systems have been installed in the past without first obtaining a building permit. Most PV systems installed prior to the mid-1990s were stand-alone systems and may not have met present code requirements. However, with the renewed interest in PV systems for buildings as a result of financial incentive programs and government initiatives, most new systems comply with local codes. And utility-interactive PV systems are more likely to be inspected by building code officials than are stand-alone systems because utilities will typically not allow an interconnect without a signed interconnect agreement that requires inspection of the system, usually by the authority having jurisdiction and sometimes by the utility as well.

5.3.2 Building Code Requirements

The most important standard affecting the structural design of PV systems is entitled *Minimum Design Loads for Buildings and Other Structures* [6]. As noted in previous chapters, it is a standard of the ASCE and is periodically updated. It provides the minimum load-carrying requirements for buildings and other structures, and these requirements apply to PV systems. It is important to note that standards have meaning only when they are adopted by an enforcing jurisdiction and become part of their code. Many local jurisdictions reference parts of this ASCE standard in their building codes. However, local building codes are less uniform than local electrical codes, almost all of which use the *National Electrical Code* [7].

The basic requirements of building codes are that the PV system and any building or other structure to which it is attached shall be designed and constructed to safely support any structural load or combination of loads to which they are subjected. In other words, the ASCE standard provides formulas for computing the structural loads from wind, snow, ice and so on. The engineer computes these loads for the conditions expected at the site of the PV system. The stresses these forces produce must not exceed the appropriately specified allowable stresses for the structural materials being used.

In addition to establishing strength requirements, building codes often limit the deformations that can occur with physical structures such as PV systems. For example, there may be restrictions on the amount of deflection or lateral drift of a structure. Such a deflection might have an adverse effect on the use of the system or attached buildings or structures. Because vibration can lead to fatigue failure, the

Structural Considerations

local building code may contain provisions that address the stiffness of the structure and the associated frequencies and amplitudes of structural vibration.

5.3.3 Aesthetics

It is important for arrays on buildings to be aesthetically pleasing. They should be designed to blend into the building lines and colors. In designing PV arrays for building applications, the architect and engineer should consider the following suggestions:

- Mounting the array parallel to the roof (if appropriate)
- Using the roof dimensions to establish the array aspect ratio
- Avoiding harsh contrasts and patterns
- Making the mounting hardware as inconspicuous as possible
- Avoiding shading, even if appearance suffers somewhat

Many of the new building-integrated PV (BIPV) products add to the architectural attractiveness of buildings. However, even in the case of stand-off mounting, the additional lag screws driven into the trusses result in a more secure fastening of the roof decking to the trusses.

5.4 FORCES ACTING ON PV ARRAYS

5.4.1 Structural Loading Considerations

Because solar energy is a relatively dilute resource, PV modules and arrays are area intensive. These large-area solar electric devices are capable of transmitting a variety of forces to themselves, their support structures, buildings and other frames or foundations to which they are mechanically attached. These forces include the dead weight of the array, the weight of array installation and maintenance people and their equipment (live weight), aerodynamic wind loading, seismic effects and the forces due to rain, snow, hail and ice. Determining these forces under prescribed conditions can be an interesting and challenging structural problem.

Of the various types of forces mentioned earlier, aerodynamic wind loading presents the most concern. At most locations around the world, the force effects due to wind loading are much higher than the other forces acting on the structure. For example, both dead and live weight loads due to a PV array on a building are usually less than 4 pounds per square foot (psf) (191 Pa). In contrast, the computed wind loads are typically between 24 psf (1149 Pa) and 55 psf (2630 Pa), but sometimes greater than 100 psf (4790 Pa), depending on location and the associated design wind speed for that location.

Because of the geographic variation in environmental conditions such as rain, snow, ice, wind speed, seismic activity, etc., local building codes tend to be much less uniform than electrical codes. Consequently, it is important for the PV systems engineer not only to become familiar with local building code requirements but also to compute the minimum structural design loads that are relevant for a given location. In addition to the previous considerations, the design engineer must select and configure the array of mounting materials so that corrosion and ultraviolet degradation are minimized. Also, if the array is mounted on a building, the structural integrity of the building must not be degraded and building penetrations must be properly sealed so that the building remains watertight over the life of the PV array.

The ASCE standard [6] provides instructions and formulas for computing the following types of loads:

- Dead loads
- Live loads
- Soil and hydrostatic pressure and flood loads
- Wind loads
- Snow loads
- Rain loads
- Earthquake loads
- Ice loads—atmospheric icing
- Combinations of the aforementioned loads

5.4.2 Dead Loads

Dead loads consist of the weight of all the materials that are supported by the structural members, including the weights of the structural members themselves. As an example, consider a PV array mounted on the roof of a building. The modules are structurally tied together in panels using steel or aluminum structural members. The weight of the modules is carried by the structural members and transmitted to the roof structure of the building. The total dead load that the roof must carry is the combined weight of the modules, structural members used to form the panels, attachment hardware and mounting brackets. These loads can be expressed in psf or Pa and assumed to act uniformly over the area covered by the array. The equations of static equilibrium can be used to determine the individual forces at each of the brackets that connect the panels to the roof. For PV systems, the dead loads are usually assumed to act uniformly over the supporting structure and are expressed in psf. Note that these are external loads—not internal stresses—even though similar units are used. Dead loads for PV systems generally fall within the range of 2 to 5 psf (95.8 to 239 Pa) and, even though they cannot be discounted, do not pose serious structural problems for the engineer.

5.4.3 Live Loads

Live loads associated with PV systems are those produced by individuals and their equipment and materials during installation, inspection and maintenance. In designing structures to carry live loads, the engineer may treat them as uniformly distributed loads, that is, in psf or Pa, as concentrated loads (lb or kg) or as a combination of uniformly distributed and concentrated loads. For PV arrays, live loads are usually assumed to be distributed uniformly and are small, on the order of 3 psf (144 Pa) or less.

5.4.4 Wind Loads

The forces from the wind acting on PV arrays are *aerodynamic forces*. As such, their magnitudes depend partly on the properties of the atmosphere. Atmospheric properties include static pressure, temperature, viscosity and density. Of these, density is one of the more important ones affecting the mechanical forces acting on PV systems. At any given location, density is usually assumed to be constant. However, density does vary with altitude—decreasing as altitude is increased. For example, the density in the mile-high city of Denver, CO, is about 86% of the atmospheric density for Florida cities, which are only slightly above sea level. The mass density of air for the standard atmosphere at sea level is 0.00256 lb-sec^2/ft^4.

The air flowing over a PV array produces two types of forces: *pressure forces* normal to the surface and *skin friction forces* along the surface. Skin friction forces are simply surface shear forces resulting from the viscosity of the air as it contacts the surface.

Most PV arrays are essentially large flat plates that are tilted at various angles to the direction of the flow of the wind. Also, the buildings and structures to which they are attached and the terrain in their vicinity may significantly affect the flow around them. This flow often produces complex velocity and pressure distributions that, in turn, produce forces and stresses in the structural members. The determination of wind loads on PV structures does not lend itself to theoretically derived solutions. Rather, aeronautical and civil engineers have together developed empirical formulas for estimating the structural loading resulting from exposure to various wind conditions.

At this point, it is useful to summarize the important variables and factors that affect the aerodynamic forces acting on PV arrays and structures. They include:

- Wind speed
- Effects of wind gusts
- Density of the air
- Orientation with respect to the wind direction
- Shape and surface area
- Elevation above the ground
- Effects of topographical and man-made features

The list is somewhat intimidating and shows that the best wind load analysis is not as accurate as some engineers would like. Nonetheless, with "ballpark accuracy" and adequate factors of safety, good design is definitely within the engineer's grasp.

Figure 5.1 shows a view of a modular array field designed by Hughes Aircraft Company for Sandia National Laboratories [8]. Figure 5.2 shows the *net* pressure distribution on the ground-mounted array and support structure for the combined wind and dead loads. The wind speed is 100 mph (161 km/h) and is directed toward the front of the array. The resultant force acts down and puts structural member BC in compression. Hence, the structure must be analyzed for buckling under this loading. From the pressure distribution, based on wind tunnel testing, the forces and stresses in all the structural members can be computed.

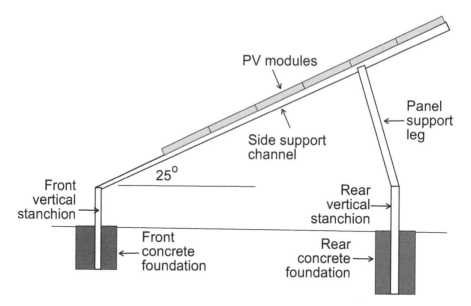

FIGURE 5.1 Sketch of the Hughes modular array field. This design was developed for Sandia National Laboratories. (From Marion, B. and Atmaran, G., *Preliminary Design of a 15–25 kWp Thin-Film Photovoltaic System*, prepared for the U.S. Department of Energy, Florida Solar Energy Center, Cape Canaveral, FL, November 13, 1987.)

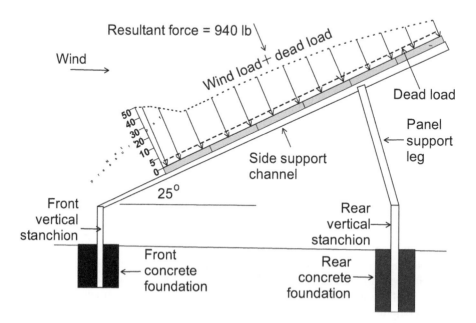

FIGURE 5.2 Net pressure on the array due to 100 mph winds toward the front of the array. Structural member BC is in compression.

Structural Considerations

Figure 5.3 shows the same array structure but with the wind directed toward the back of the array. Note that the resultant force is trying to lift the array from its supports and the structural support member BC is in tension.

Work done by Boeing showed that for single arrays, the wind forces were a minimum at array tilt angles of about 20° above the horizontal [9]. Surprisingly, maximum wind forces occur at array tilt angles of 10° to 15° and again at 90°. At 10° to 15° tilt, the array acts as a reasonably efficient airfoil and large aerodynamic lift forces result. For tilt angles of 20° and higher, the air separates from the array—similar to the flow over the wing of an aircraft when it is going into a stall [10].

Ground clearance also affects wind loading. For example, it has been found that increasing the ground clearance from 2 to 4 ft (0.61 to 1.22 m) increases the normal wind forces by 10% to 15% for an array with an 8-ft panel height [11]. For systems consisting of multiple rows of subarrays, wind tunnel tests have shown that the interior rows experience several times less wind loading than the exterior rows [12]. Consequently, wind breaks or wind fences are sometimes used to reduce the wind loading on the exterior rows.

Because of their large exposed surface areas, their elevation off the ground and their orientation with respect to wind direction, PV arrays are often subjected to exceptionally high structural forces. Not only do the modules in the array have to resist these forces, but also the attachments to roofs and/or other structures must be well secured.

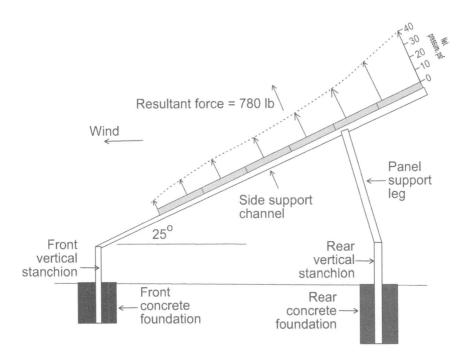

FIGURE 5.3 Net pressure on the array due to 100 mph winds toward the back of the array.

In designing rooftop PV arrays and structures to meet wind loads, the engineer must complete the following steps:

1. Determine the risk category.
2. Determine the design wind speed.
3. Determine the mean roof height.
4. Determine the exposure category.
5. Determine the module dimensions.
6. Determine the roof slope.
7. Determine the building dimensions.
8. Determine roof zones.
9. Determine the exposure coefficients.

Fortunately, the ASCE has developed standards and procedures for completing the aforementioned steps [6]. For example, the ASCE has developed maps indicating the basic wind speeds throughout the United States for each risk category (steps 1 and 2); definitions of exposure categories (step 4); definitions of roof zones based upon building dimensions (step 8) and exposure coefficients for different roof zones, roof slopes and roof classifications (hip or gable as common examples). Figure 5.4 shows wind speed contours for the eastern continental United States for risk category II buildings (ASCE 7–16 figure 26.5–1b). For a given location, the wind speed for each risk category increases as the risk increases. These additional wind maps can be found in ASCE 7–16 or 7–22. Note that interpolation is allowed between contours unless local building jurisdictions require rounding up to the next highest wind speed.

Risk categories depend upon the danger to human life if the building is damaged by external forces. Category I includes structures that represent a low risk to human life in the event of failure. Category II includes all structures not included in Categories I, III and IV. Typically, this will include a residential neighborhood. Category III includes buildings that could pose a substantial risk to human life if the structure fails which is not in Category IV. Category IV includes structures designated as essential facilities, such as hospitals, but it also includes structures that house materials that may be hazardous chemicals that might pose a significant danger to human life if the building fails. The design engineer should consult ASCE 7 for more details on Categories III and IV as a part of the design process if there is any doubt about the correct risk category.

Calculation of the wind load on a rooftop array generally begins with the calculation of the velocity pressure, q_z, using the equation:

$$q_z = 0.00256 K_z K_{zt} K_d K_e V^2 \qquad (5.3)$$

where K_z = velocity pressure exposure coefficient at height z
K_{zt} = topographical factor
K_d = wind directionality factor
K_e = ground elevation factor
V = basic wind speed in mph

Structural Considerations

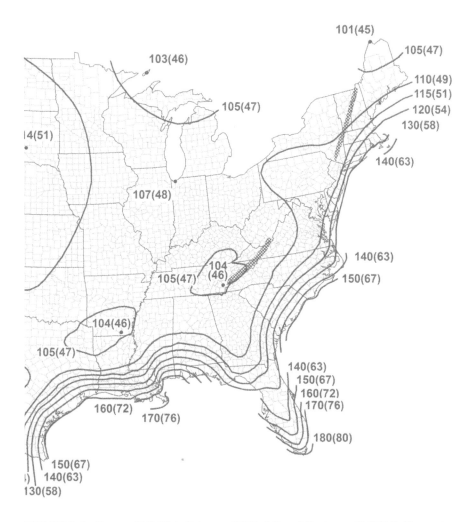

FIGURE 5.4 Eastern U.S. Risk Category II Wind Speed Contours. (ASCE 7–16, courtesy ASCE.)

Values of K_z as a function of height above ground and exposure category are tabulated in ASCE 7–16. Exposure categories depend upon the topography that surrounds the structure. Category B generally applies to buildings of a height less than 30 ft (9.14 m) that are surrounded by other buildings or trees to a distance of at least 1500 ft (457 m) on all sides. If the building height exceeds 30 ft, then the buildings and trees line extends to 2600 ft or 20 times the building height, whichever is greater. Category C applies to all cases where either Category B or Category D does not apply. Category D applies to areas where the surface in the direction of the wind is

TABLE 5.3
Tabulation of Values of K_z as a Function of Height and Exposure Category

Height in ft above Ground	Exposure B	Exposure C	Exposure D
0–15	0.70	0.85	1.03
20	0.70	0.9	1.08
25	0.70	0.94	1.12
30	0.70	0.98	1.16
40	0.76	1.04	1.22
50	0.81	1.09	1.27
60	0.85	1.13	1.31

(Adapted from ASCE 7–16)

flat or water for 5000 ft (1524 m) or 20 times the building height, whichever is larger. In addition, if the surface is B or C within a distance of 600 ft (183 m) or 20 times the building height, whichever is greater and then continues as Category D, the site is also classified as Category D. Table 5.3 shows representative values for K_z up to a height of 60 ft (18.29 m).

The topographical factor, K_{zt}, is defined by

$$K_{zt} = (1 + K_1 K_2 K_3)^2, \text{ where} \quad (5.4)$$

K_1, K_2 and K_3 are defined in ASCE 7–16 figure 26.8–1, a portion of which is shown in Figure 5.5. The figure incorporates the definitions and descriptions of an escarpment and a ridge or three-dimensional asymmetrical hill along with formulas and tables to determine K_1, K_2 and K_3. A set of multipliers applies to each case, based on the ratio of H/L_h. K_1 varies between 0.29 and 0.72 when H/L_h varies between 0.2 and 0.4 for a 2D ridge, between 0.17 and 0.43 for a 2D escarpment and between 0.21 and 0.53 for a 3D axisymmetrical hill. K_2 varies between 1.0 and 0.00 for values of x/L_h between 0.00 and 4.00 for a 2D escarpment and drops from 1.0 to 0.0 between $x/L_h = 0$ and 1.5 and remains at 0.0 for larger values of x/L_h for all other cases. K_2 also has a dependence on vertical height, z, compared to L_h for each of the 3 types of hill. By the time z reaches a height of L_h, K_2 drops from 1.0 to almost 0.0 for all 3 cases. Thus, the range of K_{zt} will be from 1 to $(1 + 0.72 \times 1.00 \times 1.00)^2 = 2.958$. Note that if a hill is involved, a careful look at this table is important, especially if the "hill" is anything but a 2D escarpment. In these cases, $K_2 = 0$ for any distance further upwind than 1.5 times the height of the "hill," which means for relatively small hills, it is likely that $K_{zt} = 1$. In fact, as the location is moved away from the hill, K_{zt} decreases to 1. Perhaps the most important feature of K_{zt} is that it is only > 1 if the exposure of the location is C, since exposure B already has so many buildings and trees as obstructions, there is insufficient clear space to locate a hill within the exposure area. In effect, the hill acts similar to a building.

The wind directionality factor, K_d, depends on the shape of the building. Only unusual shapes like chimneys and tanks have values of K_d different from 0.85. Since

Structural Considerations

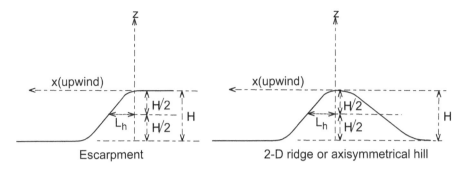

FIGURE 5.5 Defining parameters of an escarpment and a 2-D ridge or 3-D axisymmetrical hill. (Adapted from ASCE 7–16 with permission.)

it would be very difficult to install a PV system on a chimney, it is reasonable that for any location chosen for a PV system, the use of $K_d = 0.85$ will be an acceptable choice.

The ground elevation factor, K_e, takes into account the density of the air at the elevation of the location. As the altitude increases, the density of air decreases, and thus K_e decreases exponentially. The exact expression for K_e is

$$K_e = e^{-0.0000362 z_g}, \text{ where } z_g \text{ is the ground elevation above sea level in ft, or} \quad (5.5a)$$

$$K_e = e^{-0.000119 z_g}, \text{ where } z_g \text{ is the ground elevation above sea level in m.} \quad (5.5b)$$

A quick calculation shows that K_e drops to 0.8 at an altitude of 6000 ft (1.8 km) and to 0.5 at 19,000 ft (5.79 km). For a more conservative calculation, it is permissible to use $K_e = 1$ for all altitudes.

The next step after determining the value of q_z for a location is to calculate the design wind pressure from

$$P = q_h (GCp)(\gamma_E)(\gamma_a) \text{ lb/ft}^2 \text{ (or N/m}^2\text{), where} \quad (5.6)$$

$q_h = q_z$ evaluated at the mean roof height, h,
(GCp) is the external pressure coefficient determined from ASCE 7–16 30.3–2,
γ_E is defined as the array edge factor, and
γ_a is defined as the solar panel pressure equalization factor.

External pressure coefficients apply to all surfaces that are exposed to wind. Each surface is characterized by a different set of coefficients. These coefficients are based on wind tunnel tests. It has been found that wind forces are stronger near the edges and corners of a surface than in the middle of a surface. For PV modules attached to roofs, values of (GC_p) depend on the slope of the roof, the effective wind area of the module and the location on the roof. This coefficient is determined from a series of figures in chapter 30.3–2 of ASCE 7–16. As an example of the amount of detail involved in determining (GC_p), Table 5.4 lists values of (GC_p) for a hip roof with a 6:12 (26.6°) slope for roof zones 1, 2e, 2r and 3 for a module with an area of 20 ft² (1.86 m²).

γ_E = array edge factor = 1.5 for uplift loads on panels that are exposed and within a distance of 1.5(L_p) from the end of a row at an exposed edge of the array: γ_E = 1.0 elsewhere for uplift loads and for all downward loads as illustrated in ASCE 7–16 figure 29.4–7. Figure 5.6 illustrates the defining array layout. A panel is defined as exposed if d_1 to the roof edge > 0.5h and one of the following applies:

1. d_1 to the adjacent array > 4 ft, or
2. d_2 to the next adjacent panel > 4 ft.

The final coefficient is (γ_a), which is defined as the solar panel pressure equalization factor, as defined in ASCE 7–16 figure 29.4–8, which is adapted as Figure 5.7.

Example 5.1 Determine the design wind pressures for a roof in Omaha, NB. The roof is a hip roof with a slope of 26.6° and a mean roof height of 15 ft. The building is a risk category II building located in an exposure B area that is relatively flat. For this example, it will be assumed that there are no edge effects.

TABLE 5.4
Values of (GCp) for a Hip Roof with a Slope of 26.6° and 20 ft² Modules

Zone	1		2e		2r		3	
(GC_p)	+0.6	−1.2	+0.6	−1.75	+0.6	−1.75	+0.6	−1.75

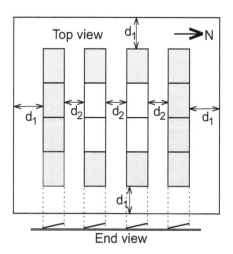

FIGURE 5.6 Defining parameters of exposed modules.
(Adapted from ASCE 7–16 with permission.)

Structural Considerations

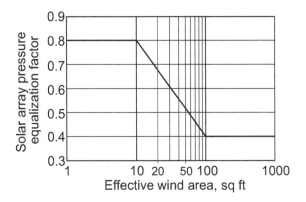

FIGURE 5.7 Solar panel equalization factor versus effective wind area. (Adapted from ASCE 7–16 with permission.)

Solution: Step 1 is to evaluate q_z, which means determining the basic wind speed, K_z, K_{zt}, K_d and K_e. From Figure 5.4, the wind speed is 114 mph (51 m/sec), $K_z = 0.7$ in exposure B up to a mean roof height of 30 ft (9.14 m), $K_{zt} = 1$ since the area is flat and $K_d = 0.85$ since the building is a "regular" building. Although 1.0 can always be used for K_e, given that the elevation of Omaha is 1089 ft (332 m), evaluating (5.5a) yields $K_e = 0.96$. Thus,

$$q_z = q_h = 0.00256 K_z K_{zt} K_d K_e V^2 = 0.00256 \times 0.7 \times 1.0 \times 0.85 \times 0.96 \times (114)^2$$
$$= 19.00 \text{ psf } (909.9 \text{ Pa}).$$

Step 2 is to evaluate (5.6) to find the design wind pressure. Since for the 20 ft² (1.86 m²) module, $\gamma_a = 0.7$, (5.6) becomes $P = q_h(GCp)(\gamma_E)(\gamma_a) = 19 \times 1.0 \times 0.7(GCp) = 13.3(GCp)$. Since (GCp) has two values for each roof zone, it is convenient to tabulate the result for P, as shown in Table 5.5. Since most modules can withstand uplift pressures of 50 psf or more, thus far, it appears that whatever module is chosen will be able to withstand the pressure in any of the roof zones.

5.4.5 Snow Loads

Snow loads on PV arrays and sloped roofs decrease as the tilt angles and slopes increase. There are three reasons for this. First, as the tilt angle increases, the component of the weight force parallel to the array surface increases relative to the component normal to the array surface, thus helping to shed the snow. Second, wind forces tend to shed the snow from the array. Third, melt water reduces the friction between the array surface and helps to shed the snow. Most PV arrays have "slippery" glass surfaces that also aid the snow-shedding process. On the other hand, obstructions and rough surfaces hinder the shedding of snow and its weight. For example, some tile roofs contain built-in protrusions or rough surfaces that prevent

TABLE 5.5
Design Wind Pressures for the Roof of Example 5.1

Zone	1		2e		2r		3	
P (psf)	7.98	−15.0	7.98	−23.3	7.98	−23.3	7.98	−23.3
P (Pa)	382	−764	382	−1115	382	−1115	382	−1115

the snow from sliding. The snow that accumulates on arrays mounted on such roofs may be obstructed from shedding [6]. An extreme case of snow loading occurs when unusually large amounts of snow accumulate on an array at a relatively small tilt angle. However, this is rare because shallow tilt angles are uncommon in colder, snow-prone parts of the world. In most cases, snow loads on PV arrays are less than 10 psf (479 Pa) unless the snow is unusually dense (wet). Of course, unlike wind loads, snow loads are downward only.

5.4.6 OTHER LOADS

The PV array and support structure may experience structural forces due to other loads, such as rain, ice, hydrostatic pressure and seismic activity. The latter is critical in earthquake-prone regions and the ASCE has detailed procedures for designing for seismic loads.

Foundations, footings or slabs that support PV arrays may experience uplift forces due to hydrostatic pressure in locations where ground water is close to the surface. If expansive soils are present, the hydrostatic pressures can be quite large. In such cases, the expansive soils are sometimes excavated to a depth of at least 2 ft (0.61 m) below ground level, followed by back filling with freely draining sands and gravel.

Regardless of the location, the PV engineer is responsible for ensuring that all of the relevant loads and load combinations have been properly accounted for in the structural system design.

5.5 ROOFTOP MOUNTING SYSTEM DESIGN

In addition to meeting structural, code and safety requirements, good design of the array mounting system should achieve the following:

- Minimizing installation costs
- Enhancing array performance
- Providing reasonable accessibility for installation and maintenance
- Making the system aesthetically appropriate for the site and application

The last two items require no further discussion. They are pretty obvious, particularly since beauty is in the eye of the beholder.

5.5.1 Minimizing Installation Costs

Life cycle costing is used in economic analyses of renewable energy technologies because these technologies realize their value over time. They are usually characterized by high capital costs and low operating costs. For PV systems on buildings, the combined array-roof configuration is a major contributor to the life cycle cost. It is important that these configurations last at least 25 years. However, this is not always achieved without unexpected added costs. For example, asphalt shingles often have to be replaced about every ten years in warm, humid climates. The life cycle cost analysis should include the added cost of removing and re-installing the PV array when re-roofing. Life cycle costing will be covered in detail in Chapter 9.

In general, installation costs for the array depend upon the type of roof, the quality of the components and the anticipated forces on the array. For example, it is more difficult to install an array mount on a tile roof than on a shingle roof. However, a tile roof will normally last longer than a shingle roof. It is also easier to attach array mounts to roof decking than to roof trusses, but a truss attachment will generally have more than twice the uplift resistance strength than a deck mount. Then there is the case of the metal seam roof, where it is often desirable to avoid any penetration of the metal, resulting in the use of clamps that clamp to the seams. The bottom line is that very few generalizations can be made about the best way to attach an array to a roof.

5.5.2 Enhancing Rooftop Array Performance

The best way to enhance the performance of an array is to design the roof with a maximum amount of roof facing south at a tilt of latitude as possible. However, since most roofs have already been designed to achieve some other purpose, good, south-facing roofs are often difficult to find. Some of the remaining options were discussed in the last chapter, such as using microinverters or optimizers to maximize production at the module level. Other relatively practical as well as achievable options include avoiding shading, using stand-off mounts to allow air to circulate beneath the array for cooling and simply using the most efficient modules as possible to maximize production of kWh for the space available.

5.5.3 Standoff Mounting

Standoff arrays are mounted above and parallel to the roof surface as illustrated in Figure 5.8. Standoff mounts work well for buildings with sloping roofs. When installing a standoff-mounted array, the PV modules are often attached to the roof using point connections—usually along the edges of the modules. As a minimum, the standoff height between the roof and the bottom of the module frame should be at least three inches. Four to five inches is preferable. Lag screws that penetrate the roof rafters 2 to 3 in (5.1 to 7.6 cm) may be used to fasten the mounting brackets to the roof. Spanner attachments, as shown in Figure 5.9, are stronger than lag bolts or lag screws and are commonly used in areas where no truss is located where the mounting bracket needs to be placed.

To promote passive cooling of standoff arrays, the engineer should consider the following designs:

- Designs that allow both lateral and vertical airflow along the back surface of the modules
- Designs that induce pressure differences between air inlet and exit regions
- Arrays with larger lateral dimensions than vertical dimensions (i.e., higher aspect ratios)

5.5.3.1 Determining Rooftop Wind Zones

Section 5.4.4 introduced the observation that wind uplift pressures are different at different points on a roof. Clearly, if uplift forces are higher in one zone than in another, it is likely that more attachment points may be necessary in zones with higher uplift forces. Although roof zones were mentioned in Section 5.4.4, now is

FIGURE 5.8 Example of standoff roof mount using two rails with and without modules mounted.

(Photo courtesy Lord & Lawrence Consulting Engineers.)

FIGURE 5.9 Example of spanner attachment.

Structural Considerations

the time to show how these zones are determined for different classes and slopes of roofs. Once the zones for a roof have been determined, assuming everything but the pressure coefficients is known, the pressures for each zone can be calculated.

A common denominator for defining roof zones for sloped roofs is the "a" factor, which is used to establish the boundaries between zones. ASCE-7 [6] defines the "a" factor as "a = 10% of the least horizontal dimension or 0.4h, whichever is smaller, but not less than either 4% of the least horizontal dimension or 3 ft, whichever is the smallest."

ASCE 7–16 figure 30.3 (Figure 5.10) shows the roof zones for flat, hip and gable roofs with slopes between 0° and 7° and for hip and gable roofs with slopes between 7° and 45°. Although a significant fraction of buildings have more complicated shapes, generally only the horizontal dimensions may cause some confusion as to which measurement to use; however, the definition of "a" does specify the *least* horizontal dimension, and no particular mention is made about how many horizontal dimensions are allowed in the process of circumnavigating the building.

Slopes and mean roof heights do not generally pose any problems; except if the roof has sections of different heights, slopes and/or hip for some and gable for others, different rules may apply to different sections. In fact, ASCE 7 has defining figures for (GCp) for gable and hip roofs with mean roof heights ≤ 60 ft (18.3 m) for slopes of $0° < \theta \leq 7°$, $7° < \theta \leq 20°$, $20° < \theta \leq 27°$ and $27° < \theta \leq 45°$ as well as for overhangs for a few roof types and a number of less common roof structures such as stepped roofs.

This new information on defining roof zones now makes it possible to return to a roof from Chapter 4 to define roof zones and calculate the wind load on the array. Although it is tempting to use the roof of Figure 4.5, the straightforward roof of Figure 4.3 will be used instead.

Example 5.2 Design a standoff roof mount for the Oklahoma City residence of Chapter 4. (Figure 4.3)

Solution: The roof is shown in Figure 5.11, but this time the wind zones can be determined from the information on the figure. Note that the mean roof height is the

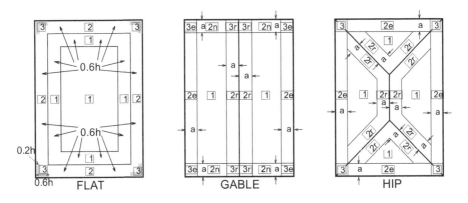

FIGURE 5.10 Roof zones for flat, hip and gable roofs with slopes between 0° and 7° and for hip and gable roofs with slopes between 7° and 45°.

(Adapted from ASCE 7–16 with permission.)

average of the height of the roof at the peak and the height at the eave. The difference between the height of the roof at the peak and the eave is $(33/2)\tan 22.6° = 6.8$ ft (2.07 m). If the eave is 9 ft off the ground, then the mean roof height will be 12.4 ft (3.78 m).

To determine "a," 10% of the least horizontal dimension is 3.3 ft and 40% of the mean roof height is 4.96 ft. Thus, a = 3.3 ft, and is included in the figure in dashed lines around the roof section. Note the locations of wind zones 1, 2e, 2r and 3. There are no edge or exposed modules in this array, and thus $\gamma_E = 1$ and $\gamma_a = 0.7$. Since the module size is 67.8 × 44.65 × 1.18 in (1722 × 1134 × 30 mm), the module area is 21.02 ft² (1.953 m²), which is close enough to 20 ft² to use this size for determining the values of (GCp). In addition, since the values of (GCp) in Table 5.5 are valid for hip roofs with slopes between 20° and 27°, they work for this roof. Thus, all that is needed is to determine the values of a few more constants and the design wind speed.

For this location, the land is relatively flat and the exposure is B. This means $K_z = 0.7$, $K_{zt} = 1$, $K_d = 0.85$, and it is acceptable to use $K_e = 1$. The wind speed map shows that Oklahoma City is nearly midway between the 104 mph and the 114 mph wind speed contours. Using 114 mph is taking the conservative approach, rather than

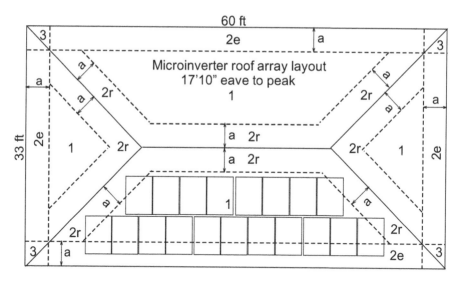

FIGURE 5.11 Array layout on hip roof showing wind zones.

TABLE 5.6
Design Wind Pressures for the Roof of Example 5.2

Zone	1		2e		2r		3	
P (psf)	8.32	−16.63	8.32	−24.26	8.32	−24.26	8.32	−24.26
P (Pa)	398	−797	398	−1162	398	−1162	398	−1162

Structural Considerations

interpolating. Thus, $q_z = 0.00256 \times 0.7 \times 1.0 \times 0.85 \times 1.0 \times (114)^2 = 19.8$ psf (948 Pa). Finally, using (5.6), the downward and upward pressures in all zones can be calculated and tabulated as shown in Table 5.6.

Finally, it is now possible to calculate the wind pressure on all the modules. A closer look at Figure 5.11 reveals that for the top row, six modules are totally in zone 1 and two of the modules have about 20% of their area in zone 2r and 80% in zone. Thus, the modules in zone 1 will experience a total uplift pressure of 21 × 16.63 psf = 349 lb (158.3 kg), and the modules with parts in both zones will have a total uplift of 21(0.2×24.26 + 0.8×16.63) = 381 lb (172.8 kg). The total uplift force on the top row of modules will be (6×349) + (2×381) = 2856 lb (1295 kg).

For the bottom row, nine of the modules have about 35% in zone 2e and 65% in zone 1. The total uplift force on each of these will be 21(0.35×24.26 + 0.65×16.63) = 405 lb (183.7 kg). The end modules have about 60% of their total area in either zone 2e or zone 2r. Since each of these zones has the same pressure, the total uplift forces on each of the two end modules of the bottom row will be 21(0.6×24.26 + 0.4×16.63) = 445 lb (202 kg). Thus, the total uplift force on the bottom row will be 4535 lb (2057 kg).

If each row is supported by two mounting rails, then the next step is to determine how many mounting points will be needed. This decision will depend upon the pull strength of each mount and the allowed distance between mounts allowed by the mounting rail specifications. The mounting rails will have maximum allowed spans for each uplift force and also maximum allowed cantilevers between the last mount and the end of a rail. The pull strength of the attachments will depend mostly upon where the attachment is attached and maybe on whether the attachment itself will have a limit on the uplift force it can withstand. Exactly where the attachment is attached will usually depend upon the type of roof, such as shingle, tile or metal seam.

The range of pull strengths of attachments is from less than 200 lb (90.7 kg) to over 1000 lb (453.6 kg) each. Thus, clearly, the type of attachment will be a major factor in deciding how many will be needed and where they will be attached. On the other hand, once the attachment type has been chosen, its pull strength will be known and the minimum number of attachments for a row of modules will be the force on the row divided by the available uplift force for the attachment.

Thus, if a deck mount is chosen that can withstand an uplift force of 300 lb (136 kg), then the top row will need at least 2856 ÷ 300 = 10 (rounded up) and the bottom row will need at least 16 (rounded up). However, if truss mounts are used, then the spacing of the attachments will be multiples of truss spacing. As a result, if the maximum spacing happened to end up at 42 in (107 cm), the attachment spacing would need to be 24 in (61 cm), since 48 in (122 cm) would be too wide. It should be noted that the rated pull strength of all attachments is generally based upon a safety factor between 2 and 4. Thus, if an attachment has been rated at 300 lb (136 kg), then it has been tested to somewhere between 600 and 1200 lb (272 and 544 kg) to achieve failure of the attachment.

Figure 5.12 illustrates a possible mounting configuration for the system with 10 attachments on the top row and 16 attachments on the bottom row. Note that due to the slightly higher average uplift forces on the modules in the bottom row, the attachment points are spaced somewhat closer than the top row attachments. Note also that the end modules on each row have higher percentages in either zone 2e or zone 2r and, as a result, the spacing of the attachments at the ends of the rows is somewhat closer to provide adequate holding power for all wind zones.

5.5.3.2 Determining the Pull Strength of Lag Screws Into Trusses

Since a high percentage of standoff-mounted PV arrays are attached to the roof using lag screws, it is interesting to take a closer look at how the pull strength is determined. The lag screw typically passes through a hole in a mounting bracket, possibly a mounting pad, shingles, waterproof membrane, sheathing and, finally, into the truss, spanner or other primary structural wood support member. The withdrawal strength of this and other attachment points is a function of the diameter of the lag screw, the length of thread embedded in the primary structural wood support, and the specific gravity of the wood. The allowable withdrawal load of lag screws and bolts per inch of embedded thread is given by [4]:

$$p = 1,800 D^{3/4} G^{3/2} \tag{5.7}$$

where p = allowable withdrawal load (lb./in.)
D = shank diameter of lag screw or bolt (in.)
G = specific gravity of oven-dry wood

Values for p computed using (5.7) include a safety factor of 4.5. This relatively high value provides added protection just in case the actual wood used may have reduced strength due to variations in specific gravity, knots, grain irregularities, holes and other defects [13].

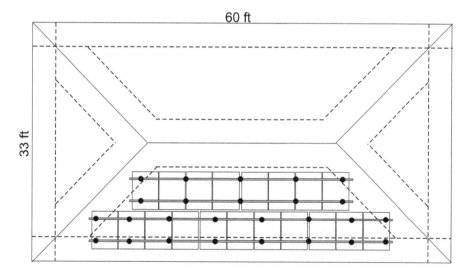

FIGURE 5.12 Array mount using attachments rated at 300 lb pull strength.

Structural Considerations

The typical diameters of lag screws used to attach PV arrays are 1/4 in (6.35 mm) and 5/16 in (7.9 mm), although they may be as large as 3/8 in (9.5 mm) or 1/2 in (12.7 mm) if the trusses consist of larger beams, rather than 2 in × 4 in (5.1 × 10.2 cm) or 2 in × 6 in (5.1 × 15.3 cm) lumber. The specific gravities of various types of lumber can be found in the *Wood Engineering and Construction Handbook* [14]. Table 5.7 [14] was developed using the previous equation and the specific gravities for various types of lumber.

5.5.3.3 Edge and Exposed Modules

A roof without any obstructions is a rare find, indeed. Most roofs have plumbing vents and any number of other protrusions that prevent the installation of a module over them. As a result, instead of the perfectly aligned, straight rows that have been dealt with thus far, most roofs have obstructions that will interrupt the alignment of modules in the row. This can result in the need to locate modules in places where they might not otherwise have been located to install the desired number of modules on a roof. If a module is mounted too close to the edge of a roof, it can experience up to 50% higher wind uplift forces. Whether or not a module is too close to the edge is determined by the distance, h, between the module and the roof. If the module is closer than 2 h from the edge, it is classified as an edge module and will require consideration for its contribution to the wind load on the array of a factor of 1.5.

On the other hand, if a module is mounted *too far* from the edge, it can also experience higher wind forces. These modules have been identified as exposed modules and the effect has been defined by the array edge factor, γ_E.

Exposed modules have at least one section of the module that is exposed to higher wind speeds. Figure 5.6 [6] illustrates the various locations that fall into the exposed category. ASCE 7–16 defines the edge factor as follows:

$\gamma_E = 1.5$ for uplift loads on panels that are exposed and within a distance of $1.5L_p$ from the end of a row at an exposed edge of the array; $\gamma_E = 1.0$ elsewhere for uplift

TABLE 5.7
Allowable Withdrawal Loads for Lag Screws

Lumber	White Oak	Southern Yellow Pine	White Spruce	Douglas Fir
Specific gravity, G	0.71	0.58	0.45	0.41
Diameter, D (in.)		Allowable withdrawal load, p (lb/in)		
1/4	381	281	192	167
5/16	450	332	227	198
3/8	516	381	260	226
7/16	579	428	292	254
1/2	640	473	323	281

Values for p were computed using (5.7) and include a safety factor of 4.5.

loads and for all downward loads, as illustrated in figure 29.4.7 (ASCE 7–16). A panel is defined as exposed if d_1 to the roof edge > 0.5h and one of the following applies:

1. d_1 to the adjacent array > max($4h_2$, 4 ft (1.2 m) or
2. d_2 to the next adjacent panel > max(($4h_2$, 4 ft (1.2 m) applies.

Figure 5.6 shows a portion of figure 29.4.7 that shows the defining parameters of the edge factor, which is sometimes referred to as the *exposure* factor to avoid confusing it with the factor for the modules that are too close to the edge of the roof.

5.5.4 Rack Mounting on Rooftops

Rack-mounted arrays are above and tilted with respect to the roof. Rack mounts work well on flat roofs and roofs with a slope of 2:12 or less (<7°). They may be mechanically attached to the building structure or may employ ballast to resist wind and other structural loads. Rack mounts are usually subjected to higher structural loads, incur higher costs for mounting hardware and are often less attractive than standoff mounts. However, for the same array area, the total energy output is sometimes higher because of better orientation and lower average operating temperatures, especially at northern latitudes, where the added slope helps to clear snow loads more quickly.

Ballasted rack assemblies offer the distinct advantages of simplicity and avoidance of roof penetrations, and should be considered if acceptable by local code jurisdictions. However, most rack mountings are firmly attached to the roof substructure. These are usually point connections, although several distributed attachment methods are sometimes used. Rack-mounted arrays typically run relatively cool compared with other mounts and they can reduce heat gain through roofs. Since the primary mechanism of rack array cooling is convection from the front and back surfaces of the modules, cooling can sometimes be enhanced by locating the rack in natural air channels and by reducing obstructions to airflow, such as screens, grates and walls.

ASCE 7 includes a wind load calculation procedure for rack-mounted arrays on roofs with slopes < 7°. The defining parameters for the determination of net pressure coefficients, (GC_{rn}), are shown in Figure 5.13, which is an excerpt from ASCE 7–16 figure 29.4–7, a more comprehensive illustration. An important distinction to be made from this figure is the difference between (GC_{rn}), which is defined as the net pressure coefficient, and $(GC_{rn})_{nom}$, which is an intermediate value obtained from graphs of $(GC_{rn})_{nom}$ that are included in the remaining sections of the ASCE 7 figure. Once the value of $(GC_{rn})_{nom}$ has been determined, the net pressure coefficient, (GC_{rn}) is calculated from

$$(GC_{rn}) = \gamma_p \gamma_c \gamma_E (GC_{rn})_{nom}, \text{ where} \tag{5.8}$$

$\gamma_p = \min(1.2, 0.9 + h_{pt}/h)$,
$\gamma_c = \max(0.6 + 0.06L_p, 0.8)$
γ_E = the exposure/edge factor as defined in Section 5.5.3.3

Structural Considerations

Graphs of $(GC_{rn})_{nom}$ are shown for the angle between the module and the roof, ω, in two ranges: $0° \leq \omega \leq 7°$ and $15° \leq \omega \leq 35°$. One set of values applies to the lower angle and another set applies to the higher angle range. For the range between 7° and 15°, linear interpolation is allowed between the values of (GC_{rn}) that correspond to the same normalized wind area and the same roof zone. Once the net pressure coefficients have been determined, the design wind pressure is determined from

$$P = q_h(GC_{rn}) \text{ in units defined by the units of } q_h, \qquad (5.9)$$

A closer look at Figure 5.13 indicates that the L_p dimension of the module, which is measured from the lowest point of the module from the roof to the highest point from the roof, is limited to 6.7 ft (2.04 m). Coincidentally, at the time of this writing, the length of most commercial modules with rated power less than 500 W is less than 6.7 ft. However, L_p is not necessarily intended to represent the longest dimension of the module. In fact, if the actual length of the module is $< L_p$, then the module can be mounted in a portrait position. However, if the actual length of the module is $> L_p$, then it is still highly likely that the module can be mounted in landscape instead of portrait, since in this position, the *width* of the module would need to be $< L_p$. Limits on other defining parameters include $h_1 \leq 2$ ft (0.61 m) and $h_2 \leq 4$ ft (1.22 m).

Example 5.3 Design of a 24-module array with a 5° tilt on a flat, rectangular roof.

Solution: The first step is to list the module dimensions. Using the "pick one and see how it fits" method, assume the module dimensions are 1×2 m (3.28 × 6.56 ft). Since 2 m < 2.04 m, this dimension can be used as L_p and as long as the constraint on h_1 and h_2 is met, then the modules can be installed with a portrait orientation. From Figure 5.13, it can be seen that since $\sin\omega = (h_2 - h_1)/L_p$, then $(h_2 - h_1) = L_p \sin\omega = 2\times 0.0872 = 0.1744$ m (0.5722 ft), which shows that the $h_2 - h_1$ criterion is met as long as $h_1 < 2$ ft (0.61 m). If a mounting system that results in $h_1 = 4$ in (0.1016 m), then h_2 will be 0.276 m (10.9 in).

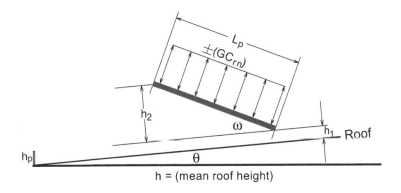

FIGURE 5.13 Defining parameters of racked modules on low slope roof.
(Adapted from ASCE 7–16 with permission.)

The next step is to use the module area, 2 m² (21.536 ft²), to determine the normalized net pressure coefficient, $(GC_{rn})_{nom}$. From Figure 5.13 for $\omega = 5°$ and 21.536 ft² module area, $(GC_{rn})_{nom}$ is found to be 0.9 in roof zone 1, 1.3 in roof zone 2 and 1.4 in roof zone 3. If the roof height, h, is 10 ft, then zone 1 is defined as > 2h from the roof edge, zone 2 is defined as < 2 h from the roof edge and zone 3 is at the corners where zone 2 from two adjacent sides overlap. Thus, zone 2 is 20 ft wide, zone 3 is 4 squares, 20 ft on a side, and zone 1 is the leftovers in the middle.

The next step is to determine γ_p, γ_c and γ_E to use (5.8) to evaluate (5.9). For this particular example, there is no parapet. Thus $h_{pt} = 0$ and $\gamma_p = 0.9$. Using $L_p = 6.56$ ft, $0.06 L_p = 0.394$ results in $\gamma_c = 0.8$ and for the worst case where there are exposed modules, $\gamma_E = 1.5$. Thus, $(GC_{rn}) = 0.9 \times 0.8 \times 1.5 \times (GC_{rn})_{nom}$. Since $(GC_{rn})_{nom}$ depends upon the roof zone, (GC_{rn}) will be 0.97 in zone 1, 1.40 in zone 2 and 1.52 in zone 3.

The building exposure, wind speed and four more constants are now needed to determine q_h from (5.3). If the building is a relatively "normal" building in Miami, FL, where the basic wind speed is 175 mph and all exposures except D are classified as C, $K_z K_{zt} K_d K_e = 0.85 \times 1.0 \times 0.85 \times 1.0 = 0.72$. Thus, $q_h = 0.00256 \times 0.72 \times (175)^2 = 56.45$ psf (2703 Pa). Finally, the pressure on the array is 54.76 psf in zone 1, 79 psf in zone 2 and 85.8 psf in zone 3.

If the array consists of three rows of eight modules each, then the area of each row is 172.29 ft² and the overall pressure will be the sum of the individual module pressures, depending upon the zone or zones occupied by each module. It is left as an exercise for the reader to determine these pressures (Problem 5.9).

5.5.5 Integrated Mounting

For integrated arrays, also referred to as BIPVs or integral mounting, the array replaces conventional roofing or glazing materials as illustrated in Figure 5.14. Because integrated arrays replace conventional roofing and glazing materials, they become a significant architectural feature of a building and can be aesthetically very pleasing. Dimensional tolerances may be tighter for some integrated arrays than for standoff or rack mounts. If commercial curtain-wall glazing techniques are used, larger modules are preferable to smaller modules. As in most modules and mounting systems, technical challenges faced by integrated mounting include efficiency of about half the efficiency of crystalline silicon as well as lack of cooling beneath the modules and the resulting increase in array temperature. However, these drawbacks can be acceptable if the modules can be mounted in places where using standoff mounts is not practical.

5.6 LARGE-SCALE GROUND MOUNT ARRAYS

5.6.1 Introduction

Ground-mounted arrays are supported by racks, poles or tracking stands. These arrays are secured to the ground to resist uplifting caused by wind loads. All ground-mounted arrays run relatively cool because good airflow is possible over both the

Structural Considerations 179

FIGURE 5.14 Integrated array mounting.

front and back surfaces of the modules. This cooling can be enhanced by minimizing obstructions to airflow such as shrubbery and fences. Although it may seem that installing a ground mount would be easier than installing a roof mount, it is important to note that it is usually straightforward to determine what is underneath a roof, but it can be a bit more complicated to determine what is underground. Since it is possible that ground screws or other anchors may need to be driven several feet or meters into the ground, it is important to know whether any natural obstructions, piping or electrical equipment is buried. Furthermore, the type of soil is also important to know. Fortunately, regional soil maps are available with soil information. Often a geotechnical engineer will be employed to assess the site to determine whether there are any surprises.

Fixed tilt arrays are exactly that—fixed in place like a rooftop array, but not with lag screws.

On the other hand, some ground-mounted arrays are designed to track the sun. There are several types of tracking arrays, two of which will be considered to the extent of creating an appreciation of the methodology needed to complete the designs. Because tracking arrays receive more sunlight than stationary arrays, especially in areas with high percentages of direct sunlight, each individual module produces more energy. Whether to use a tracking array depends on the tradeoff between additional energy produced and added cost and complexity and space needed between trackers to avoid one row shading another.

Active trackers use drive mechanisms, such as electric motors and gears, to point the array toward the sun. They may track about either one or two axes. The tracking motion may be controlled by a computer or by sun-seeking sensors. A more sophisticated version of tracking arrays is the backtracker. These arrays detect inter-row shading and begin rotating in the opposite direction to eliminate inter-row shading as the sun moves lower in the sky.

5.6.2 Large Fixed Tilt Arrays

5.6.2.1 Introduction

This section makes the distinction between small and large arrays. The distinction is relatively simple. Small arrays do not have enough modules such that the outer modules in the array will provide shielding from wind for the inner modules, similar to the rooftop situation where modules at the edges experience more wind force than modules at the center. Figure 5.15 is an adaptation of ASCE 7–22 figure 29.4–9, which is a new figure, the first to appear in ASCE 7–22. This figure shows the locations of defining measurements used in determining wind zones for this array. Figure 5.16 shows how the wind zones associated with the array are defined. Figure 5.17 shows an end view of 3 rows of the array. The parameters are defined as follows:

L_C is the panel chord length.
W_g is the shortest row length in the array.
ω is the angle between the solar panels and the ground surface in degrees.
h is the mean height of a panel.
S is the center-to-center row spacing.
s_p is the gap between adjacent panels in both directions.
S_L is the horizontal distance in the longitudinal direction of the open area within a single row.
S_T is the horizontal distance in the transverse direction of the open area between adjacent rows.
d_s is the offset of the ends of 2 rows in the longitudinal direction as shown in Figure 5.15.
d_p is the horizontal distance from the row end, defined as $d_p = 4L_C$ or 30 ft, whichever is smaller.
A = effective wind area of an element (unbroken row) of the array

The wind load calculation is subject to limits on some of these parameters as follows:

6 ft (1.8m) $\leq L_C \leq$ 14 ft (4.4 m),
$(W_g/L_C) \geq 7$,
$0° \leq \omega \leq 60°$,
$0.5 \leq (h/L_C) \leq 0.8$,
$0.20 \leq (L_C/S) \leq 0.60$,
$S_p \leq 0.014 L_C$
$S_L \leq 0.25 L_C$ and
$S_T \leq 2S$.

In addition, there must be three or more rows in the array and the ratio of the area blocked by support framing to the total area below the lowest edge of panels must be $\leq 8\%$ over any length of $4L_C$.

Structural Considerations 181

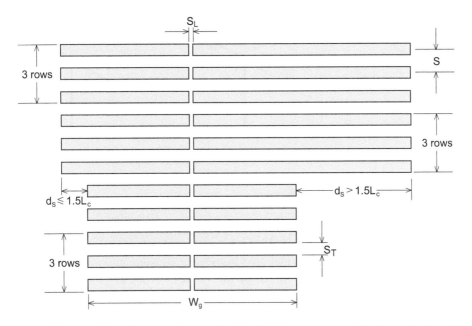

FIGURE 5.15 Ground-mounted array showing defining parameters for determination of wind load zones for the array.

(Adapted from ASCE 7–22 with permission.)

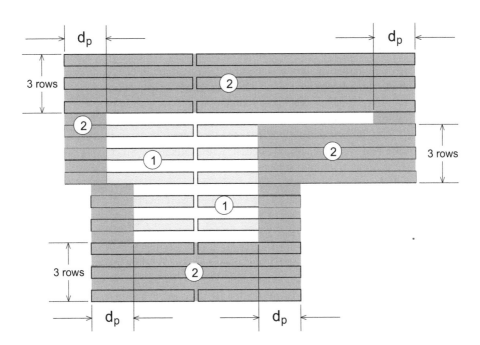

FIGURE 5.16 Wind load zones for the array of Figure 5.15.

(Adapted from ASCE 7–22 with permission.)

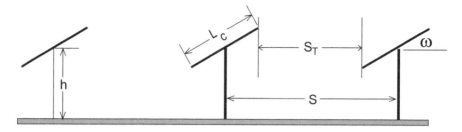

FIGURE 5.17 End view of the array of Figures 5.15 and 5.16 showing additional defining parameters of the array.

(Adapted from ASCE 7–22 with permission.)

5.6.2.2 Calculations of Zones, Pressures and Moments

Whether or not there will be a region of lower wind pressure on the array (zone 1) is made quite clear by Figure 5.16. Given that in the direction perpendicular to the array rows, Zone 2 has a width of three rows, this means that considering that the wind perpendicular to the array can come from two directions, there will be three rows in each of these directions that are perpendicular to the rows or a total of six rows. Thus, any array with six or fewer rows of modules will not have a zone 1.

At this point, it is also interesting to note that up to this point with rooftop arrays, only *forces* on the arrays have been considered. Since ground mount arrays are often attached to a single axis down their middle, as suggested by Figure 5.17, the ground mount analysis includes a determination of *torque* (moment) in addition to wind pressure. In fact, in addition to the determination of moments, this analysis allows for the calculation of dynamic effects, another consideration not used for rooftop arrays. Why? Keep in mind that standoff rooftop arrays have span limits as well as cantilever limits, both of which minimize any possible vibration modes.

Thus, assuming the array has more than 6 rows, for discussion purposes, suppose it has 11 rows, as in Figures 5.15 and 5.16. In fact, to make life as simple as possible, suppose the array is shaped exactly like these two figures. To determine the wind load in each of the two zones, it will be necessary to evaluate d_p. Since d_p is defined by ASCE 7–22 as the horizontal distance on zone 2 from the row end as determined by $d_p = 4L_C$ or 30 ft (9.1 m), whichever is smaller, this means it is necessary to determine L_C. In fact, it makes sense to list all the various measurements indicated on Figures 5.15, 5.16 and 5.17 in anticipation of them all being useful at some point.

Keeping in mind the limitations of the various parameters, a set of values valid for the analysis can be listed. Starting at the top of Figure 5.15, S_L is the spacing between two racks, which, allowing for a person to walk along this opening, suggests that 4 ft (1.22 m) may be a reasonable starting point. Next, S is the spacing between the centers of racks. To minimize shading between rows, S is generally determined after module length (L_C) tilt (ω) and row-to-row shading conditions have been established.

Assuming a module with a length of 6.56 ft (2 m) is used in a two-up configuration, with a spacing of ½″ (0.042 ft) between the short sides of the modules (S_p),

Structural Considerations

results in the chord length, $L_C = 6.56 + 6.56 + 0.042 = 13.16$ ft (4.01 m). Next, assuming a tilt angle of 30°, S_T becomes $S - L_C\cos\omega = S - 11.4$ ft.

The vertical distance between the top of one panel and the bottom of the panel in the next row is found from $H = L_C\sin\omega = 6.8$ ft (2.07 m). Assuming a shading analysis has been done using these values of H and S_T, and the result is that adequate exposure will result if $S = 25$ ft (7.62 m), S is now known. At this point, since L_C and 30 are both known, since $4L_C = 52.64$ ft (16.045 m), $d_p = 30$ ft, which defines the boundary between wind zones 1 and 2.

At this point, all that is needed is to determine that all the other constraints on array dimensions are satisfied. Thus

If $W_g > 7L_C = 92$ ft (28.04 m), then this criterion is met.
If $0.5L_C < h < 0.8L_C$, or 6.58 ft $< h <$ 10.53 ft, or 2.01 m $< h <$ 3.22 m, then this criterion is met.
If $0.20 \le (L_C/S) \le 0.60$, this criterion will be met. Since $L_C/S = 0.5264$, this criterion is met.
If $S_p \le 0.014L_C$, this criterion will be met, and since $S_p/L_C = 0.042/13.16 = 0.0032$, this criterion is met.
It is left as an exercise for the reader to determine whether $S_L \le 0.25L_C$ and $S_T \le 2S$. (Hint: One of these criteria is not met.)

Now that all the design criteria are known and valid, the wind forces and moments can be calculated for both zones. The equations for these calculations are

$$F_n = q_h K_d[\pm(GC_{gn})]A \quad \text{(lb or N as preferred) and} \quad (5.10)$$
$$M_c = q_h K_d[\pm(GC_{gm})]AL_C \quad \text{(lb/ft or N/m as preferred)} \quad (5.11)$$

In (5.10) and (5.11) note that all dimensions must be consistent in either English or metric units. It is also important to note that since this example is from ASCE 7–22, the definition of q_h is slightly different from its definition in ASCE 7–16. The difference is that K_d is considered separately in ASCE 7–22 instead of as a part of the coefficient of q_h as in ASCE 7–16. This means that the formula for q_h to use in this example is the value of q_z at the mean chord height h, which, in this case, is defined in Figure 5.17. To be sure everything is calculated correctly, it is a good idea to begin with the calculation of q_z using the definition of ASCE 7–22, which is

$$q_z = 0.00256 K_z K_{zt} K_e V^2 \text{ (lb/ft}^2\text{) with V in mi/h, or} \quad (5.12a)$$
$$q_z = 0.0613 K_z K_{zt} K_e V^2 \text{ (N/m}^2\text{) with V in m/s} \quad (5.12b)$$

Note that K_d is not included in these expressions as it was in (5.3), the origin of which is ASCE 7–16. In ASCE 7–22, K_d is found in Table 26.6–1, which shows that K_d depends upon the structure of interest. For most structures with square corners, $K_d = 0.85$. The other coefficients have the same definitions as in the previous edition. Note that K_z depends upon the exposure category and the mean height above ground. For a large ground-mount PV array, the exposure will necessarily be C before the array is installed. Furthermore, it is highly likely that the mean height of the array will be less than 15 ft (4.57 m). Under these conditions, $K_z = 0.85$ [15].

If the array field is flat, then $K_{zt} = 1.0$. If it is rolling hills, then K_{zt} will need to be evaluated in terms of K_1, K_2 and K_3 and may be > 1.0.

K_e is defined by (5.5a & 5.5b). This is the final coefficient needed to evaluate q_h in (5.10) or (5.11), since q_h is q_z evaluated at the mean array height, h. The only items remaining for the calculation are to look up the wind speed on wind speed maps for Risk Category I structures and to look up the values of GC_{gn} and GC_{gm}. For a larger safety factor, Category II could be used for the wind speed.

At this point, the reader may have begun wondering what all these GC_{xy} things are and where they come from. In fact, they are the result of a significant amount of empirical wind tunnel research. A look at ASCE 7 will show a significant number of very detailed equations associated with these various coefficients. All it takes is one look at these equations for one to appreciate being able to look up the coefficient values on convenient graphs, a few of which are shown in Figures 5.15 and 5.18. A significant observation is that the coefficients apply to two tilt angle ranges: 0° – 5° and 15° – 60°. These coefficients are dependent upon effective wind areas of array sections. As long as the modules are close enough to each other in each row, as defined by $S_p \leq 0.014 L_C$, the effective wind area of a section of the array will simply be the product of the number of modules mounted horizontally times the area of a single module. Thus, for two rows of 25 modules each, one on top of the other and modules separated horizontally and vertically by S_p, each module having a width of 3.33 ft (1.016 m) and $L_C = 13.16$ ft (4.01 m), the effective wind area, A, will be 13.16 × (3.33 + 0.042)×25 = 1110 ft².

For a tilt angle of 30°, this results in $GC_{gn} = 1.4$ for zone 1 and 2.3 for zone 2. For zone 1, $GC_{gm} = 0.22$ and for zone 2, $GC_{gm} = 0.32$. If h = 10 ft (3.048 m), then $K_z = 0.85$. If the field is flat, then $K_{zt} = 1$. If the location is near Wichita, KS, the elevation will be approximately 1300 ft (396 m), and thus $K_e = 0.96$ and V = 105 mph (46.94 m/s) for risk category I. Thus, using $K_d = 0.85$,

$q_h = 0.00256 \times 0.85 \times 1.0 \times 0.96 \times 105^2 = 23.0$ lb/ft² (1101.25 N/m²).
$F_n = \pm 23.0 \times 0.85 \times 2.3 \times 1110 = \pm 49,911$ lb or 44.97 lb/ft² in zone 2 and 27.37 lb/ft² in zone 1.
$M_c = \pm 23.0 \times 0.85 \times 0.32 \times 1110 \times 13.16 = 91,385$ lb/ft or 134,045 N/m.

5.6.3 Single Axis, East–West Trackers

There once was a time when PV modules were very expensive, and to squeeze every drop of electricity out of them, they were mounted on tracking stands that kept them facing the sun. Today, with modules on the order of 300 times less expensive, the cost of the fancy trackers discouraged designers from using them. As a result, the tracking community began working on less expensive ways to increase the array annual energy production. One method that had some advantages was the single-axis east–west tracker. If the rows in Figure 5.17 were installed from north to south,

Structural Considerations

(a)

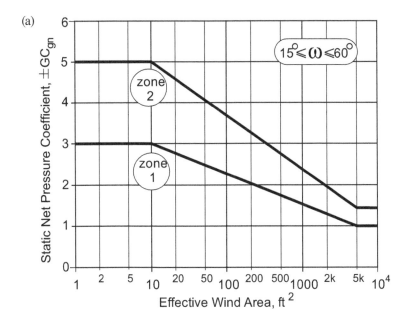

(b)

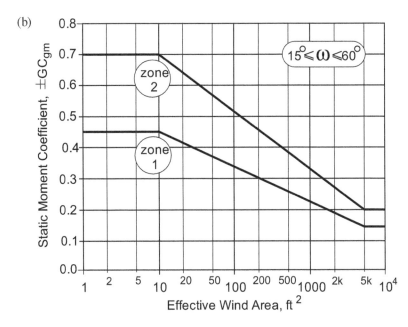

FIGURE 5.18 Static net pressure and static moment coefficients for a large ground-mounted array with tilt angle between 15° and 60°.

(Adapted from ASCE-7–22 with permission.)

the modules could then rotate from east to west, enabling them to more efficiently capture the sun's rays earlier in the morning and later in the afternoon. Later in the afternoon was particularly desirable, since late afternoon is generally utility peaking time—a time when utilities can use all the electricity they can get their wires on. Of course, to generate power late in the afternoon, it is important for the rows of the array *NOT* to shade each other as the solar altitude decreases and the shadows get longer. Unfortunately, this is what happens at a particular critical sun altitude where adjacent rows begin shading each other. The wider the row spacing, the lower the shading angle, but this means fewer modules per acre, and this might also be a consideration.

Then, along came the backtracker, which was first introduced as one way of capturing sunlight back in Chapter 2 (Figure 2.10, repeated as Figure 5.19). It was also shown in Chapter 2 that, over the year, a backtracker in the right location can produce more energy than a fixed array.

The same analysis that was used for the fixed array can also be used for the backtracking array. The challenging part is to determine row spacing and when the backtracking begins in the morning and afternoon. Then, even more fun is determining which tilt angle and corresponding values of S_T will result in the highest wind loads. Fortunately, computers can become very useful in this level of analysis, and the manufacturers have lots of data, some proprietary, for use in determining wind loads. At this point, the reader should have developed some appreciation of the ongoing effort to more accurately determine wind loads and optimal design parameters for large ground-mounted arrays.

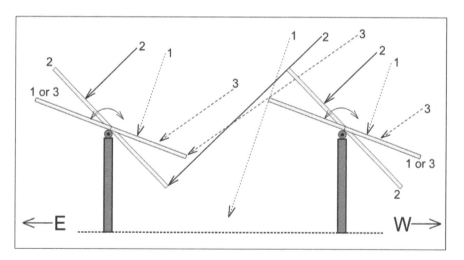

FIGURE 5.19 Illustration of operation of a backtracking version of a single-axis tracker with a horizontal N-S axis.

Structural Considerations

5.6.4 DYNAMIC EFFECTS

If any resonance effects, due to rapid changes in wind speeds, are possible, they will be more possible with ground-mounted arrays when the individual chords are attached only at the middle of the chord. The reason for estimating moments also holds for conducting a dynamic analysis. ASCE 7–22 includes an additional eight graphs, similar to those in Figure 5.18. The dynamic net pressure coefficients and the dynamic moment coefficients also depend on the same tilt angle ranges of the static coefficients, but the horizontal axes now show a dependence on reduced frequency, N_s, and the data is also separated into two different areas, A_1 and A_2. Different sets of dynamic net pressure and dynamic moment coefficients are shown for $A \le A_1$ and for $A \ge A_2$, where A_1 and A_2 are defined in terms of L_C^2. A damping coefficient, β, is also incorporated into the analysis.

The purpose of this section, however, is simply to create an awareness and appreciation of the need to farm out a large array of designs to specialists, which may well be the manufacturers. Those readers who are feeling the need to investigate this material further are encouraged to visit ASCE 7–22, section 29.4.5.4. Meanwhile, readers looking forward to learning about energy storage might decide to move on to Chapter 6.

Homework Problems

Note that for most problems that reference ASCE 7, it will be acceptable to use either 7–16 or 7–22 as references. For problems on large ground mount arrays, it will be necessary to use ASCE 7–22. To a large extent the results from either edition will be the same or at least almost the same.

5.1 Why are building codes that affect the mechanical and structural design of PV systems much less uniform throughout the United States than electrical codes? Explain.

5.2 Give two examples of cases when the design loading for the PV array and structural supports is not dominated by wind loading.

5.3 List four objectives in designing an array mounting system.

5.4 Six lag screws of 5/16 in. diameter and 3.5 in. length are used to attach a 6 ft by 13 ft standoff array to an asphalt shingle roof. The total thread length is 2.5 in. and the combined thickness of the mounting bracket, pad, shingles, roof membrane and plywood sheathing is 1.0 in. If the lag screw penetrates the roof truss made of Southern yellow pine, what is the allowable withdrawal load on each lag screw? What is the maximum uplift loading in psf that the array can withstand?

5.5 A ten module PV array has an area of 210 sq. ft. and will be secured using 32 mounting brackets. Each bracket uses two 5/16 in. lag screws to secure the standoff-mounted array. The roof trusses are made of white spruce. The design wind load is 55 psf. What is the minimum thread penetration for each fastener?

5.6 An array of 10 PV modules, each with dimensions of 1m x 1.6 m., is to be standoff-mounted in a single row above and parallel to a standing seam metal roof using clamps attached to the seams. The modules themselves can withstand an uplift load of 75 psf. The spacing of the seams is 16 inches. If each clamp has an allowable withdrawal resistance of 240 lb, what is the minimum number of clamps that must be used?

5.7 For more effective cooling of standoff-mounted PV arrays, would you recommend higher or lower aspect ratios, that is, portrait or landscape? Explain.

5.8 A residence has a rectangular shape of width = 30 ft and length = 50 ft. It has a gable roof, with two identical sections. The ridge is parallel to the length of the structure and is 54 ft long (2 ft overhang at each end). The eaves of the roof have the same dimensions. The sections of the roof have a 4:12 slope and a 2 ft overhang (direct view) at the eaves. The distance between the ground and the edge of the eaves of the roof is 8 ft. It is located in exposure B.

   a. Determine the elevation of the peak of the roof.
   b. Determine the actual dimensions of each roof section. (Not the plan view.)
   c. Determine the mean roof height.
   d. Determine "a."
   e. Draw the roof sections to scale and then determine the roof zones.
   f. Look up and tabulate the values of (GCp) for each roof zone, using either ASCE 7–16 or ASCE 7–22.

5.9 Assume the dimensions of the building of Example 5.3 are 150 ft x 100 ft. Using the end result for pressures and zones obtained in the example, lay out the array in three rows of eight modules. Then, depending upon the zone occupied by each module, determine the pressures on each module and each row. Then find an attachment designed to be attached to the roof with tapcons, determine the allowable uplift on the attachments and determine a layout for the attachments and mounting rails of the array mounts. Note that the use of tapcons implies the roof is concrete and flat.

5.10 Referring to Figure 4.12(a), which is repeated as Figure p5.1,
   a. Show that the array as shown must be treated as two arrays for determining wind zones 1 and 2.
   b. Check to verify that each half satisfies all the conditions required to divide each half into two zones.
   c. Determine the wind zones for each half of the array.
   d. Determine the maximum spacing between the two array "halves" (S_L) that would be allowed for the total array to be considered as one array for wind load calculation purposes. Then determine the wind zones if S_L meets the spacing requirements.
   e. How would you expect the wind load on the inverters to compare with the wind loads on zone 1 and zone 2 of the array?

Structural Considerations

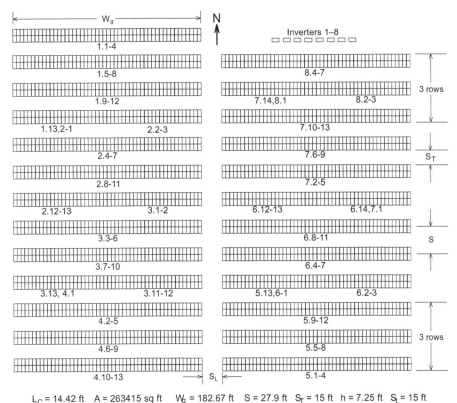

FIGURE P5.1 Proposed layout for 25% of the 5-MW array showing wind load parameters.

REFERENCES

[1] Singer, F. L., *Strength of Materials*, 2nd ed., Harper & Brothers Publishers, New York, 1962.
[2] Dieter, G. E., *Engineering Design: A Materials and Processing Approach*, 2nd ed., McGraw-Hill, New York, 1991.
[3] Clauser, H. R., Editor-in-Chief, *The Encyclopedia of Engineering Materials and Processes*, Reinhold Publishing, New York, 1963.
[4] Avallone, E. A. and Baumeister, T., Eds., *Marks' Standard Handbook for Mechanical Engineers*, 10th ed., McGraw-Hill, New York, 1996.
[5] Kutz, M., *Mechanical Engineers' Handbook*, 2nd ed., John Wiley & Sons, New York, 1998.
[6] *Minimum Design Loads for Buildings and Other Structures*, ASCE Standard, SEI/ASCE 7–16, American Society of Civil Engineers, Reston, VA, 2017.
[7] *NFPA 70 National Electrical Code*, 2020 ed., National Fire Protection Association, Quincy, MA, 2019.
[8] Modular Photovoltaic Array Field, "Prepared for Sandia National Laboratories," *SAND83-7028*, Hughes Aircraft Company, 1984.

[9] Miller, R. D. and Zimmerman, D. K., "Wind Loads on Flat Plate Photovoltaic Array Fields, Phase II Final Report, Prepared for Jet Propulsion Laboratory," *Contract No. NAS-7-100-954833*, Boeing Engineering and Construction Company, Seattle, WA, 1979.
[10] Marion, B. and Atmaram, G., *Preliminary Design of a 15–25 kWp Thin-Film Photovoltaic System*, Prepared for the U.S. Department of Energy, Florida Solar Energy Center, Cape Canaveral, FL, November 13, 1987.
[11] *Wind Design of Flat Panel Photovoltaic Array Structures*, Prepared for Sandia National Laboratories, SAND79–7057, Bechtel National, Inc., June 1980.
[12] Miller, R. D. and Zimmerman, D. K., "Wind Loads on Flat Plate Photovoltaic Array Fields, Phase III and Phase IV Final Reports, Prepared for Jet Propulsion Laboratory," *Contract No. NAS-7-100-954833*, Boeing Engineering and Construction Company, Seattle, WA, April 1981.
[13] *Wood Handbook—Wood as an Engineering Material*, U.S. Department of Agriculture, Forest Service, Forest Products Laboratory, Madison, WI, 1999.
[14] Faherty, K. F. and Williamson, T. G., *Wood Engineering and Construction Handbook*, 3rd ed., McGraw-Hill, New York, 1997.
[15] *Minimum Design Loads for Buildings and Other Structures*, ASCE Standard, SEI/ASCE 7–22, American Society of Civil Engineers, Reston, VA, 2022. https://www.asce.org/publications-and-news/civil-engineering-source/article/2021/12/02/updated-asce-7-22-standard-now-available

SUGGESTED READING

Epstein, S. G., *Aluminum and Its Alloys*, Aluminum Association, Inc., presented at the Annual Liberty Bell Corrosion Course, sponsored by the National Association of Corrosion Engineers, Washington, DC, September 1978.

Shigley, J. E. and Mischke, C. R., *Mechanical Engineering Design*, 5th ed., McGraw-Hill, New York, 1989.

Smith, W. F., *Principles of Materials Science and Engineering*, 2nd ed., McGraw-Hill, New York, 1990.

Ventre, G. G., *A Program Plan for Photovoltaic Buildings in Florida*, prepared for the Florida Energy Office/Department of Community Affairs and Sandia National Laboratories, Florida Solar Energy Center, Cocoa, FL, FSEC–PD–25–99, January 1999.

6 Energy Storage

6.1 INTRODUCTION

Most electrical engineers have some familiarity with batteries, especially the ones used in flashlights, toys, laptop computers, watches, calculators, cameras, automobiles and golf carts. The extent of familiarity normally extends to the decision as to whether to purchase C–Zn batteries or whether to buy the more expensive but longer-lasting, alkaline or lithium cells. Remembering to get the correct size, whether a 9-volt or a 1.5-volt AA battery, is another of life's energy storage dilemmas. In this section, considerations in battery selection for PV systems are presented. Alternatives to battery energy storage are also presented, so the PV design engineer will be able to make educated choices, based on sound economic and engineering principles, as to which type of energy storage to incorporate into a specific PV system design.

Many different types of rechargeable batteries suitable for PV applications are currently available. Although several rather exotic technologies are now available, lithium batteries are now the most common for relatively economical storage of useful quantities of electrical energy and will likely remain so for at least the next decade. *Ni–Cd* batteries have been in common use in applications that require sealed batteries capable of operating in any position, and still require high energy density. But they are more expensive per kWh stored than other units. Nickel metal hydride (NIMH) and lead-acid batteries were popular for a while, but at this time it appears they have taken a back seat to lithium. And, for utility-sized battery backup, flow batteries are moving beyond the experimental stage.

But, then, batteries are not the only way to store energy. It is also possible to store energy by producing hydrogen, pumping water uphill, spinning a flywheel, compressing air or charging a large chemical capacitor. This section explores the applications, advantages and disadvantages of some of the storage methods currently in use as well as some of the new ideas that have been proposed.

As discussion of various technologies follows, two important characteristics of batteries are their charging rates and discharging rates. Some are designed to discharge quickly and others are designed to discharge slowly. Some can do both. The charging or discharging process is measured in terms of how long it will take to either fully charge or fully discharge a battery. This rate is measured in terms of C/x, where C is the battery capacity and x is the time in hours that it takes to fully charge or discharge a battery. Thus, a 100 Ah battery, discharging at a C/10 rate, will take 10 hours to fully discharge. The actual discharge can be measured as a current (100 Ah ÷ 10 h) = 10 A. If the battery voltage enters the picture, then the discharge rate will be measured in W. Thus, if the 100 Ah battery is a 48-V battery, then the discharge power will be 10 A×48 V = 480 W.

Each battery has its own limits on the amount of charge it can hold (kWh) and the maximum rates at which it can be charged or discharged (kW). Each limit serves different purposes. The kWh rating is of most concern when determining how many and how long various loads can be operated if the battery energy needs to replace grid energy. The kW rating determines the maximum load in kW that can be served instantaneously by the battery. Depending upon the purpose of the storage, it may be specified in terms of either or both of these quantities.

All batteries have some internal resistance, and, as a result, since energy is lost in the form of heat as current flows through the internal resistance, faster charging or discharging times result in a decrease in useable battery energy during discharge or an increase in energy required to charge the batteries. Battery temperature can also affect the rate of chemical reactions in the batteries and reduce performance at colder temperatures.

Another important quantity associated with energy storage systems is the *round-trip efficiency*. This is defined as the percentage of energy used to store the energy that can be recovered when the stored energy is released. Among currently popular storage mechanisms, round-trip efficiency ranges between about 25% and 95%.

6.2 LITHIUM BATTERIES

6.2.1 LFP Batteries

Since lithium batteries are the current dominant battery storage technology for use with PV systems, it makes sense that the designer should have some knowledge of lithium battery chemistry and technology. Thanks to a renewed interest in electric and hybrid vehicles as well as the need for low-weight, high-energy-density storage for computing and communications devices, lithium technology has made significant advances over the past 30 years. Lithium-ion batteries are typically one-third the weight of a lead acid unit with the same storage capacity. They are about half the volume of a comparable lead-acid unit, have significantly better low-temperature capacity, have a relatively steady voltage during discharge at a constant current and have a lower internal impedance than comparable lead-acid units, which allows them to be charged and discharged at faster rates due to lower internal I^2R losses. They have minimal self-discharge, require almost no maintenance and have nearly double the expected lifetime of their lead-acid counterparts.

Six different lithium technologies have been developed since 1991 to meet the needs of specific loads. Lithium cobalt oxide ($LiCoO_2$) units have high energy density and are used mostly in cell phones, laptops and cameras. The reader may have noticed that they are not cheap. Lithium manganese oxide ($LiMn_2O_4$), lithium iron phosphate ($LiFePO_4$) and lithium nickel manganese cobalt oxide ($LiNiMnCoO_2$) are popular for power tools, electric vehicles, medical and hobbyist use. Lithium nickel cobalt aluminum oxide ($LiNiCoAlO_2$) and lithium titanate ($Li_4Ti_4O_{12}$) batteries are used mostly for electric vehicles and grid storage [1].

Lithium batteries have three basic cell configurations (building blocks)—cylindrical, prismatic and pouch, depending on the particular lithium technology [2]. The cylindrical geometry is generally used with lithium iron phosphate (LFP) technology, which is the current preferred technology for PV energy storage systems.

One might expect that large batteries with capacities in the 100s or 1000s of kWh would have relatively large basic building blocks. Presently, however, the largest basic cells will store up to about 5 Ah with a terminal voltage of approximately 3.5 V (17.5 Wh). Thus, a 10 kWh storage system at 48 V will require approximately $10,000 \div 17.5 = 571$ individual cells in a series-parallel configuration. More specifically, since $48V \div 3.5V = 13.7$, 14 cells in series will produce 49 V. This means a total of $571 \div 14 = 40.79$, or 41 series sets will need to be connected in parallel to achieve the desired 10 kWh unit. At an energy density of 125 Wh/kg, this 10 kWh unit can be expected to weigh approximately 80 kg (176 lb). Of course, this is only the weight of the energy storage part. With packaging, battery management and additional electronics for conversion of AC to DC and DC to AC, the overall weight may end up over 136 kg (300 lb).

The operational challenge for this topology is thus to be sure each and every cell is charged and discharged equally such that at any time, every cell will be at the same state of charge (SOC). In addition to maintaining uniform SOC (cell balance), charging rates and discharging rates must be limited, charging must be slowed at about 80% and halted when the SOC of the unit reaches 100% and discharging must be halted as the SOC approaches 0%. It is also desirable to track cell temperature, since most of the cell operating parameters are temperature dependent. Since terminal voltage is temperature dependent, the SOC of individual cells is essentially monitored by counting charge carriers by integrating current over time for each cell. All these functions are the responsibility of the battery management system (BMS), which happens to be ever so slightly more complicated than the typical lead-acid battery charge controller. Thus, a microcontroller will likely be the heart of any large-scale BMS.

Figure 6.1 shows a block diagram of a typical BMS [3]. This particular system is designed to manage four cells, so one only needs to imagine how it will need to be scaled up to manage over 500 cells. Note that the functional capabilities of the BMS as previously mentioned are shown in the individual blocks. Not mentioned previously is the fact that all the electronics happen to need their own uninterruptible power supply, derived from the overall storage package and shown as the BMS local power supply in the figure. This enables all the electronic functions to carry out their mission, whether it be current sensing, watchdogging or being the failsafe Microcontroller Unit (MCU). For those not familiar with electronic watchdogs, their purpose is to monitor whether or not a system is operating normally by watching the clock signals from the MCU. The Failsafe MCU, as one might infer by simply looking at all of its inputs, looks important. And, indeed, it is. This microcontroller is designed to identify certain conditions in the system, such as high temperature, high or low voltage or other conditions that could result in damage to the overall system and then implement control commands to alleviate the potential for catastrophic damage resulting from the irregularities.

Other components of the BMS include a means of monitoring the current, voltage and charge of the individual cells (cell monitor and balancer), a temperature sensor or a collection of them, an overcurrent protection module, an overall current sensor that measures either the total charging or total discharging current, a fuse as a last resort for shutdown and an electronic charge/discharge switch that limits charging and discharging and responds to the system needs as will be discussed later.

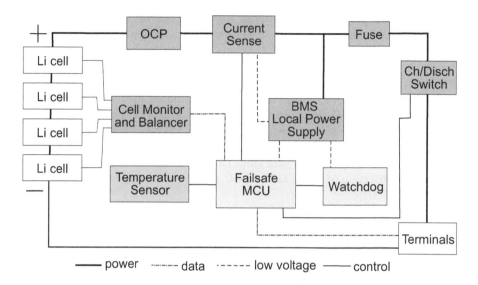

FIGURE 6.1 Block diagram of a typical small battery management system.

(Adapted from Sanino, Enrico, *Introduction to Battery Management Systems*, All About Circuits, Feb 8, 2021 For educational purposes only.)

Given the importance of lithium technology, plus the assumption that the reader loves chemistry, it is time for an explanation of the chemistry of the LFP cell [4]. The reaction is described by

$$x\text{Li}^+ + xe^- + 6\text{C} \leftrightarrow \text{Li}_x\text{C}_6 \qquad \text{LiFe(III)PO}_4 \leftrightarrow \text{Li}_1-x\text{Fe(II)PO}_4 + x\text{Li}^+ + xe^- \qquad (6.1a)$$

    Anode   Electrolyte LiPF_6 in   Cathode
                 ethylene carbonate/
                 dimethylcarbonate (DMC)

Combining these two reactions results in an overall reaction that can be expressed as

$$\text{LiFePO}_4 + 6\text{C} \leftrightarrow \text{LiC}_6 + \text{FePO}_4 \qquad (6.1b)$$

where discharging is in the left-to-right direction and charging is in the right-to-left direction. Note that, during the charging process, electrons leave the anode via the external circuitry and enter the cathode and during the discharge process, electrons from the cathode enter the anode from the external circuitry.

This reaction differs from many other batteries, since the space between the positive and negative electrodes, in addition to the electrolyte, incorporates a separator membrane that has a permeability that will only allow penetration by lithium ions. This forces the electrons to flow through the external circuit, either in the charging direction (anode to cathode) or in the discharging direction (cathode to anode).

Energy Storage

If the cell is 100% charged, then the composition of the anode is carbon in the form of graphite and the composition of the cathode is LiFePO$_4$. If the cell is 100% discharged, then the composition of the anode is LiC$_6$ and the composition of the cathode is FePO$_4$. The nominal operating voltage of the LFP cell is 3.2 V, the fully charged voltage is 3.65 V and the fully discharged voltage is 2.5 V [5].

6.2.2 NCA BATTERIES

Another lithium technology that is becoming of interest to the grid storage community is lithium nickel cobalt aluminum oxide (LiNiCoAlO$_2$), conveniently shortened to NCA [6]. NCA cells have high energy density (250–300 Wh/kg) [7, 8] and can be rapidly charged and discharged, which means they have high efficiency and very low internal resistance. The primary interest in NCA technology in 2023 is electric vehicle batteries, but their properties also make them very attractive for grid storage. At present, the downside of the equation is the cost and availability of materials. Nickel and cobalt are expensive and recently have been subject to supply chain interruptions. Given the possible grid applications, it will be interesting to follow the story of the development of this technology.

6.3 NICKEL-BASED BATTERY SYSTEMS

6.3.1 NICKEL–CADMIUM (NI–CD)

Ni–Cd batteries are more robust than lead-acid batteries. They can survive freezing and high temperatures, and they can be fully discharged and they are less affected by overcharging. As a result, in some applications, Ni–Cd batteries may be a better choice because their robust nature may enable the elimination of the system charge controller. If the batteries are to be used in a location where access for maintenance is difficult, the higher cost of these batteries can also often be justified.

The overall discharge reaction at the electrodes is given by

$$2NiOOH + 2H_2O + Cd \rightarrow 2Ni(OH)_2 + Cd(OH)_2. \tag{6.2}$$

The reaction is reversed in the charging direction. The voltage of the fully charged cell is 1.29 V. Unlike the lead-acid system where the specific gravity of the electrolyte changes measurably during discharge or charge, the KOH electrolyte of the Ni–Cd system changes very little during battery operation. In some batteries, LiOH is also added to the electrolyte to improve cycle life and also to improve higher temperature operation.

These batteries are in use in industrial, military and space applications where the previously mentioned properties are important. Capacities range from small sizes up to more than 1200 Ah. Three different thicknesses are used for electrodes, depending on whether the battery is designed for high, medium or low discharge rates. Energy densities range from 20 Wh/kg to more than 50 Wh/kg, depending on cell composition. Unlike the lead-acid system, which loses capacity under conditions of heavy discharge, the Ni–Cd system can be discharged at rates up to C/1.0 over a wide temperature range, while still providing more than 90% of its capacity to the load. This is partially attributable to the very low internal resistance of the cells, which, depending on cell area, can be less than a milliohm.

If Ni–Cd batteries are charged and then left unused, they will lose charge at the rate of approximately 2% per day for the first few days, but then stabilize to a relatively low loss rate. Over a six-month period, the total loss is typically about 20%, depending on whether the battery is a high, medium or low discharge rate battery. The higher the discharge rating of the battery, the greater the loss of charge over time. Loss of charge is also temperature dependent. The loss rate is greater at higher temperatures, but at a temperature of $-20°C$, there is almost no loss at all.

The lifetime of a Ni–Cd battery depends on how it is used but is less dependent on the depth of discharge than that of lead-acid batteries. A lifetime of at least 2000 cycles can be expected for a battery when it is not used extensively at elevated temperatures. As a result, under certain applications and operating conditions, a battery may last as long as 25 years, while under more frequent cycling, the lifetime may be reduced to eight years. A Ni–Cd battery can last twice as long as its lead-acid counterpart.

Disadvantages of Ni–Cd batteries include difficulty in determining the SOC of the batteries and the toxicity of the cadmium, which creates an environmental concern during production and disposal. They are also more expensive than most other rechargeable battery technologies. As electric and hybrid automobiles become more popular, the demand for high-energy-density batteries will likely lead to the replacement of Ni–Cd technology with lower-cost, higher-energy-density technologies.

6.3.2 Nickel–Zinc Batteries

The *nickel–zinc* battery is a combination of the Ni–Cd system and the Cu–Zn system, which provides some attractive features, including long life and a capacity advantage. The specific energy of the Ni–Zn system is double the specific energy of the Ni–Cd system. The overall discharge chemical reaction is given by

$$2NiOOH + Zn + 2H_2O \rightarrow 2Ni(OH)_2 + Zn(OH)_2. \tag{6.3}$$

If the battery is allowed to overcharge, it dissociates the water in the KOH electrolyte, resulting in gassing and the release of hydrogen and oxygen. Hence, charging must be controlled to avoid overcharging. Fully charged cell voltage is approximately 1.73 V. Ni–Zn battery systems can be made in sizes in excess of 200 Ah. At normal operating temperatures ($\cong 20°C$) and reasonable discharge rates ($\cong C/3$) the cell voltage remains quite stable at about 1.6 volts over most of the discharge range [10].

6.3.3 NIMH Batteries

Another technology that was very popular, particularly in smaller applications such as camcorders, is the NIMH battery [11]. This battery replaces the cadmium cathode with an environmentally benign metal hydride cathode, allowing for higher energy density at the cathode and a correspondingly longer lifetime or higher capacity, depending on the design goal. The anode is the same as in the Ni–Cd cell and KOH is used as the electrolyte. The overall discharge reaction is

$$MH + NiOOH \rightarrow M + Ni(OH)_2. \qquad (6.4)$$

The NIMH cell requires clever use of an oxygen recombination system to prevent the loss of oxygen during the charging cycle. Electrodes are also carefully sized to ensure that the useful capacity of the battery is determined by the anode electrode to ensure that the cathode will moderate any overcharge or overdischarge condition.

6.4 FLOW BATTERIES

6.4.1 Introduction

Flow batteries have been around for over 40 years, but until recently, they have remained in the shadows. A number of technologies exist, each of which has its advantages and disadvantages. Attention has returned to flow technologies because they are capable of storing megawatt-hours of energy and are capable of releasing it at rates up to megawatts. Non-flow technologies can be optimized to release their energy quickly or slowly, that is, to provide high power, such as needed for starting an internal combustion engine, or to provide lower power, such as would be used in a typical PV application. These applications are generally classified as shallow discharge and deep discharge. Once the device has been configured, the charge and discharge limits are established. On the other hand, true redox flow batteries can be optimized for both power and energy, which can mean optimizing the unit to meet the load conditions.

In general, the main difference between flow batteries and previously discussed technologies is that in lithium-ion, nickel and most other technologies, the energy is stored at the electrodes. In true redox flow batteries (RFBs), such as vanadium–vanadium and iron–chromium systems, the energy is stored in the electrolytes. In hybrid redox flow batteries, such as zinc–bromine and zinc–chlorine systems, some energy is stored in the electrolytes and additional energy is stored at the electrodes. To store energy in electrolytes, separate electrolytes are needed for the anode side and the cathode side of the battery. The electrolytes are pumped from storage containers through the electrochemical stack, which houses the electrodes and a membrane that separates the two electrolytes but allows ion transfer through the membrane as a part of the charging or discharging process. Disadvantages of flow batteries include lower volumetric energy storage and higher energy losses between charge and discharge than non-flow units, and some flow technologies use toxic chemicals.

Recently, work has been progressing on organic flow batteries that use non-toxic, organic electrolytes. Reasonable energy density has been achieved and the researchers believe that this technology may result in a more cost-effective technology for utility-scale energy storage [12, 13].

6.4.2 Sodium Sulfur

Unlike batteries previously discussed that have liquid electrolytes and operate at ambient temperatures, sodium sulfur batteries are high temperature (>300°C) and use a solid electrolyte. The overall cell reaction is

$$2Na + 3S = Na_2S_3 \tag{6.5}$$

During discharge, $Na \rightarrow Na^+ + e^-$ and the sodium ions then migrate toward the sulfur anode via the electrolyte. The freed-up electrons pass through the external circuit toward the anode. As the sodium ions and the electrons meet at the sulfur anode, the sodium and sulfur are combined into Na_2S_3. During charging, the sodium moves back to the cathode until the anode is free of sodium. Cell voltage is approximately 2 V, depending upon the specific cell chemical reaction, battery efficiency is close to 90% and cell-specific energy is somewhere between about 90 and 170 Wh/kg. A significant portion of the development work with sodium-sulfur technology has been carried out in Japan. One installation in northern Japan can store 245 MWh and deliver 34 MW for grid stabilization [14].

6.4.3 Zinc Bromine

Important advantages of zinc bromide flow batteries include 100% discharge capability on a daily basis, cycle life in excess of 5000 cycles, non-flammable electrolytes, no cooling systems needed, relatively lower cost, readily available materials and straightforward end-of-life recycling. In addition, they are relatively easily scalable to high storage capacity by increasing the size of the tanks and stacks. However, the downside is lower energy density (60–85 kWh/kg), lower round-trip efficiency (70%–80%), the need to be fully discharged every few days and lower charge and discharge rates [15].

Battery operation is based upon two tanks of ZnBr salt solution. During charging, metallic zinc is plated onto the cathode surfaces. At the anode, bromide ions are converted to bromine and stored in a safe, organic phase. During discharge, the zinc plated onto the cathode dissolves again into the electrolyte. It is interesting that many batteries become nearly useless if they are drained to zero SOC, but these batteries are happy in that condition, since the cathode prefers not to have the zinc attached. To accomplish the charging or discharging process, the electrolytes are pumped from one tank to the other via a set of reactor stacks. During the operation of the battery, measures need to be taken to prevent the bromine from coming into contact with the zinc. If this happens, the battery is essentially short-circuited internally and it will self-discharge. Thus, a membrane impervious to the transfer of the bromine needs to be installed between the two electrodes.

The chemical reactions are

$$\text{Cathode: } Zn \leftrightarrow Zn^{2+} + 2e^-, \text{ anode: } Br_2 + 2e^- \leftrightarrow 2Br^-, \text{ overall: } Zn_{(s)} + Br_{2(aq)} \leftrightarrow 2Br^-_{(aq)} + Zn^{2+}_{(aq)} \tag{6.6}$$

This reaction produces a cell voltage of 1.67 V.

Work on improving this technology is ongoing to the extent that one manufacturer in 2023 has claimed an order backlog of 2.2 GWh [16].

6.4.4 Vanadium

Vanadium redox flow battery characteristics include round-trip efficiency up to 90%, between 12,000 and 14,000 lifetime 100% cycles, 10–20 Wh/kg and a cell terminal voltage of 1.15–1.5 V. The charge carriers in the electrolytes are vanadium ions. Electrons produce the current in the external circuit and the current between the anode and the cathode that balances the electron current is via protons, provided that the membrane adequately opposes the flow of any larger ions that might want to flow through it [17].

The interesting thing about vanadium ions is that there are 4 possible oxidation states with them, including V^{2+}, V^{3+}, VO_2^+ and VO^{2+}. Two are used in the positive half-cell electrolyte (VO_2^+ and VO^{2+}) and the other two are used in the negative half-cell electrolyte. In somewhat simpler terms, the two ions in the positive electrolyte are designated as V^{5+} and V^{4+} and the two ions in the negative electrolyte are designated as V^{3+} and V^{2+}. Each electrolyte has its own storage tank and circulating pump. The electrolytes are circulated between their respective storage tanks and the half-cells associated with each electrolyte. The two half-cells are separated by a special membrane that allows only the passage of hydrogen ions (protons) and electrons but does not allow the passage of any of the vanadium ions. The half-reactions, in terms of V^{2+}, V^{3+}, V^{4+} and V^{5+} are $V^{5+} \leftrightarrow V^{4+}$ in the positive electrolyte and $V^{3+} \leftrightarrow V^{2+}$ in the negative electrolyte.

It's interesting to figure out which represents charging and which represents discharging. The first thing to note is that only electrons can flow in the external circuit. Thus, if an electron leaves the positive electrode (charging), the positive electrode becomes more positive. Thus $V^{4+} \rightarrow V^{5+}$ during a charge cycle. If an electron leaves the positive electrode, it must enter the negative electrode. Thus $V^{3+} \rightarrow V^{2+}$. During the discharge cycle, the electron flow in the external circuit reverses and thus, the two half-reactions also reverse. In each case, the abandoned hydrogen ions (protons) pass through the membrane to balance the charge, flowing in the direction within the cell opposite the electron flow external to the cell. In terms of the extra oxygen atoms involved in the vanadium SOC, the half-reactions can be represented as

$$VO_2^+ + 2H^+ + e^- \rightarrow VO^{2+} + H_2O \text{ and } V^{3+} + e^- \rightarrow V^{2+} \quad (6.7)$$

The pioneering research work on the vanadium redox flow battery can be attributed to the University of New South Wales back in the mid-1980s [18]. Since then, with additional improvement, these batteries have become popular whenever either rapid response (100% load change in half a millisecond) or stabilization of renewable source output and long-term storage with very little maintenance are desired. Rapid response is particularly useful for utility frequency stabilization and uninterruptible power supplies and longer response capabilities are important for frequency regulation and load shifting. As of 2022, the largest battery in operation was a 400 MWh,

100 MW unit in China, with plans to expand it to twice the size [19]. Other countries that have shown interest in the large-scale deployment of vanadium redox flow batteries since 2005 include Japan, Germany and the United States.

6.4.5 ALL IRON REDOX

The two tanks with the half-cell design are also used for the Iron Redox Flow Battery. The electrolytes contain dissolved iron ions. Once again, the membrane that separates the half cells is the critical component of the assembly, since it must only allow certain ionic species to pass through it, depending upon the overall system chemistry. One of the important functions of the membrane is to prevent any iron in any form from penetration. The battery stack consists of single cells connected in series [20].

The discharge processes in each half cell and overall are

$$\text{Positive: } 2Fe^{2+}_{(aq)} \rightarrow 2Fe^{3+}_{(aq)} + 2e^-, \text{ Negative: } Fe^{2+}_{(aq)} + 2e^- \rightarrow Fe^0_{(s)} \text{ and overall}$$
$$3Fe^{2+}_{(aq)} \rightarrow 2Fe^{3+}_{(aq)} + Fe^0_{(s)} \quad (6.8)$$

The overall basic cell voltage is 1.21 V, round-trip efficiency is about 70% and energy density is about 20 Wh/kg in present commercially available units. Work at USC Dornsife [21] suggests a theoretical value of 170 Wh/kg might be achieved with improvements to the cell.

Because the available power is dependent upon the stack size and the available energy is dependent on the electrolyte volume, these batteries can be scaled for desired power and energy levels. The components are generally more available at less cost, and they can have extended cycle lifetime as long as a number of possible degradation or failure modes are dealt with. Work continues on mitigating issues such as hydrogen release and capture, maintaining sufficiently low electrolyte pH and preventing any rust from forming. This technology has been commercialized, with capacities of up to 600 kWh and 90 kW at the time of this writing.

6.5 EMERGING BATTERY TECHNOLOGIES

A number of additional battery types are currently under investigation for PV and similar uses, such as electric vehicles and large-scale grid storage. A partial listing of these batteries includes zinc/silver oxide, metal/air, iron/air, zinc/air, aluminum/air, lithium/air, zinc/bromine, lithium-aluminum/iron sulfide, lithium-aluminum/iron disulfide, sodium/sulfur and sodium/metal chloride and a collection of hybrid flow batteries and organic storage systems. The young engineer can expect to see progress in these new technologies during his/her path toward becoming an older engineer, especially now that public demand for electric vehicles, hybrid vehicles, photovoltaic (PV) systems with battery storage and large-scale utility grid backup storage is rapidly increasing.

In addition to battery storage, which is presently the most common method of energy storage for PV systems, a number of other energy storage methods are either

Energy Storage

6.6 HYDROGEN STORAGE

Aside from using batteries to store energy produced by PV systems, it is also possible to use hydrogen as a storage medium [22]. The advantage of using hydrogen for energy storage is when it is desired to recover the stored energy, the hydrogen can be reacted with oxygen to form water, which is an exothermic reaction that does not have any carbon byproducts. If the hydrogen is burned in the air to produce steam to turn a turbine to generate electricity, water is still produced as the byproduct, with a relatively minimal amount of nitrogen oxides. In a fuel cell, hydrogen combines with oxygen to produce water, electricity and heat. Of course, the hydrogen does not need to be used to produce electricity. It can also be used as an engine fuel to power a vehicle, such as an automobile, bus or space shuttle. Indirectly, the result is a solar-powered vehicle that stores energy in hydrogen rather than in batteries. The engineering community has not yet developed a battery-powered, space shuttle booster engine, although there are now some lithium battery-powered aircraft mainly being used by bush pilots.

Perhaps the most important advantage of hydrogen storage is the energy density of hydrogen. One of the highest energy densities in conventional fuels is that of gasoline, which contains approximately 1,047,000 BTUs/ft^3 (39,010,173 kJ/m^3 or 434 kWh/ft^3). To store the same amount of energy in high-capacity storage batteries would require approximately 120 batteries having approximately 2000 BTUs/ft^3 (74,518 kJ/m^3 or 0.586 kWh/ft^3) storage capacity. The storage capacity of liquid hydrogen is close to 240,000 BTUs/ft^3 (8,942,160 kJ/m^3 or 99.48 kWh/ft^3), and gaseous hydrogen can store approximately 47,000 BTUs/ft^3 (1,751,173 kJ/m^3 or 19.48 kWh/ft^3). Since water cannot be decomposed into gasoline, however, and since gasoline burning produces CO_2, among other undesirable combustion products, clean-burning hydrogen is a more attractive alternative for combustion.

In fact, since the energy stored in gasoline is generally obtained through combustion, and since combustion engines running to produce electricity are subject to the limited efficiency of heat engines, which is generally less than 40%, less than 40% of the energy stored in gasoline can be converted to electricity. Aside from liquid and gaseous hydrogen storage, other means of obtaining relatively high energy density, including various hydrides, are also available.

Electrolysis of water is a common method of producing hydrogen. The output of a PV system is conveniently DC at a level consistent with that needed for relatively efficient electrolysis. Production of hydrogen by electrolysis may seem to be simply a matter of passing a current through water to produce hydrogen and oxygen, just as many readers have already done in chemistry lab. The challenge is to produce hydrogen efficiently with a system whose power output varies significantly whenever a cloud passes over the PV array.

Once the hydrogen is produced, it can be used on-site or it can be transported to other locations. Hydrogen proponents have suggested that hydrogen be produced in tropical latitudes having greater annual insolation and shipped to regions having less sun. Thus, after the production of hydrogen, storage and transport of the gas also present engineering challenges. Recent interest in fuel cells for use in automotive applications has once again piqued interest in the efficient and convenient production of hydrogen. The same technology that may emerge for automotive fuel cell applications may very well spin off into PV energy storage applications as well.

6.7 THE FUEL CELL

Fuel cells provide a convenient means of reacting hydrogen and oxygen directly to produce electricity and water that is not subject to the limiting efficiency of combustion. They utilize essentially the inverse of the electrolysis process. The first fuel cell was built in 1839 by Sir William Grove [23]. The high cost of fuel cells, however, had kept them out of practical use until NASA decided to use them on space flights to provide power and water.

Fuel cells are being constructed with a wide range of power output capabilities. The smaller cells are sized to power video cameras and laptop computers. On a larger scale, The Southern California Gas Company had ten 200-kW plants on line in 1993 and an 8 MW unit is planned for operation in 2024 [24]. The fuel cell reaction is exothermic, so if cleverly applied, the cells can supply both heat and electricity to a building. By capturing the heat, the overall system efficiency is increased, provided that the heat can be put to good use, such as for district heating or other space or water heating applications.

The alkaline cell (AFC) contains a 30% solution of KOH as the electrolyte. The proton exchange (or polymer electrolyte) membrane cell (PEMFC) incorporates a semipermeable membrane that will pass hydrogen ions, resulting in a proton exchange. The phosphoric acid cell (PAFC) is based on a 103% phosphoric acid solution. The molten carbonate cell (MCFC) is based on a eutectic mixture of molten lithium carbonate and potassium carbonate. The solid oxide cell (SOFC) is based on a stabilized zirconium oxide electrolyte [25].

The low-temperature cells are approximately twice as efficient as conventional heat engines in the conversion of energy stored in hydrogen to electricity. This makes fuel cells a very attractive possibility for energy storage and conversion to electricity in an electric vehicle, since electric motors can be operated at very high efficiency, especially with modern solid-state controllers. The present disadvantage, of course, is that fuel cells remain quite costly and will require further development to reduce their cost. Fuel cells also require a source of hydrogen, so a means for providing fuel for vehicular applications of fuel cells will need to be created.

6.8 MECHANICAL, THERMAL AND OTHER STORAGE OPTIONS

6.8.1 Introduction

Perhaps one of the simplest means of storing energy from the sun is to fill a water storage tank so the water can be used after sundown. Maybe the use will be for drinking or irrigation, or maybe the water will be used to turn a turbine to generate electricity before it is used for some other purpose. It is not difficult to calculate the

Energy Storage

amount of potential energy stored in a gallon of water raised to a height of 10 feet above ground.

Other possible storage mechanisms include compressed air, flywheels, superconducting magnets and chemical capacitors. All of these can be shown to be useful for certain end uses, but, in general, the cost per kWh stored is quite high at present. Engineers are generally familiar with various forms of potential and kinetic energy expressions, such as $E = mgh$, $E = \frac{1}{2} mv^2$, $E = \frac{1}{2} J\omega^2$, $E = \frac{1}{2} CV^2$ and $E = \frac{1}{2} LI^2$. Note that in this mix of electrical and mechanical quantities, there is a possibility of confusing the symbols for electric current and moment of inertia. After flipping a coin and also noting that electrical current appears more often than moment of inertia in this publication, the authors decided to use J for moment of inertia and I for electrical current. Many of the other storage options make use of one or more of these electrical or mechanical storage means, or, perhaps, some version of thermal storage.

6.8.2 Pumped Hydro

With all the excitement about batteries, it is often overlooked that pumped hydro is by far the largest storage means worldwide [26]. In 2022, a total of 137.06 GW of pumped hydro capacity were in operation worldwide, compared to 99.76 GW in 2010 [27].

The first thing to notice about this number is that it is in *GW, not GWh*. Thus, it represents available instantaneous power as opposed to cumulative energy. For hydro, this works fine, since the same rotational machinery either pumps the water uphill or generates electricity as it comes back down again. If one is interested in the associated kWh, then one needs to ask how many hours can the system sustain its capacity while in the generation mode. For pumped hydro, this number is generally measured in days. Much of the reason for this large number is due to the inherent instability of the nuclear electrical generation process.

Controlling the fission process is very delicate, and it takes more than a few hours to stabilize the output of a nuclear electrical generating plant to the rated value. At that point, the plant is pretty much insistent on generating a fixed amount of power. Of course, the electrical load on the electrical grid is nowhere near constant. As long as the load exceeds the nuclear capacity, other generation methods can make up the difference. But when the load drops below the nuclear capacity, this electricity needs somewhere to go to keep the nuclear generator happy. That's when the experts realized that pumping water uphill at night to use the excess nuclear load could be a good solution to the problem. All that was needed was water, a hill, some pumping/generation equipment and supportive neighbors.

It's interesting to calculate the amount of energy that can be stored with pumped water. Since a cubic foot of water weighs 62.4 lb (28.3 kg) and since 1 kWh = 2,655,200 ft lb (3,599,968 n-m), all one needs to do to calculate the number of cubic ft of water needed to store a kWh if it is pumped uphill, if the hill is 100 ft (30.48 m) high. The result is (2,655,200 ft lb) ÷ (62.4 lb/ft³) ÷ (100 ft) = 425.5128205 ft³ (12.04 m³). To obtain 1 MW as the water runs back downhill means the generation equipment needs to be able to generate 1 MW. Since 425.5 ft³/hr would be 1 kW, then 425,500 ft³/hr would equate to 1 MW. If the flow rate per minute is desired, it would be 7092 ft³/min (0.163 Acre ft/min, 200.8 m³/min), which is a bit higher flow rate than a typical shower, which explains the huge piping associated with a pumped hydro facility. Of course, the round-trip efficiency of this process is not without loss. A well-designed system may achieve a round-trip efficiency of about 80%.

In any case, pumped hydro is no longer a pump at night and is generated during the day operation. With near Gigawatts of daytime PV, it is possible that daytime generation at some times may exceed the needs of the grid. Or, maybe it might be excess wind power. As a result, now the pumped hydro stations of the world are important for renewables as well as for nuclear to help avoid curtailment of generation when the fuel is free. The time to switch from pumping uphill to generation is less than 30 minutes.

6.8.3 Compressed Air

Compressing air is yet another method of storing energy that can be effective at any hour of the day [28]. It is likely that some readers have watched as their scuba tank was filled with compressed air. In particular, it may have been noticed that the tank is first immersed in a barrel of water. Then it is filled by the air compressor. As the air is compressed, the tank gets hot and is cooled by the water in the barrel.

Probably the most interesting part of compressed air energy storage (CAES) is finding a suitable location for storing the compressed air. Scuba tanks were ruled out for use as energy storage systems a while ago. Even larger tanks have generally been ruled out in favor of storage in underground caverns. These caverns may be depleted natural gas reservoirs or salt caverns or other stable geologic formations. Different areas have different types of formations. A major advantage of underground storage is the significantly reduced above-ground footprint of the facility. Just as all the other storage mechanisms can be explained by kinetic or potential or chemical energy relationships, CAES is based upon the familiar ideal gas law, $PV = nRT$. Note that if P has the units of lb/ft² and V is in ft³, then the units of PV will be ft-lb. Thus, if somehow temperature can be kept constant (good luck on that one!), then it might be possible to at least estimate to a first approximation what size cavern and what pressure might be necessary to store 100 MWh of compressed air energy.

First, using the now familiar 1 kWh = 2,655,200 ft lb relationship, 100 MWh = 100,000 kWh = 100,000×2,655,200 = 2.655×10^{11} ft lb. Now all that is needed is to pick a pair of P and V that will yield this number. If a starting point might be that the typical fill pressure of a scuba tank is 3000 psi = 432,000 psf, then all one needs to do is determine an associated volume, which, in this case, will be (2.655×10^{11} ft lb) ÷ (432,000 lb/ft²) = 614,583 ft³. If this volume represents a cube, the linear dimensions will be about 85 ft on a side, which is a surprisingly small cube that is highly pressurized. Clearly, someone with the geotechnical data for the site will need to confirm the available volume and a reasonable pressure for the cavern.

The compressor and the generator do not necessarily need to be the same size. In any case, it is now interesting to estimate the approximate motor size to fill the cavern in half a day. Since 1 hp = 550 ft lb/sec, one method would be to calculate the HP it will take to deliver (2.655×10^{11} ft lb) in 12 hours (43,200 sec). Thus, the power needed is (2.655×10^{11} ft lb) ÷ 43,200 sec = 6,145,833 ft-lb/sec = 11,174 hp. Of

Energy Storage

course, this has assumed 100% efficiency for the overall process. The existing CAES system's overall efficiency is reported to be in the 60% range.

The estimate on compression hp also leads to an estimation of expansion hp assuming the stored energy is used in a 12-hour period. Again, at 100% efficiency, energy out = energy in. Thus, equal compression and expansion times suggest approximately the same size for both ends of the cycle. The engineering challenge is to optimize the performance of the system taking into account site conditions. Generally, it takes less than 12 minutes to switch directions for the equipment, similar to pumped hydro [29]

6.8.4 Molten Salts

Different salts have different melting temperatures. Every material has its own thermal mass. Perhaps these properties are best known for water, with its freezing point of 0°C (32°F), its boiling point of 100°C (212°F) and its specific heat capacity of 4182 J/kg/°C (1 BTU/lb/°F). Water has served well over the years for producing steam to turn turbines. Of course, to produce the steam, energy needs to be added to the water at a temperature that exceeds the boiling point. This is where molten salts enter the picture. The idea is to find a salt or other suitable material that can remain in the liquid state over a wide temperature range while being heated either thermally or electrically by renewable or nuclear energy, then be stored with minimal heat loss until the heat energy is needed and then passing through a heat exchanger that will produce steam to generate electricity while maintaining the molten state of the salt. When the sun shines or the wind blows, the cycle repeats itself.

The ideal high-temperature liquid (molten salt) will have a wide range of temperatures where it is in the liquid state and will be abundant, nontoxic, nonflammable, non-corrosive and have a high specific heat capacity. One of the more popular liquid salts is a mixture of approximately 60% sodium nitrate ($NaNO_3$) and 40% potassium nitrate (KNO_3). The melting point of the mixture is between 213°C and 242°C and its boiling point is between 450°C and 590°C. Increasing the percentage of $NaNO_3$ increases both the melting point and boiling point. Operation is generally between 260°C and 550°C. The specific heat capacity of this mixture is 1530J/kg/°C (4.25×10^{-4} kWh/kg/°C). Thus, 1 kg of the molten salt will deliver or absorb 4.25×10^{-4} kWh for each degree C it is cooled or heated [30].

Putting this into perspective, assuming operation between 260°C and 550°C, each kg of the molten salt can deliver 0.123 kWh of heat when cooling through this range. The available power, of course, depends upon how quickly the salt is cooled. For example, if the cooling takes place in 0.1 hr, the 0.123 kWh is delivered in 0.1 hr, which amounts to 1.23 kW. Of course, ideally this power will be delivered with 100% efficiency, but, in reality, it will be less than that. In any case, the numbers are now available to make ballpark estimates of quantities and flow rates needed to produce kWh and kW in the MWh and MW range.

An interesting variation on Molten Salt energy storage has been under investigation by a research team at Pacific Northwest National Laboratory [31]. Rather than using the sun to melt a salt and then have the salt run a turbine when the sun is not shining, the PNNL project stores renewable energy in the molten salt at a temperature of 180°C, allows the molten salt to solidify and then recovers the energy (discharges) at a later time when the battery is heated.

The process involves using renewable-generated electrical current to move ions through the battery to effectively store energy in an electric field created by the separation of the ions during charging. Then, if the electrolyte is cooled back to solid, the ions are frozen in place at ambient temperatures and can remain that way for a long time, not too different from an old-fashioned flashlight battery. Then, several months later, such as during the winter in Alaska, the batteries can be reheated to enable the batteries to discharge to provide grid support.

6.8.5 Flywheels

Flywheel operation is well-known (presumably) to the mechanical engineers who may be digesting this material. The formula is simply $E = \frac{1}{2}I\omega^2$, where I is the moment of inertia and ω is the rotational speed in radians/sec. Using this formula as a starting point, it is interesting to estimate the size of the rotor needed to store 1 MWh of energy at a rotation speed, ω = 150,000 rad/min = 2500 rad/sec. Using $I = 2E/\omega^2$ gives

$$I = 2\times 10^6 \text{ (Wh)} \div 2500^2 \text{ (sec}^{-2}) = 0.32 \text{ (Wh)(sec}^2).$$

Now the fun part is making sense of these units. One way is to convert Wh to joules, using 1 Wh = 3600 joules, to get I = 1152 joule-sec². Next, noting that 1 Newton = 1 kg-msec^{-2} and 1 joule = 1 Newton-m, the moment of inertia can be expressed as J = 1152 kg-m².

Next, assuming the use of a flywheel for which $I = MR^2$, where M is the mass in kg and R is the radius in m of the flywheel, one simply needs to assign a value to either the radius to determine the mass or vice versa. If a radius of 1 m is assumed, that gives the obvious result that M = 1152 kg (2534 lb).

6.8.6 Weight Lifting—Lifted Weight Storage

Water is only one form of weight that can be lifted. For that matter, there are limited hills that can be used for pumping water uphill. For a lifted weight storage system, the principal requirements are gravity and weights and a way to convert potential energy into electrical energy and vice versa. Once again, as in pumped hydro, the simple relationship, E = mgh is used. Initially, concrete was used in experimental designs, but since concrete historically has been very energy intensive, other less energy-intensive materials are now being used, such as dirt or solid waste that does not use cement as binding material. The technology originated in Europe, and, in 2022, one popular deployment consisted of lifting 35-ton bricks. The up-down efficiency is in the range of 80–85% and is limited by the efficiencies of the motors and generators that are needed to lift the bricks and then to extract the energy as they descend back to ground level.

Energypost.eu reported in 2022 that a single vault with 10,000 bricks will have an annual output of 27 GWh, which converts to about 75 MWh per daily cycling,

Energy Storage

allowing downtime for maintenance during the year [32]. The rate at which the bricks are allowed to descend controls the desired power available to the grid. Knowing that reader curiosity will prompt the reader to want to explore heights and weights needed for various power and energy levels, a few problems await the interested reader at the end of this chapter.

6.8.7 Supercapacitors and Superconducting Magnets

Most students remember clearly sitting at lunch after their electric circuits class repeating over and over again: "The voltage across a capacitor cannot change instantaneously, the current in a capacitor leads the voltage, the energy in a capacitor is stored in an electric field" and "The current through an inductor cannot change instantaneously, the current in an inductor lags the voltage, the energy in an inductor is stored in a magnetic field." These important features of these two important components offer useful stabilizing opportunities for the grid. In addition, capacitors and inductors are also energy storage elements. For capacitors, $E = \frac{1}{2}CV^2$ and for inductors, $E = \frac{1}{2}LI^2$ when C is in farads, L is in henrys, V is in volts, I is in amperes and E is in joules. As a result, if C, V, L and/or I are large enough, perhaps some energy can be stored for a while. Note that 1 kWh = 3,600,000 joules (2,655,200 ft-lb).

Recall that the time constant associated with an RC circuit is given by $\tau = RC$ and the time constant associated with an RL circuit is L/R. In other words, ohms × farads = seconds and Henries ÷ ohms = seconds. In these expressions, R will usually represent the impedance of the utility line if the capacitor is connected across the line or if the inductor is connected in series with the line.

A check of the internet reveals capacitors that will store 15 kWh at 48 V. Since 15 kWh = 54,001,500 joules, this suggests the capacitance will be 2×54,001,500 ÷ (48^2) = 46,876 farads. This is an interesting result in a world where 50 years ago a microfarad was a lot of capacitance.

Probably the biggest advantage of the capacitor is that it can store energy with no chemical changes. It simply stores the energy in the form of charges that are separated, creating an electric field between the two charged surfaces. As a result, charging and discharging can be very rapid, since very little resistance is generally present to limit the current flow. Since the voltage across a capacitor cannot change instantaneously, this means that, in addition to being candidates for energy storage, capacitors can also play an important role in voltage stabilization. In particular, the loss of a source can result in transient voltages on a line and capacitors can help to minimize these transients.

Conceptually, energy storage in a magnetic field is straightforward, but in practical terms, it is still not among the economically attractive solutions [33]. While it is possible to wrap long lengths of superconducting mercury, vanadium or niobium-titanium wire around a magnetic core, it is not an easy task to keep these superconductors at temperatures below their critical temperature. It appears that this technology may take a while longer in the lab until it can be commercialized. However, if resistance is allowed in the system, then copper wire wound on a laminated iron core can be useful in reducing current transients just as capacitors can reduce voltage transients, since inductors store energy in a magnetic field created by the current and the current cannot change instantaneously.

6.9 AC AND DC BATTERIES

Some of the previously discussed energy storage mechanisms are based on DC batteries and some are based upon using energy stored in a mechanical or thermal form using AC and/or DC for the conversion process, either during the storage cycle or during the discharge cycle. Ultimately, with very few exceptions, the electrical output of all the storage mechanisms is AC. Even if the energy is stored in a DC battery, it is usually converted to AC before it is used, especially in grid-connected applications.

It is often taken for granted that an Energy Storage System that is shown with a single input/output port where AC power can enter for charging or leave during the discharge process will easily work in either direction. In fact, that may appear to be the case, but it is more complicated than that. Consider a lithium battery in the charging cycle that is approaching full charge. Assume the input to the battery is the output of an inverter. This means the inverter output needs to be converted efficiently to DC for charging the DC battery cells. As the BMS measures full charge, it needs to disconnect the battery from the charging power. This may seem pretty straightforward to simply shut off the battery. However, the interesting part is if the battery is disconnected, how then can it discharge?

It seems reasonable to assume that if PV power is still available after the batteries reach full charge, the PV power can be used to supply other loads, provided that they are connected. But what if the utility shuts down or what if it asks the batteries to send it some power? How do the batteries turn on again and how do they protect themselves from the possibility of the charging mechanism trying to continue charging them?

One answer is a smart internal transfer switch, which, in fact, is a part of the BMS and is shown in Figure 6.2. The electronic automatic transfer switch is controlled by the microcontroller in the BMS on the basis of inputs from its collection of sensors. As the storage element reaches full charge, the switch is switched from charge to off and waits for further instructions from the microcontroller. Since the charging voltage needs to exceed the cell voltage to drive current into the storage element, when the DC charging voltage is disconnected, the terminal voltage drops. Since the microcontroller of the BMS is smart enough to know when the storage element will need charging again, since it is keeping track of the SOC of the system, it will not turn the charging back on again until charging is needed.

On the discharge cycle, discharging is also determined by how the microcontroller is programmed. Several options are available. In one case, the storage remains charged until grid power is lost and then the controller immediately switches the system to discharge. This enables the inverter and AC power are supplied to the backup loads. Since the inverter is a one-way connection to the storage, even if the PV system is producing power, it supplements the power required by the loads but does not charge the storage until the controller gives it the instruction to do so.

Energy Storage

Another possibility for enabling discharge is if the system is programmed for grid support. In this case, a signal from the grid can request a specific amount of power for grid support, from the PV system, the storage system, or both. These options will be discussed as various energy storage systems are designed in the next chapter.

If the SOC of the storage drops below the minimum allowable charge level, the ATS is switched from discharge to charge. At this point, the battery waits patiently for a charging source, which could be either the grid or the sun (or the wind or . . .). After reaching an acceptable charge level, the storage once again is ready to turn on if the controller directs it to do so.

The final fine point of interest is the fact that storage voltage drops slightly, depending upon the load and the specific storage discharge Thevenin equivalent resistance. During charge, because of this internal resistance, terminal voltage exceeds internal voltage as mentioned earlier. During discharge, the terminal voltage drops below the internal voltage level. What this all boils down to is the need for separate voltage ranges for charging and discharging. Figure 6.3 illustrates this situation. Fortunately, the BMS or other controller will be able to respond to this hysteresis loop to prevent rapid charge-discharge cycling.

One variation on the AC-in-AC-out is used in certain DC-coupled PV systems with energy storage. These systems will be discussed in the next chapter. The point is that sometimes it is desired to store energy using DC for charging and discharging. The same rules need to apply to maintain a healthy SOC level in the storage. The difference is that the charge control/management function needs to be achieved either with an external charge controller, which is typical of stand-alone PV systems, or by

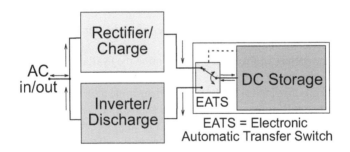

FIGURE 6.2 DC storage with AC in and out.

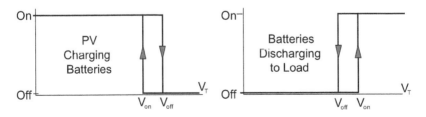

FIGURE 6.3 Hysteresis loops in charge controller using regenerative comparator for voltage sensing.

the inverter that uses the storage as one of its DC inputs, with the PV system being the alternate inverter power source.

For higher power levels, if the inverter is operating at the DC storage voltage and if this voltage is 48 V, very high currents will be required to deliver kilowatts of power. Usually, these systems operate with PV array voltages in the neighborhood of 400 V, which is directly fed into the MPPT input of the inverter. If a DC:DC converter is a part of the storage system, then it can track the PV voltage and convert it down to the storage voltage, whatever that may be. When the inverter needs power from storage, for any of the reasons previously mentioned, its controller can switch the direction from charge to discharge and convert the storage voltage up to the desired inverter input operating voltage. Since 10 kW at 400 V requires a mere 25 A, which can be supplied by #10 copper wire, this is a lot less than the 208 A that would be needed at 48 V. Of course, there might be up to almost 2% loss in the round-trip conversion process, but sometimes this is considered to be acceptable for an overall system. Both types of systems will be discussed in the next chapter.

In general, the field of energy storage is emerging rapidly. The reader is encouraged to explore the internet for information and photos of existing and emerging technologies.

Homework Problems

6.1 A 10 kWh Li battery is to be charged at a C/1 rate between 10% and 80% charge and at a C/3 rate between 80% and 100%. If the charging begins when the battery is at 20% charge level, how long will it be before the battery is 100% charged?

6.2 For the battery of Problem 6.1, determine the charging current for the two different charging rates.

6.3 For the battery of Problem 6.1, determine the available power if the maximum discharge rate is C/2.

6.4 If Li cells are rated at 5 Ah @ 3.5 V, determine a series-parallel combination that will store a nominal 5 kWh @ nominal 24 V.

6.5 Look up the typical voltage of a BMS local power supply on the internet.

6.6 Compile a list of possible irregularities in the charging or discharging of a lithium battery pack that would cause the Failsafe MCU to shut down the system.

6.7 For the 10 kWh @ 48 V lithium storage system of Section 6.2.1, determine the terminal voltage when the system is fully charged, at the nominal operating voltage and fully discharged if the fully charged cell voltage is 3.65 V, the nominal operating voltage is 3.2 V and the fully discharged voltage is 2.5 V.

6.8 Rewrite the chemical equations of the Ni–Cd and NIMH systems showing the half-reactions at the anodes and at the cathodes. Determine the charge transfer mechanism.

6.9 Assuming the efficiency of a gasoline engine to be 30%, what would the efficiency of a hydrogen engine need to be for an equal volume of liquid hydrogen to do the same amount of mechanical work as a gasoline engine?

6.10 How many gallons would have to be pumped into a tank raised 10 feet from the ground to be able to recover 1 kWh of electricity at a conversion efficiency of 100%, assuming the water is allowed to fall the entire 10 feet to operate an electrical generator?

6.11 Look up the performance specifications of one of the two CAES plants in operation in 2023. Can you find the volume of the storage reservoir, the storage pressure, the storage capacity in MWh, the horsepower needed to compress the air to fill the reservoir and the available MW from discharge? What about operating voltages?

6.12 Determine the size and rotation speed of a flywheel designed to store 20 kWh of energy. Note that there is no unique solution to this problem. Use reasonable assumptions.

6.13 Determine the capacitance of a capacitor needed to store 1 kWh if it is charged to 500 V.

6.14 Determine the inductance needed to store 1 kWh if the inductor is carrying a reasonable current, to be determined by the problem solver.

6.15 Check to see whether the local utility that serves your hometown has provisions for a home or business or electric vehicle owner to provide grid support at the request of the utility, in the form of either providing power to the grid or storing excess power from the grid. Do they have a rate schedule available?

REFERENCES

[1] Richmond, R., *Home Power Magazine*, February/March 2013. https://www.homepower.com/archive-browse

[2] For Information on Li Battery Cell Configurations: www.power-sonic.com/blog/lithium-battery-configurations/

[3] Sanino, Enrico, "Introduction to Battery Management Systems," *All About Circuits*, February 8, 2021.

[4] LFP Charging and Discharging Chemistry: https://en.wikipedia.org/wiki/Lithium_iron_phosphate_battery

[5] LFP Cell Voltage vs SOC Information: www.google.com/url?sa=t&rct=j&q=&esrc=s&source=web&cd=&cad=rja&uact=8&ved=2ahUKEwjQle2C0MAAxVVRDABHTabAjkQFnoECA0QAQ&url=https%3A%2F%2Fen.wikipedia.org%2Fwiki%2FLithium_iron_phosphate_battery&usg=AOvVaw0KEMBN5SzTyW1x1lf9eCnp&opi=89978449

[6] LiNiCoAlO2 Battery Details: https://skill-lync.com/student-projects/week-1-understanding-different-battery-chemistry-136

[7] More Information on NCA Batteries: https://skill-lync.com/student-projects/week-1-understanding-different-battery-chemistry-136

[8] Information on Use of NCA Batteries: www.petro-online.com/news/measurement-and-testing/14/breaking-news/how-are-lithium-nca-batteries-used/59149

[9] Linden, D., ed., *Handbook of Batteries*, 2nd ed., McGraw-Hill, New York, 1994.

[10] Sunden, Bengt, *Hydrogen, Batteries and Fuel Cells*, Section 4.5: www.sciencedirect.com/topics/engineering/nickel-zinc-battery

[11] Information on Nickel-metal Hydride Batteries: www.sciencedirect.com/topics/chemistry/nickel-metal-hydride-battery

[12] Information on Organic Flow Battery: www.sciencedaily.com/releases/2023/07/230710180520.htm

[13] More Information on Organic Flow Batteries: https://onlinelibrary.wiley.com/doi/pdf/10.1002/anie.201604925
[14] Information on Sodium-Sulfur Flow Batteries: www.sciencedirect.com/topics/earth-and-planetary-sciences/sodium-sulfur-batteries
[15] Information on Zinc Bromide Batteries: https://en.wikipedia.org/wiki/Zinc%E2%80%93bromine_battery
[16] Improvement Work on Zinc Bromide Batteries: "Eos Energy Enterprises Reports First Quarter 2023 Financial Results." https://finance.yahoo.com/news/eos-energy-enterprises-announces-first-203000553.html
[17] Information on Vanadium Redox Flow Batteries: https://en.wikipedia.org/wiki/Vanadium_redox_battery
[18] Univ New South Wales Vanadium Redox Research: https://research.unsw.edu.au/projects/vanadium-redox-battery
[19] Announcement of World's Largest Vanadium Flow Battery in China: www.energy-storage.news/first-phase-of-800mwh-world-biggest-flow-battery-commissioned-in-china/
[20] Information on All-Iron Redox Flow Battery: https://en.wikipedia.org/wiki/Iron_redox_flow_battery
[21] USC Dornsife All-Iron Flow Battery Research: https://dornsife.usc.edu/labs/narayan/all-iron-redox-flow-battery/
[22] Hancock, O. G., Jr., *Hydrogen Storage: The Hydrogen Alternative*, Photovoltaics Int., 1986.
[23] Skerrett, P. J., "Fuel UPD," *Popular Science*, 1993.
[24] Information on Planned 8 MW Fuel Cell Installation: https://seekingalpha.com/news/3978155-plug-power-to-supply-8-mw-of-hydrogen-fuel-cells-to-energy-vault
[25] Roland, B., Nitsch, J. and Wendt, H., *Hydrogen and Fuel Cells—The Clean Energy System*, Elsevier Sequoia, Oxford, 1992.
[26] Information on Pumped Hydro Systems: www.iea.org/energy-system/electricity/grid-scale-storage
[27] Worldwide Pumped Hydro Statistics: www.statista.com/statistics/1304113/pumped-storage-hydropower-capacity-worldwide/
[28] Information on Compressed Air Energy Storage: www.pge.com/en_US/about-pge/environment/what-we-are-doing/compressed-air-energy-storage/compressed-air-energy-storage.page
[29] Information on Operation Details of CAES: www.ctc-n.org/technologies/compressed-air-energy-storage-caes
[30] Johnson, Samuel C., Webber, Michael E., et al., "Selecting Favorable Energy Storage Technologies for Nuclear Power," in *Storage and Hybridization of Nuclear Energy*, Academic Press, 2019.
[31] Blaustein, Anna, *Rechargeable Molten Salt Battery Freezes Energy in Place for Long-Term Storage*, Scientific American, May 6, 2022.
[32] Lifted Weight Storage Information: Lifting and Lowering Tons of Bricks: The Best Storage Solution for Wind and Solar Intermittency? February 16, 2022 by James Conca.
[33] Johnson, Samuel C., Webber, Michael E., et al., *Storage and Hybridization of Nuclear Energy*, 2019.

SUGGESTED READING

Federal Consortium for Advanced Batteries, *Executive Summary National Blueprint for Lithium Batteries 2021–2030*, June 2021. Agency is US Department of Energy. https://www.energy.gov/mesc/federal-consortium-advanced-batteries-fcab

7 Grid-Connected PV Systems With Energy Storage (ESS)

7.1 INTRODUCTION

This chapter focuses on the design of grid-connected PV systems with energy storage. All one needs to do to discover an interest in these systems is to spend a week or two after a hurricane, an ice storm or some other natural disaster without electric power. And, believe it or not, there are places in the world where the grid is not available on a 24-hour-per-day basis. Those who have purchased fossil-fueled generators for backup power generally appreciate the idea of a quiet, reliable, pollution-free, non-fuel-consuming source of electricity. Rather than wait in long lines for gasoline for their generator, they can sit by the swimming pool drinking cold beer. With enough PV, a solar water heater, or a hybrid water heater, they will likely have hot showers as well. In fact, they will probably be the most popular family on the block.

Battery backup grid-connected PV system components still must comply with UL 1741 [1]. In the event of a utility disturbance or outage, the system must shut down its grid connection for the allotted five-minute interval, monitor the grid and reconnect again only if the grid has been stable in voltage and frequency for the five minutes. This means that the system must be a multipurpose system. It supplies power to the grid as a current source that continuously monitors the grid whenever the grid is energized. It also acts as an alternate means of charging the system batteries in the event that the PV system for any reason has not kept them charged. Finally, it acts as a stand-alone voltage source to power designated standby loads in the occupancy while the utility is not available, making this switchover in a few ms.

Recently, ac-coupled battery backup PV systems have dominated the market. For those who have installed non-battery-backup systems and then found they would have preferred a battery backup system, it is possible to add a few extra components and a standby distribution panel to achieve backup operation. There is also an argument that this type of system is inherently more efficient than the dc-coupled systems that have been popular, since they immediately invert the PV array output to useful ac power, rather than have batteries as an intermediate stop. These systems are also adaptable to multiple ac sources, such as wind generators and low-head hydro systems.

Both types of systems will be discussed in this chapter. The reader might want to note that "battery" and "Energy Storage System" (ESS) tend to be used interchangeably. Technically, there is a subtle difference. A battery is simply a device that is able to store energy, usually chemical, and then deliver it as DC power when demanded by a load. The battery may or may not be rechargeable.

An ESS is more than just a battery. Some sort of rechargeable battery will normally be a part of an ESS, but the ESS will likely also include some form of charge control or battery management as well as provisions for AC:DC conversion, DC:AC conversion or DC:DC conversion, along with its very own microcontroller that is responsible for making sure everything works exactly the way it has been set up to work. The other subtle difference between a battery backup system and an ESS is that the battery backup system only serves as a backup system if the utility is down. The ESS serves as backup power when the utility is down, but is also available for utility stabilization if the utility is reaching peak load or has excess generation and needs help. Note, however, that pumped hydro also constitutes an ESS and does not use any DC batteries. If the storage system involves AC and batteries, it will likely be described as a battery energy storage system (BESS).

7.2 BESS DESIGN BASICS

7.2.1 AC-Coupled Systems

Figure 7.1 is a simplified block diagram representation of an AC-coupled PV system with energy storage. Note that the PV array output is immediately converted to AC, either by a standard direct grid-connected inverter or, perhaps, by microinverters under the array that serve the multiple roles of inverter, MPPT, GFDI, AFCI and RSS. Because of the AC output prior to any other connections, including loads, controllers or energy storage, this system earns the title AC-coupled.

Following the path of the power flow, the standby loads are the first to benefit, with minimal losses between array and load, much the same as in a straight grid-connect system as discussed in Chapter 4. Next, note that any additional power source, as long as it is in compliance with IEEE 1547 and UL 1741, can also be fed to the standby load panel. The power flow up to the standby loads is in only one direction.

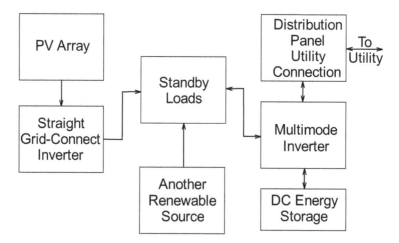

FIGURE 7.1 Block diagram of AC-coupled battery backup grid-connected PV system. Grounding conductors and disconnects are not shown.

The heart of the system is the battery backup inverter, which is becoming better known as a *multimode inverter or hybrid inverter*. This inverter monitors the grid and when the grid is available, it passes through any excess PV power that has not been used by the standby loads to the main distribution panel (MDP), where it is then available for use by loads in that panel. If the needs of all the loads in the MDP are also satisfied by the PV output, any excess is delivered to the grid, just as in a direct grid-connect system.

Power for the operation of the multimode inverter comes from the storage batteries. Another job performed by this inverter is to control the charging and discharging of the batteries. If the batteries need a charge and PV power is available, it will be used to charge the batteries. If the batteries need charging at night when the PV system is down, the inverter can charge them with grid power, provided that the grid is available. If the grid goes down, it is the responsibility of the multimode inverter to disconnect from the grid, meaning the connection between the inverter and the main distribution panel is opened up.

Immediately upon sensing the absence of the grid, the inverter switches from the current source to the voltage source with stable amplitude and frequency. In effect, it is now simulating the grid, and, as a result, the straight grid-connect inverter doesn't have a clue that the grid is down. It continues generating power because it is convinced that the multimode inverter is really the grid. Since the connection to the grid is open, however, the PV power is only available to the standby loads and to the multimode inverter for battery charging, if needed.

There are effectively five modes of operation of this system:

1. Grid is up and sun is up. This is the first normal mode of operation. As long as the batteries are charged, the system operates essentially the same as a straight grid-connected system, with the multimode inverter simply acting as a controller that incorporates a straight busbar connection to the grid. All of the PV energy is used for the backup loads, the non-backup loads or the grid.
2. Grid is up and sun is down. This also happens often. When this happens, all the loads, including the batteries if needed, are supplied by the grid. When the sun comes up, the system goes back to square one.
3. Grid is down and sun is up. Ideally, this never happens, but nothing is completely ideal. In any case, the multimode inverter disconnects from the grid, becomes a voltage source and the PV system generates electricity for use by the standby loads. If not all available power is needed by the standby loads or batteries, then the multimode inverter needs to get the message to the PV system to slow down. One way to do this is to vary its output frequency as a message to the straight grid-connect inverter to slow down its output to the level needed by the backup loads. During this condition, the multimode inverter also needs to be sure no further charge will be delivered to the storage system.
4. Grid is down and sun is down. Power to the backup loads is provided by the multimode inverter using energy from the ESS. This is all well and good as long as the ESS has enough energy to last through the night. If this is

the case, then the PV system wakes up again to the tune of the multimode inverter's pretend grid output and does its best to meet the needs of the backup loads and to recharge the batteries with any leftovers. The fun part is when the batteries do not have enough charge to make it through the night. Fortunately, the multimode inverter is also responsible for monitoring the ESS state of charge. If it drops too far, the inverter shuts down and no further power is available to the loads. As the reader might imagine, this is a very interesting situation for the multimode inverter. If the straight grid connect inverter is completely loyal to UL 1741, it needs to see a stable grid for five minutes before it can start back up again. This suggests that one way or another, the multimode inverter needs to communicate with the straight grid-connect inverter to determine whether PV power is available yet without draining the batteries while undertaking this task.
5. This option has only been around since 2018. In 2018, IEEE 1547–2018, as mentioned in a previous chapter, provided for the PV system owner to enter an agreement with the utility for the utility to communicate with the ESS to help to stabilize the grid. Prior to this time, the ESS would only spring into action when the grid went down. Now the ESS can help the grid to NOT go down.

With all the previous talk about the multimode inverter, one might conclude that it is an indispensable device for this system. In fact, in a number of commercial systems currently available, the function of the multimode inverter is carried out by a combination of ESS and a system controller. As long as a controller and some means of storage are available to perform all the functions of the multimode inverter, who would argue against such a possibility? This situation will be seen in several of the examples that follow.

One more important point should be mentioned before exploring DC-coupled systems. It's called load shedding. Load shedding can be done automatically or manually. The idea is simple. Just establish a priority list for the operation of all the backup loads. In effect, this means which load is shut down first and which is shut down last. In many cases it will be the refrigerator that stays on as long as possible, but in other cases it may be a piece of essential medical equipment. As the state of charge of the ESS drops, certain loads are disconnected automatically to save energy for more essential loads. It is now possible to acquire distribution panels that serve exactly this function. They can be useful during grid outages but also can be useful when the grid is up but needs a boost.

7.2.2 DC-Coupled Systems

Figure 7.2 shows a DC-coupled PV System with energy storage. This system actually became popular before the AC-coupled systems were introduced. They were an extension of stand-alone with good sine waves to grid-connected with good sine waves. The older systems had separate source circuit combiners, charge controllers and MPPT. The newer systems, as shown in Figure 7.2, incorporate all of these features into a single enclosure.

Grid-Connected PV Systems With Energy Storage (ESS)

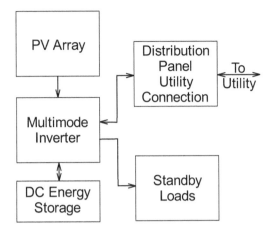

FIGURE 7.2 Block diagram of DC-coupled battery backup grid-connected PV system. Grounding conductors and disconnects are not shown.

The basic idea is that the PV array for smaller systems produces its DC power with strings that operate in the 100–500 V range at a current in the 10–14 A range. These strings feed the array input section of the Multimode inverter, where there may be several different string-level MPPT inputs for connection options. The MPPT outputs are combined into a single, lower-voltage DC output at the specified inverter working voltage. This could be almost anything between 24 and 400 V as will be seen in later examples.

Note that, once the DC is at the inverter operating voltage, the flow becomes bidirectional, depending upon whether and how the ESS may be charging or discharging. The multimode inverter incorporates the same functions as the one in the AC-coupled system. The difference in the DC-coupled system is that the PV comes in at the DC input of the multimode inverter and needs to be inverted by this inverter before it can proceed to loads or the utility. Thus far, multimode inverters have efficiencies in the 95% range, whereas straight grid-connect inverters advertise efficiencies over 98%. Since during normal operation, when the batteries are charged, the array output has the choice of charging batteries or feeding the inverter, this means this system has more losses before the PV power reaches the desired load. And most of the time, operation is normal operation.

To some extent, the overall performance of an AC or DC-coupled system depends upon when the loads want the power. If the load is predominantly daytime, then the 2% loss rate of the AC-coupled system is less than the 5% loss rate of the DC-coupled system. At night, however, since all loads are served by the multimode inverter when the grid is down for both AC and DC-coupled systems, if the grid is down a lot, then the difference in performance is not as large.

Before leaving this discussion, it is interesting to note that when the multimode inverter is connected ONLY to the standby loads, power only flows out of the inverter to the loads and one could easily be led to believe the power can only flow out of the

AC out port, when, in fact, this port is bidirectional since it connects to an internal busbar in the inverter.

7.2.3 Load Determination

Twenty years ago, when PV was still relatively expensive, it was important to determine the minimum-size PV system that could supply the backup loads. This meant a fairly careful study needed to be made of the backup loads. In 2023, backup loads are determined more by a "guesstimate" process, since most modern energy storage elements have both energy and power ratings. For example, if a storage system has an output current limited to 16 A at 240 V AC, this says a lot about what can be backed up. It's likely not going to run the electric dryer or the air conditioner, unless the air conditioner is a small one. However, if one wants to be a purist, it is not that difficult to make a list of desired backup loads and then make an educated guess at how long they will operate every day and come up with desired minimum kW and kWh ratings for the ESS. The listing might look like Table 7.1.

At first look, one might wonder immediately about the load claimed for the refrigerator. Doesn't the NEC require refrigerators to be listed at 1500 W? The simple answer is that 1500 W is used unless one has better information. In fact, a quick web search suggests that 230 W is likely a better estimate. A smaller unit could be as low as 150 W.

The other load powers in the table are either nameplate values or estimates of the power of the new LED lighting.

Again, load current is simply from I = P/V, with the exception of anything with a motor, which will have a higher current due to its lagging power factor. The load kWh/day numbers are the guesstimates of the table, since one needs to essentially guess how long the various items will operate each day. The refrigerator's annual energy star kWh divided by 365 gives a reasonable estimate of the refrigerator's daily kWh. At this point, whether or not these numbers are used, at least they yield ballpark figures just in case they might be needed.

TABLE 7.1
Summary of Standby Loads for Battery-Backup System

Load Description	Load Power, W	Load Current, A	Load kWh/day
Refrigerator	230	2.0	1.4
Microwave oven	1200	10	0.9
Washing machine	1500	12.5	0.4
Computer	200	1.7	1.0
Lighting ckt 1	200	1.7	1.2
Lighting ckt 2	200	1.7	1.2
Ceiling fans	150	1.3	1.8
Mini-split A/C	500	5.2	5.0
Totals	4180	36.1	12.9

7.2.4 Inverter Selection

The load power column of Table 7.1 suggests the minimum inverter AC output power should be at least 4180 W, unless some sort of load management is employed. In any case, 4180 W will cover all the standby loads simultaneously. This number will also be useful in the selection of the energy storage, since ESS are both current-limited and power limited in addition to having a specified kWh capacity. This selection can also be important, since the air conditioner will have a starting current requirement that will need to be met.

It should be noted that, in some cases, the inverter limits the size of the array and sometimes the storage limits the size of the array. And, in fact, sometimes the storage limits the size of the inverter. Newer systems have overcome most of these issues, but it is still important that the designer be aware of these possibilities. It's all a matter of how the system controller can control the array output if the utility is down. If nothing else, there may be a limit to the amount of PV or storage current that can be controlled by the controller.

7.2.5 Storage Selection

Table 7.1 pretty much eliminates any further guessing that may be involved with the storage. However, one issue is the daily variation in PV system energy output. This can involve a trade-off between array size and storage size. Also, the variation over the day when loads are actually used may affect the storage needed. For example, if only 50% of the total storage is needed at night, then, provided that the PV system can produce adequate output on a daily basis, the storage size might be reduced a bit. In any case, it appears that for this collection of standby loads, storage in the range of 10–15 kWh would be adequate.

7.2.6 PV Array Selection

The minimum array size should be capable of producing the necessary energy for the load on the days of the year it will most likely be needed. This could be during hurricane season, ice storm season or any other time of the year when the likelihood of losing the utility will be the greatest. Larger arrays will generally be preferable if there is room for them, since most of the time, the utility will be there to absorb any excess power produced by the array. When the utility is down, the array needs to be able to power the daytime standby loads while simultaneously recharging the batteries so they can fill in for the PV when the PV output decreases. It also needs to be remembered that the energy out of the batteries will be less than the energy into the batteries, due to the round-trip efficiency issue. Since lithium batteries will be used, about 5% loss should be allowed for the round-trip losses.

7.2.7 Fossil Fuel Generators

In some cases, emergency power may be needed for medical or other purposes. Thus, in case the sun and batteries both give out early one stormy morning, it may be desirable to incorporate a fossil fuel generator into the system as a backup for the sun and the storage. Of course, unless the fossil system consists of a large natural gas generator, the fossil system has its own storage, namely gasoline, so even a fossil generator is not necessarily the perfect backup system. It likely gives a reason for increasing the size of the array and storage to minimize the need for the generator.

7.3 A MICROINVERTER-BASED 120/240 VOLT AC-COUPLED PARTIAL HOME BATTERY BACKUP SYSTEM

7.3.1 INTRODUCTION

As has already been determined in Chapters 4 and 5, a 10-kW PV array is possible for a large majority of residential installations. Since the inverter needs to have a capacity of at least 4180 W, a 10 kW array with a DC:AC ratio of about 1.3 will result in an inverter size of 7.7 kW, which exceeds the needs of the loads. It has already been determined that the annual system kWh is dependent upon the array size as long as the inverter size is within reasonable DC:AC ratio limits. Thus, the first check might be the annual array output of the proposed 10 kW array, which, as it happens, would be implemented by 5-kW facing SE at a tilt of 22.6° and 5 kW facing SW at a tilt of 22.6°. The system will be in Albany, NY, where winter ice storms can produce power outages.

An NREL SAM analysis for the array gives the results shown in Figure 7.3. The annual output of the SE array is estimated to be 5988 kWh/yr and the output of the SW array is estimated to be 6099 kWh/yr. Thus, total annual production is 12,087 kWh, or 1208.7 kWh/kW/yr. The interesting part from a backup load perspective is the winter months, where monthly production drops to about 500 kWh for December, 630 kWh for January and about 800 kWh for February, vs nearly 1350 kWh for July. Thus, the average daily kWh for December is 16.13, the lowest daily average for the year during any month. Allowing a round-trip efficiency of 95%, this suggests the storage input kWh to deliver 12.9 kWh, will need to be 13.6 kWh. This is 2.53 extra kWh for the average day, which suggests that the not-so-average day might produce about 80% of this amount, while other better days might produce an extra 3 kWh. In any case, despite this array size, there may be days in December when the array output may not meet the needs of the backup loads on its own.

Before confirming modules and inverters, at this point, it perhaps makes some sense to explore some of the system options for increasing the system availability in December. To explore options, a closer look can be taken at the backup loads. For example, in December, it is not likely that 1.8 kWh will be used by the ceiling fans and it is possible that the washing machine could use a vacation day, removing another 0.4 kWh from the load. And the mini-split could be the big one. If it is just a straight cool unit, then forget it. But if it is a heat pump, then it could come in very handy with a few extra blankets. Thus, one option could be to curtail the load as needed.

Another option could be to install a few more modules if there is room for them. That would increase array production proportionately to make more kWh available for the loads and storage.

Yet another option would be to increase the amount of storage to have a bit left for a "rainy/icy day." Which option to use? It's either a matter of cost or convenience.

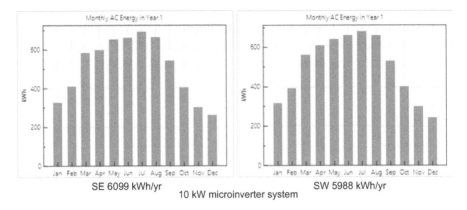

FIGURE 7.3 Albany, NY, NREL SAM analysis of output of 10 kW system split into SE and SW parts.

But the options are there for discussion. In the meantime, during the days the utility is connected, the array can perform at its optimal level, supplying house loads and selling the leftovers back to the utility. And, for that matter, if, on a hot summer afternoon, the utility is looking for extra kW, the storage should be able to provide about 5 kW to the grid at the request of the utility.

7.3.2 ESS Selection

Assuming that the sun will provide some of the daytime load requirements, it is reasonable to assume a range of storage might be between 10 and 15 kWh, hence, a trip to the data sheets. Since the array output is AC, due to the microinverters, the storage will need to be AC in and AC out. The microinverter manufacturer happens to sell storage in 5 kWh increments, with some 3 kWh units also available. Thus, 5, 10, 13 and 15 kWh are available. The 3 kWh unit is normally used along with additional units, such as to reach 13 kWh. The 5 kW is designed to deliver higher power and requires a special controller. Then there's this other manufacturer that makes a 13.5 kWh unit that works with pretty much any straight grid-connected inverter, including microinverters, as long as its own controller is a part of the system. Once again, checking out prices, availability, flipping coins and whatever else needs to be done to make the decision is in order. Before making that decision, it makes sense to look at some of the performance characteristics of the storage options.

The first step is to take another look at Table 7.1 to determine the desired performance characteristics and then compare the desired results with the storage options data, which are shown in Table 7.2. In particular, given the total load current for the standby loads, it appears that operating all loads simultaneously would require either two of the 5 kW units or two of the 10 kW units to come close. But a second look at the loads shows that all the loads are 120 V loads. This means the load currents can be split as evenly as possible between L_1 and L_2. If the microwave oven is connected to L_1 and the washing machine is connected to L_2, only 13.6 A remains to be divided

TABLE 7.2
Summary of Operating Parameters of Four Different Storage Options

Option	Max Output Current	Max Output kVA	Notes
3 kWh storage	5.3 A	1.28	8 A for 10 sec
5 kWh storage	16 A	3.84	32 A for 3 sec 25 A for 10 sec
10 kWh storage	16 A	3.84	5.76 kVA for 10 sec
13.5 kWh storage	24	5.0	120/240 V 88-106 A locked rotor

up. One such division might be to assign L_1 to the microwave oven, the computer, the ceiling fans and the mini-split, for a total of 18.2 A. The remaining loads on L_2 add up to 17.9 A. This new revelation now suggests that a combination of a 10 kWh and a 3 kWh storage unit might be a reasonable choice, since this combination will result in a total of 21.3 A on both lines, 13 kWh of storage and a reasonable amount of surge current for motor starting of the mini-split.

7.3.3 Module and Inverter Selection

The original array performance analysis was based upon an assumption of a DC:AC ratio of 1.3. If a 400 W module is selected, this would result in a microinverter size of 308 W. The choices are either a 290-W or a 325-W unit. The output current of the 290 W unit is 1.21 A and the output current of the 325 W unit is 1.35 A. Since the current per string is limited to 16 A that feeds a 20 A circuit breaker, this means that a string can accommodate 13 of the 290 W units and 11 of the 325 W units.

If 400 W modules are used, it will require 25 of them to achieve a 10-kW rating. With the 1.21 A microinverters, this can be achieved with a string of 13 and a string of 12. With the 1.35 A microinverters, it will take three strings, probably of 8, 8 and 9 modules. Noting that a NREL SAM [2] simulation with the DC:AC ratio of 1.38 gives nearly the same result as with a ratio of 1.3, it begins to make sense to use the lower power unit with two strings, since the overall system performance will not be significantly affected. Thus, in effect, the inverter and module have been selected simultaneously. More than likely, it will make sense to use the most recent version of each.

7.3.4 System Controller Selection

Communication is essential for this system. All the system electronic components need to work together as a team. This is why the selection is important. It is necessary to select a controller that will be compatible with the energy storage, the inverters, and, likely, the utility. Of course, such a controller does exist, since all the equipment will be supplied by the same manufacturer. The controller actually takes the place of the multimode inverter in this design. In fact, the controller does not incorporate an inverter because the inverters in the storage elements are designed as multimode inverters. As a result, as soon as the grid is lost, they immediately spring to life and begin delivering power. The controller incorporates the microgrid interface device, which is simply a fancy name for a contactor that senses the presence or absence of the grid and immediately disconnects if grid power disappears. This device is at the utility end of a 200 A busbar inside the controller.

Although the storage springs to life immediately after the utility goes down, the other microinverters in the system may or may not come to life immediately. It all depends upon whether the grid-forming model or the non-grid-forming models are used. The grid-forming models turn on right away and synchronize with the signal from the storage elements. The older models wait five minutes to verify the stability of the "grid" and then turn it on. Another important difference is that the AC output power of the older models is limited to 150% of the rated storage output power, while the grid-forming models work with the system controller to ensure that the loads are met, the storage gets charged and the inverters slow down if they are generating more power than the system needs at the moment.

The controller can handle up to 64 A of continuous PV current (@240VAC) and another 64 A of energy storage current, also @240VAC. The overcurrent protection for these two sources is calculated at 125% of maximum current and, thus, have a maximum value of 80 A each.

An important part of the controller is its neutral-forming autotransformer. Figure 7.4 shows how it is connected across L_1 and L_2 such that its center tap can be connected to the neutral wire coming in from the utility. This creates a 120/240 V split phase distribution source that mimics those in a single phase, 120/240 V occupancies. Note that the neutral only carries the unbalance between the loads on L_1 and L_2. The output of the microinverters is straight 240VAC with no neutral.

Another option offered by the controller is the provision for connection of an external fossil fuel generator for emergency backup. The controller can provide a starting signal and then isolate the generator from the PV and the utility so it can operate backup loads, charge the batteries or both.

For this system, the controller will be connected in a partial backup configuration, which means simply that it will only provide power to the backup load panel when the grid is down. Any other larger loads, such as the water heater, electric range, three-ton air conditioner and heat, will be disconnected, as will be evident as we proceed to finalize the system design.

7.3.5 BOS Selection and Completion of the Design

The next system component that has communications capability is the string combiner, which should be familiar from Chapter 4. The difference here is that the

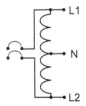

FIGURE 7.4 Neutral forming autotransformer.

string combiner data and communications in this system get their operating power, albeit only a few watts, from the system controller. Communication between the system combiner, microinverters and system controller is via a power line carrier (PLC). This involves superimposing a signal in the 100 kHz range upon the 60 Hz power connections between the microinverters, combiner and system controller. Communication between the system controller and storage elements is via a wireless channel. Selection of the combiner is straightforward—use what the controller and microinverters say to use in their installation manuals. The fun part here is keeping track of which components happen to be either upward or downward compatible with earlier versions.

Another small, but important, part of the system is the rooftop junction box or boxes. This is where the open wiring of the array is transitioned to wiring in the conduit that delivers the array power to the string combiner. The junction boxes are often equipped with terminal strips to make it easier to keep track of which wire is which when eight or ten wires may be transitioning through the junction box.

The system controller will also be equipped with an external rapid shutdown (RS) switch that will open contactors in the PV supply and the storage supply lines, thus shutting down the entire system in the event of an emergency. This switch is typically located near the service entrance of the building where it is readily available to emergency workers.

Depending on any local rules, it may be required to install separate lockable disconnects for the PV and the storage. Just as the PV strings are combined at the string combiner, a battery combiner panel is also often required for the batteries so only a single feed will run from the storage to the system controller storage input. Then there's the wiring from the system controller to the main distribution panel and from the system controller to the backup load panel. And that pretty much accounts for all the wiring and the various enclosures. Wiring details will be calculated after the array mount design is completed.

7.3.6 Array Mounting Equipment

Figure 7.5 shows the roof of the occupancy populated with 25 modules. Module dimensions are 6.258 ft × 3.458 ft (1.907 m × 1.054 m). Module thickness is 1.6 in (40.64mm). There is quite a bit of remaining roof space in case the owner decides to add more modules another time. ASCE 7–16 shows the design wind speed for Albany to be 110 mph for a category II building. The building exposure is B, and both roof sections are hip roofs. The property is relatively level, and the structure itself has a roof slope of 22.6° and a mean roof height of 13 ft. The "a" factor is either 40% of the mean roof height or 10% of the least horizontal dimension, except this applies to rectangular shapes. A reasonable compromise would be to use 38 ft as the least horizontal dimension. Thus, $0.4 \times 13 = 5.2$ ft and $0.1 \times 38 = 3.8$ ft. This distance is shown at a few locations on the roofs with modules, simply to give an idea of the location of this boundary on the roof planes. Also note that to avoid any "edge" modules, all the modules have been located at least 12″ from the edge of the roof.

Since $q_z = 0.00256 K_z K_{zt} K_d K_e V^2$ lb/ft², with V measured in mph, all that's left is to determine the values of the coefficients and multiply. The velocity pressure exposure

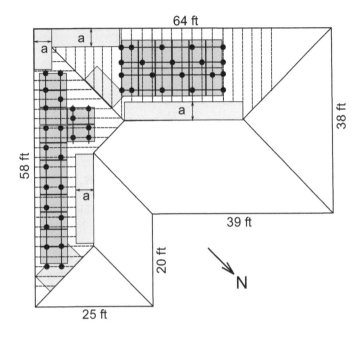

FIGURE 7.5 House with 25 400-W modules.

coefficient, K_z, is found to be 0.7 from ASCE7–16 Table 26.10–1 [4]. Since the surrounding property is essentially flat, the topographic factor is 1.0. ASCE7–16 table 26.6–1 shows the wind directionality factor, K_d, is 0.85. Since Albany is not on a mountaintop, it is acceptable to use the ground elevation factor, $K_e = 1.0$. Thus, $q_h = 18.43$ lb/ft² since q_z was evaluated at the mean roof height.

The next step is to determine the design wind pressures by first determining a few additional coefficients to calculate $p = q_h(GCp)$. The results for GC_p for hip roofs with a slope between 20° and 27° that are listed in ASCE-7–16 Figure 30.3–2G are tabulated in Table 7.3. Since the results are in lb/ft², the next step is to determine how many square feet of the array are in each wind zone, which means a return to the roof drawing.

The area of a single module is 21.64 ft² (2.01 m²). Thus, the area of 25 modules is 541 ft² (50.26 m²). The roof drawing shows the equivalent of 16.75 modules in zone 1 and 8.25 modules in either zone 2e or 2r. The overall uplift force on the array is thus 362.5ft²×22.1 lb/ft² + 178.5ft²×32.3 lb/ft² = 13,775 lb (6248 kg). The zone 1 contribution is 8016 lb (3636 kg) and the zone 2 contribution is 5758 lb (2612 kg).

A very popular array mounting foot has a pull strength of 1004 lb (455 kg) when anchored into a truss [5]. Thus, if even half this strength is used, then the zone 1 portions of the modules would need a total of 8016 ÷ 500 = 16 attachments. The zone 2 portions of the modules would need 5758 ÷ 500 = 11.5 attachments. Thus, a minimum of 28 attachments would be needed. To determine the exact number of attachments, the designer should check the maximum allowed span lengths for the

TABLE 7.3
Values of GCp for Hip Roof of Figure 7.7 [4]

Roof zone	1	2e	2r	3
GCp (up)	1.2	1.75	1.75	1.75
GCp (down)	0.6	0.6	0.6	0.6
P (uplift, lb/ft^2)	22.1	32.3	32.3	32.3
P (down, lb/ft^2)	11.1	11.1	11.1	11.1

rails used, which will likely exceed 72 in (182.9 cm) for zone 1 and 48 in (121.9 cm) for zone 2. The last item is that technically, it could be argued that the 2 modules in the top row on the SE roof meet the definition of Exposed, which adds another factor of 1.5 to the uplift. Thus, a few extra attachments for those two modules would be welcomed.

The next step is to try it out on the proposed layout. As is evident, using the 48″ (121.3 cm) constraint for zone 2 attachments results in a total of more than 28 attachments. In fact, the count as shown is 45 attachments. A rule of thumb check on this number is that the ratio of attachments to modules is 2.3. Usually, this number for a roof with portrait-mounted modules attached to the trusses will be quite secure with approximately twice as many attachments as modules.

7.3.7 Wire and Circuit Breaker Sizing

As in previous microinverter system examples, the wiring of the microinverters uses special #12 listed open air cables available from the manufacturer. The microinverters use snap-on connectors to connect to the modules and also to the AC cables. These connectors are listed as disconnects for the input and the output of the microinverter.

Because of the power line carrier communication, special terminators need to be installed at the loose cable ends to prevent reflection of the PLC signal. If the cables can be fed from the center of the cable, then the cumulative voltage drop problem can be minimized.

Since the open wiring cables are listed only for use under the modules with suitable supports, a transition to conduit needs to be made by means of a rooftop junction box. NEC 314.16(B) requires a volume of 2.25 in^3 for each #12 wire that enters the box and 2.5 in^3 for each #10 wire that enters the box and it is not unusual to have 10 of each entering the box for splicing. Also, since the terminal strip used for splicing might have a volume of 5 in^3, this means the volume of the junction box will likely be at least 52.5 in^3. Since PV rooftop junction boxes are selling like hotcakes these days, manufacturers have developed special designs for attachment either to the racking or directly to the roof. Figure 7.6 shows a few examples. Note that a 6″x6″x2″ box has a volume of 72 in^3, so this is a reasonable estimate of the minimum size of the junction box.

The wiring from junction box to combiner is usually #10, whether or not #12 might work. However, it must be remembered that whenever there are more than three current-carrying conductors in a raceway (conduit) the conductor ampacity must be derated. Neutrals and grounds do not normally count as current-carrying,

Grid-Connected PV Systems With Energy Storage (ESS)

FIGURE 7.6 Typical rooftop junction boxes.

except for a few special cases with three-phase circuits where the neutral waveform is distorted, resulting in harmonics. This does not apply in any PV circuits since all the AC waveforms are required to be of utility grade and are usually better. For up to 3 current-carrying conductors in a conduit, there is no derating. Between 4 and 6 the derating factor is 0.8; between 7 and 9, the factor is 0.7 and between 10 and 20, the factor is 0.5 [3].

The other derating factor that needs to be considered is the ambient temperature factor. Temperature derating factors depend upon the rating of the wire insulation. If insulation has a hotter rating, then it does not require as much derating. One important consideration is that as long as a conduit is more than 7/8″ above the roof surface, no additional amount due to proximity to the roof needs to be added to the local highest ambient temperature. Otherwise, *NEC* 310.15(B)(2) requires adding an extra 33°C to the ambient. Ambient derating factors can be found in NEC 310.15(B)(1) [3].

Thus, for this system with its 4 current-carrying wires from the two strings, a conduit fill derating of 0.8 will be needed. For Albany, NY, the ASHRAE (American Society of Heating, Refrigeration and Air Conditioning Engineers) high temperature is 28.9°C (84°F), using 75°C terminals, the ambient temperature derating factor is 1.00. The derating process can go two different directions. The first is to start with the rated current and then divide by each successive derating factor. The second is to start with a 30°C wire ampacity and then multiply by the derating factors.

Thus, since the highest rated string current is 13×1.21 A = 15.73 A, the minimum ampacity of the wire to be used would be 15.73 ÷ 0.8 ÷ 1.0 = 19.66 A. Since the rating of #12 Cu at 30°C is 25 A, this means #12 would be adequate. Using the multiplying approach, assuming #10 Cu with 75°C insulation, which has a 30°C ampacity of 35 A, multiplying by 0.8 and 1.0 results in 28 A. Since 28 > 15.73, #10 is also acceptable. Whether #12 or #10 is used in this case, the overcurrent protection needed will still be 20 A. Each string will backfeed a 20 A circuit breaker in the combiner box.

One subtle item should be mentioned at this point—the "either-or" rule (NEC690.8), which states that the wiring used must be either 125% of the rated current or the rated current derated for conduit fill and temperature, whichever is larger.

In this case, since the ambient temperature derate is 1.0, both results come out the same. But for Dallas, TX, where the high temperature is listed as 96.2°F, the ambient derating factor would be 0.88. In this case, applying both derating factors will exceed 125% of the rated current.

Since the conduit from junction box to combiner will have 5 #10 wires, including the equipment grounding conductor, it will need to be either ½″ or ¾″, depending upon the conduit used. To be on the safe side, it makes sense to specify ¾″. Most installers prefer the larger size, since it makes wire pulling a bit easier. The next wire run is between the combiner and the system controller. This wiring carries the combined output currents of the two strings, or 25×1.21 = 30.25 A. The wire size and overcurrent protection size must be rated at 125% of this value, due to continuous duty. Thus, 37.8 A is the basis for selection of wire and overcurrent protection. A look at the tables shows that #8 Cu and a 40 A circuit breaker will be appropriate for this connection. In addition to the wiring for the PV output power, the conduit between combiner and controller also needs to include 3 #12 Cu (L_1, L_2, N) to provide power to the communications module in the combiner. Thus, the conduit between combiner and controller will carry 2 #8 Cu (L_1, L_2), a #10 Cu G and 3 #12 Cu. This will require a minimum of ¾″ conduit. Note that the control/communication wiring is not considered current-carrying for conduit fill derating purposes.

The next wiring is between the controller and the main distribution panel. It is possible the controller may impose a limit on the minimum wire size for this connection. In any case, since 125 A is a reasonable choice for the backup load panel, it would normally be fed with #1 wire, which might be the smallest that can be connected to the controller lugs. As far as the controller is concerned, it has 200 A busbars, so anything up to 200 A could be acceptable. Thus, it is reasonable to run 3 #1 Cu (L_1, L_2, N) and a #6 Cu ground wire between the controller and the main distribution panel. In fact, since this connection will feed through to the backup load panel, the same wiring will also apply to that connection. This will require a 1 ½″ conduit. The circuit breaker used at the main distribution panel end of the connection to the controller is limited to no more than 125 A, simply because of the physical size of the circuit breakers that can be attached to the busbars of the main distribution panel.

ESS wiring is straightforward. Since the total available current is 21.3 A, each storage element can be connected via two #12 or two #10 Cu to the storage combiner panel. The 10-kWh unit will backfeed a 20 A circuit breaker and the 3 kWh unit will backfeed either a 15 or 20 A circuit breaker. Conduit size for 2 #10 Cu and a #10 G can be ½″. The combined output is then fed to a backfed 30-A circuit breaker in the controller with 3 #10 Cu (L_1, L_2, G) in ½″ conduit. Note that there are no neutral connections to these storage units, since the neutral is derived in the system controller.

7.3.8 Wiring of Standby Loads

The standby loads indicated in Table 7.1 deserve some discussion, especially if the installation is to be a retrofit in an existing occupancy. There are essentially two types of 120 V circuits in an occupancy. The simplest is when a hot wire and a neutral wire, generally run as a piece of "romex" (Type NM) cable, run from an existing electric panel to a load or, perhaps to as many as ten 120 V outlets connected in parallel. Normally these circuits are connected to single pole 15 A or 20 A circuit breakers, or, sometimes to fuses. In these cases, the circuits can be transferred to a separate distribution panel. That is, except for one detail.

The *NEC*, requires nearly every circuit in a dwelling unit to be protected by an arc fault circuit interrupter, a ground fault circuit interrupter or both. The arc fault circuit interrupter is capable of detecting arcing at current levels below the circuit breaker trip rating and interrupting the circuit. The ground fault circuit interrupter is capable of detecting a difference of as little as 5 mA between the line and neutral conductors of a circuit and interrupting the circuit if such a fault should occur. It is highly likely that a local inspector may interpret the installation of the PV system and subsequent transfer of circuits from the main panel to a standby panel to be an upgrade. In this case, even though the old circuit may have been protected by a standard circuit breaker, the moved circuit may be required to be fed by an arc fault circuit breaker or by a ground fault circuit breaker or both. Before specifying the wiring or installing the wiring, it is advisable to discuss the situation with the local electrical inspector to determine whether the circuits will need to be upgraded.

When a circuit is transferred to another distribution panel, the line and the neutral conductors associated with the circuit are simply disconnected from the old panel, run to the new panel, and reconnected at the new panel. The circuits that must be carefully checked, however, are the multiwire branch circuits. A multiwire branch circuit uses a common neutral as the return for wiring from opposite sides of a 120/240 V single-phase supply or for two or three lines that are connected to two or three phases of a three-phase system. The 120/240 V configuration is illustrated in Figure 7.7.

In the single-phase case, the neutral carries the *difference* between the two line currents (I_1–I_2), so the current on the neutral never exceeds the current on either line. For the three-phase case, the neutral carries the *sum* of the currents, but since the voltages in a three-phase system are 120° out of phase, if the currents on the three lines are balanced, the neutral current is zero. If only two lines are used, and if the line currents are equal, then the neutral current will be equal in magnitude to the line current. If the line currents are unbalanced, the neutral current will be less than the larger of the two line currents (see Problem 7.8). If a circuit is just a two-wire circuit, as shown for I_4 and I_5, then the total load current flows out and back again in the two wires. Again, under worst-case unbalanced load situations, the current in the neutral

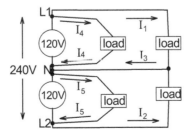

FIGURE 7.7 Example of a multiwire branch circuit.

does not exceed the current in any of the lines unless a significant third harmonic component is present (see Problem 7.9).

The problem with multiwire branch circuits is that they have been commonly used to feed bedroom circuits. A single three-conductor cable is run to a convenient location in or near two bedrooms, and then one line is fed to one bedroom and the other line is fed to the other bedroom. It is probably not too difficult to imagine one of the circuits going to the daughter's bedroom and the other going to the son's bedroom. Thus, when the time comes to determine who gets power if the utility goes out, the answer is neither or both. This is because multiwire branch circuits must be fed by a two-pole circuit breaker that disconnects both circuits when one needs to be shut down or when one trips due to an overload. This means that both lines plus the neutral associated with a multiwire branch circuit need to be transferred to the new distribution panel along with whatever means are used to provide any necessary ground-fault or arc-fault protection.

7.3.9 Equipment Grounding Conductor and Grounding Electrode Conductor Sizing

Equipment grounding conductors are sized in accordance with *NEC* Article 250.122 [3]. If the storage elements are in a metal enclosure, then the grounding conductor for the battery enclosure will need to be sized according to the 30 A ESS disconnect, which means using a #10 (13.3 mm^2) equipment grounding conductor. In fact, since all the circuits associated with the PV or energy storage part of the system involve circuit breakers less than 60 A, #10 Cu can be used for equipment grounding up to the system controller.

For the 125 A circuits, #6 equipment grounding conductors will be required. Finally, if the PV power conditioning equipment is located more than 50 ft (15.2 m) from the point of the utility connection, it may be advisable to install a separate grounding electrode for the PV equipment. For example, if the PV equipment is in an outbuilding or installed as a ground mount, then a separate grounding electrode is required. The grounding electrode conductor is then sized according to *NEC* table 250.66, noting exceptions for when the grounding electrode is a rod or a plate. This completes the design of the system. The schematic diagram, for those who have not looked at it yet, is shown in Figure 7.8.

7.4 A WHOLE HOUSE AC-COUPLED BACKUP SYSTEM USING A STRING INVERTER

7.4.1 Introduction

Every system has its advantages and disadvantages. In the case of microinverter systems, since the operating voltage on the roof is 240 V, the voltage drop can become an issue with long wire runs. Some of the string inverter systems presently available operate at DC voltages in the range of 200–800 V, depending on the rating of the building service voltage. In the residential case, the system will likely operate at about 400 V on the DC side.

Grid-Connected PV Systems With Energy Storage (ESS) 231

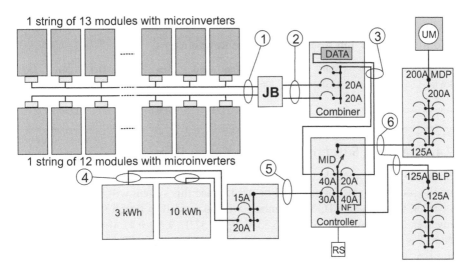

FIGURE 7.8 Single-line schematic of the 10-kW PV system with 13-kWh storage system.

Historically, multimode inverters generally fed standby loads, and generally no more than 60 A needed to be connected on the utility side of the inverter to feed the standby loads when the grid was active. Today, when modules are relatively inexpensive, a desirable feature of ESS systems is for them to be capable of whole-house backup. This means the MID switch must be capable of handling the entire building load, which, in the residential case, is often 200 A or less. This, for example, is why the busbar of the controller in the previous example is rated at 200 A. It can be used for partial backup or full backup.

In reality, full backup generally means "almost full backup." For full backup, one might expect the array and storage must be able to meet the needs of every load in the occupancy at any time. Once again, the best guesstimate method is often used in estimating whether the system can accomplish full backup. The array is designed to meet the expected daily load, and the ESS is designed to meet the nighttime loads of the occupancy and, perhaps, some of the daytime load in case of cloudy weather.

A reasonable way to begin the process is to analyze the electric bills for the past 12 months. The number to look for is the monthly kWh consumption. Depending on the location, maximum consumption may occur in the summer or the winter. To be able to begin the design, the basic residential structure used in the last example will be used. For this example, it will be located in Boise, Idaho, with the back of the house facing south. The elevation of Boise is 858 ft, the latitude is 43.56°N, the winter heating design temperature is −12.7°C and the summer cooling design temperature is 33.9°C. Table 7.4 summarizes 12 months of electric bill information. Note the low of 938 kWh in April, high of 1313 kWh in August, annual total of 13,571 kWh and average monthly kWh of 1131 kWh.

TABLE 7.4
Monthly Electric Bill Analysis for Boise Residence

Month	Jan	Feb	Mar	Apr	May	Jun	Jul	Aug	Sep	Oct	Nov	Dec
kWh	1225	1094	1031	938	1063	1219	1281	1313	1156	1000	1063	1188

Annual total = 13,569   Monthly average = 1131   High = 1281   Low = 938

7.4.2 Module Selection and Source Circuit Design

Since the 400 W module of the last example seemed to result in an acceptable fit on the roof, the same module will be used. Since this system will be using a string inverter, the module STC test numbers will be needed to match the strings to the inverter. For this module, V_{OC} = 45.33 V, V_{mp} = 37.07 V, I_{SC} = 11.31 A, I_{mp} = 10.79 A, NOCT = 44°C, TC(V_{OC}) = −0.259%/°C and TC(V_{mp}) = −0.40%/°C. Its dimensions are L = 1907 mm (6.256 ft), W = 1054 mm (3.458 ft) and t = 40 mm (1.57″). The area is 21.7 ft² (2.016 m²). V_{OC} at −12.7°C is 49.76 V and V_{mp} at 33.9°C ambient at 1000 W/m² is 31.3 V. These numbers will be needed for determining maximum and minimum string lengths. The roof slope is the usual 22.6°.

Before matching the strings to the inverter, it would be a good idea to simulate the array performance to determine how many modules will be needed to produce 13,571 kWh. Roof planes available for modules are the east, south and west roofs. Perhaps the most important observation from the simulations is that the east-facing and west-facing roof planes only produce about 64% of the annual energy produced by the south-facing array. In addition, the ratio between maximum (summer) production and minimum (winter) production exceeds 5 for the east and west roofs but only 3.13 for the south-facing roof. Clearly, the best solution is to install as many south-facing modules as possible to have the best ratio for seasonal performance, especially when the summer and winter peak kWh consumption is so close.

Figure 7.9 shows a first attempt at locating modules to meet the annual production. The 26 modules on this roof can produce a total of 15,933 kWh/yr, which is 17% higher than the original estimate. For this array, the lowest monthly production will be 10.4 kW×60 kWh/mo/kW = 624 kWh and the highest will be 1955 kWh. Table 7.5 compares estimated monthly PV production with observed monthly consumption. Since there are four months during the year that PV production is less than monthly consumption, this suggests that, unless more PV is added, some curtailment of use may be necessary during a power outage during any of these months. This, of course, leads to exploring alternatives.

One undesirable alternative would be to tilt the array more to produce more output during the winter. The problem is that this would be an aesthetic nightmare but also would involve spacing the rows, which would eliminate some modules on the south roof. Another alternative would be additional ESS to provide stored energy for the loads during bad sun days. Rather than making any of these decisions at this point, it makes sense to explore ESS options. But before that it makes sense to explore inverter selection.

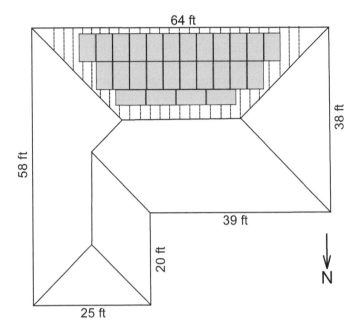

FIGURE 7.9 Proposed array layout for Boise example.

TABLE 7.5
Comparison of Monthly kWh Consumption Versus Monthly kWh Production for Boise Residence

Month	Jan	Feb	Mar	Apr	May	Jun	Jul	Aug	Sep	Oct	Nov	Dec
used	1225	1094	1031	938	1063	1219	1281	1313	1156	1000	1063	1188
gen	610	890	1310	1570	1720	1810	1930	1830	1530	1210	780	610
diff	−615	−204	279	632	657	591	649	517	374	210	−283	−578

7.4.3 Inverter Selection and String Sizing

Even though all modules face the same direction, some sort of shading is still often the case for residential properties. In this case, minimal shading will be present. Since string inverters generally incorporate several MPPT inputs to allow for parts of the array on different roof planes, if the shading patterns are known, the strings can be wired to take advantage of the different MPPT inputs such that in the worst case, the smallest number of modules will be affected.

Of the numerous string inverters on the market, only a limited number are designed to operate as a part of a system that includes a system controller that can implement the functions required for whole-house backup. One such inverter is a 7600 W unit that has four independent MPPT DC inputs with a maximum limit on array DC voltage of 600 V and an MPPT operating range of 60–480 V. It's rated output current at 240VAC is 32 A and it allows a DC:AC ratio of 1.4. Fortunately, 1.4×7600 =

10,640 W > 10,400 W. It is compatible with a controller and all the other functional components that are needed to achieve the desired full backup design. The inverter uses special RS devices that are installed after every third module in a string. The shutdown devices are activated upon loss of a grid connection by an external RS switch that can be installed at a location convenient for emergency responders.

At this point, enough information is available to make a decision on string lengths. Since the maximum DC input voltage is 600 V, the maximum string length is found from $600 \div 49.76 = 12$ (rounded down) and the minimum string length is based on the minimum MPPT inverter input voltage (60 V). Thus $60 \div 31.3 = 2$ (rounding up). Since there are 26 modules, all facing the same direction, and since the maximum string length has been determined to be 12, one string configuration could be 9, 9 and 8 modules per string. Each string will require 3 RS devices. Figure 7.10 shows a single-line representation of the string connections up to the rooftop junction box, which will need to have room for transition wiring for at least three strings. Note that the RS devices need to be installed such that one of the devices will be either at the beginning or at the end of each string.

7.4.4 Determination of Standby Loads and Storage Selection

Since this system is being designed as a whole house backup, there is no need to list the backup loads, since, presumably, everything will be backed up. Table 7.5 shows the maximum monthly kWh consumption to be 1313 kWh. On a daily basis, this will be 43.76 kWh/day. On that cold, January day, the average backup needed will be 40.8 kWh.

The ESS elements for this system come in increments of 13.5 kWh. Thus, three of these units will provide 40.5 kWh. The system can incorporate a maximum of four ESS devices. Thus, starting with 3 is a reasonable design choice, given that another unit could be added to give a bit more system availability.

The ESS devices are split-phase, 120/240 V units, and, therefore, they are the source of the neutral for the system when the utility is down. Maximum continuous output power is 5 kW (20.83 A) per unit, with a motor start capability of 106 A per unit. If the units are operated at less than the unity power factor, their maximum kVA output increases to 5.8 kVA (24.17 A). Recommended OCP is 30 A. Thus, the three combined units can provide 15 kW or 17.4 kVA continuously at a maximum current of 72.5 A. Given the flexibility in load management of the system loads, this can be a very robust system.

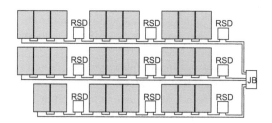

FIGURE 7.10 String connections up to junction box for Boise system.

Grid-Connected PV Systems With Energy Storage (ESS)

7.4.5 Selection of Remaining Major System Components

At this point, the only remaining major equipment includes the system controller, which has connections to the inverter, batteries, house loads, main disconnect and utility meter. Presumably the house meter and main disconnect will not be changed, but to incorporate the system controller into the system, the feeder conductors between the main disconnect and main service panel will need to be rerouted through the system controller. The original system incorporated a 150 A service, and since no loads are being added, the existing service will not be changed.

The system controller performs nearly the same functions as the system controller in the previous example, except that it does not need to create a neutral for the system. This is done at the ESS units. The inverters in the ESS units are grid-forming and thus switch over to synchronizing voltage sources when the grid is down. If the grid is up and the ESS is simultaneously delivering power, the ESS inverters act as multimode units and switch their operating mode over to the current source synchronized by the grid. The system controller enclosure is relatively large for it to be able to have room for inputs from 4 ESS units, an optional busbar for backup loads and the MID switch that disconnects the system from the grid.

7.4.6 Wire, Circuit Breaker and Conduit Sizing

Wiring between the rooftop junction box and inverter will involve the usual 3 circuits of #10 for the DC coming down from the roof. The important issue here is that because it is DC wiring, the conduit needs to be metallic and it needs to have labels every 10 ft that designate it as carrying DC circuits. The three circuits will include three positive, three negative and a ground wire in a ¾" conduit. The array itself needs to be ungrounded since the inverter is transformerless.

Wiring from inverter to system controller will need to handle the 32 A maximum inverter output and thus needs to be sized at 125%, or 40 A. Thus, the inverter will be connected to a 40-A backfed circuit breaker in the system controller via 2 #8 Cu (L_1, L_2) and a #10G in ¾" conduit. Since the neutral is derived at the ESS units, no neutral is required at the inverter.

Since the line side of the system controller will be connected directly to the 150 A main disconnect switch, it will need to be able to handle the entire system load. This will require 3 #1 Cu (L_1, L_2, N) and a #6 Cu system ground wire in a 1¼" conduit. It is significant to note that the N and G are bonded together in the main disconnect enclosure. Since this is the single point ground for the entire electrical system, the N and G wiring must be kept from any further connections throughout the rest of the system. This is why the N and G are separated in the feed from the main disconnect to the system controller. The feed from the utility to the meter enclosure will consist of three #1 Cu (L_1, L_2, N) with no equipment grounding conductor, since the G originates at the main disconnect.

The load side of the system controller will be feeding a 150 A main distribution panel that houses all of the occupancy loads. Wiring from the system controller will be the same as from the main disconnect to the system controller, with a 150 A circuit breaker located in the main service panel to protect the panel and the feeders to the panel.

The only remaining connections are the BESS units. Each will be connected to a 30-A backfed circuit breaker in the system controller via 3 #10 Cu (L_1, L_2, N) and a #10G in ½″ or ¾″ conduit. Note that all the AC wiring is allowed to be in PVC conduit if the installer so chooses. This is why it is important to include the equipment grounding conductor in all the conduits to metallic enclosures.

This completes the wiring of the system. The only remaining activity is the installation and the commissioning. This is why the manufacturers issue 100-page installation manuals to be sure everything is connected in the right place and everything is programmed for optimal system performance. Figure 7.11 shows a single-line schematic of the entire system. Note that when the utility is present, the MID switch is closed and, in effect, the utility is connected directly to the MDP. During this time, when the sun is shining, some or all of the power from the utility is replaced by solar power. Any solar power not used by the system loads is either used for BESS charging or sold back to the utility. The BESS units communicate with the utility and if the utility needs a boost, up to 15 kW can be available. Any energy delivered to the utility via the BESS will be replaced by the PV, the utility, or both. If the utility is down, up to 15 kW is available for occupancy loads on an instantaneous basis from the ESS, plus any additional power that may be available from the PV system. The numbered ovals in the drawing reference the conduits and their contents as described earlier.

And one should not forget that important final note that the BESS units are to be protected from vehicular traffic and are NOT to be installed on either the inside or the outside of a bedroom wall.

7.5 A 10-KW DC-COUPLED PV/ESS PARTIAL BACKUP SYSTEM

7.5.1 Introduction

Now that several AC-coupled system designs have been completed, before moving on to larger systems, it is interesting to look at a DC-coupled system where the

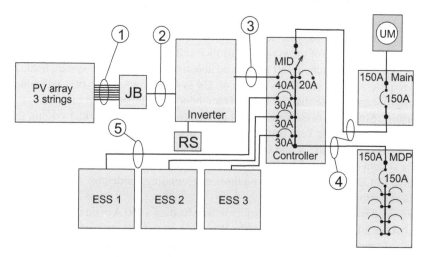

FIGURE 7.11 Final schematic drawing of the Boise system.

Grid-Connected PV Systems With Energy Storage (ESS)

batteries operate at the Inverter PV input voltage. Figure 7.12 shows a block diagram of the system. The approach with this example will be to explore the individual system components and how they work together as well as to fill in the blanks whenever more information may be needed.

The system is built around a 7.6-kW inverter that controls optimizers at the module level to ensure MPPT and RS at the module level. It has been assumed that an array with two strings of (13) 385 W modules will be acceptable to the inverter array input specifications, but this will need to be verified, along with the selection of suitable optimizers.

7.5.2 Array and Optimizers

The modules have $V_{OC} = 45.03$ V, $V_{mp} = 36.93$ V, $I_{SC} = 10.97$ A and $I_{mp} = 10.42$ A. One of several possible optimizer choices allows a maximum input voltage of 60 V, maximum rated module output power of 430 W, maximum module $I_{SC} = 12.5$ A and maximum optimizer output current = 15 A.

The minimum string length is determined by the optimizer maximum output voltage and the inverter maximum input voltage. Inverter specifications can be found on either a two-page data sheet or in the 117-page installation manual. The inverter maximum PV input voltage is 500 V and the maximum PV input power is 10,250 W with a maximum input current of 23 A. Nominal DC operating voltage is 400 V. When connected to the grid, the maximum output power to the grid is 7600 W, but, in backup mode, because of battery output power limitations, the inverter output power is limited to 5 kW. The maximum output current to the grid is 32 A.

The minimum string length can now be determined by dividing the inverter nominal operating voltage by the optimizer maximum output voltage. The result after rounding up is 7. The maximum string length per input is determined from the total

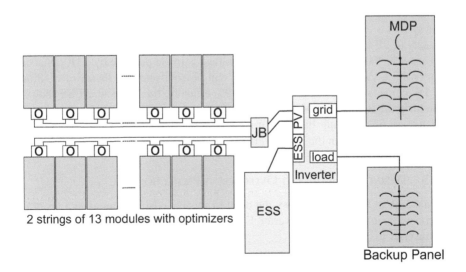

FIGURE 7.12 DC-coupled partial backup system with optimizers.

rated array PV power allowed per string. At 400 V nominal operating voltage and 15 A maximum string current, this adds up to 6000 W. However, one further refinement is needed, since the maximum PV size allowed by the inverter is 10,250 W. For 2 strings of equal length, this divides into 5,125 W per string.

Thus, the maximum string length is determined by the module power rating. Note that if the strings are of different lengths, as long as the total array power is less than 10,250 W and as long as no string is shorter than seven modules, this is also acceptable. Thus, for the system of this example, 5,125 W ÷ 385 W/module = 13.31, which must be rounded down to 13. Thus, the array configuration is acceptable and the optimizers have been selected. As a final note, it is interesting to note that the array could have been designed with higher wattage modules with matched optimizers as long as the string length and total array power are checked carefully.

7.5.3 Inverter

Not all of the inverter features are shown in Figure 7.12. Actually, the inverter incorporates a separate section that connects directly to the inverter that provides a DC disconnect switch that simultaneously disconnects the BESS and PV, has fuses on the battery inputs, a circuit breaker on the AC output to the grid, an autotransformer for deriving a neutral under backup operation, a manual inverter bypass switch, revenue grade data and lots of communications capability that provides for communications between owner and system, manufacturer and system, utility and system and repair person and system. The inverter can be programmed for backup only or also for grid support. When the grid goes down, the inverter handles the RS signal to the optimizers and switches the system over to backup operation by isolating it from the grid.

7.5.4 BESS

This system is capable of combining the outputs of two BESS units. Each unit has a capacity of 9.7 kWh, an output power of 5000 W and an operating voltage between 350 and 450 VDC. This variable output voltage is made possible by a DC:DC converter with a variable voltage output that can be controlled by the inverter. Since DC:DC conversion is generally close to 99% efficient, this particular ESS shows a round-trip efficiency of > 94.5% when battery losses are included. Another important feature of operating the ESS at the PV array voltage allows for significantly smaller wiring between the ESS and the inverter.

7.5.5 Wiring, Conduit and Overcurrent Protection

A final look at Figure 7.12 shows the need to specify wiring from array to rooftop junction box, junction box to inverter, ESS to inverter, inverter to main service panel and inverter to backup panel.

As in all previous design examples, the array and optimizer wiring will be with #12 cables supplied by the manufacturer. The transition to #10 wiring will be at the rooftop JB. The conduit from JB to the inverter will be delivering 400VDC to the inverter when the system is operational and thus must be metallic and labeled. The conduit can be ½" since it will be metallic and will carry 5 #10 Cu conductors.

Wiring between ESS and inverter needs to carry 5000 W ÷ 400 V = 12.5 A. This is consistent with the built-in 20 A fuses at the inverter and either #12 or #10 Cu can be used, depending on contractor preference.

Since the maximum current between the inverter and the grid is 32 A, this means the wiring and overcurrent protection need to be sized at 125%, or 40 A. This means a total of 3 #8 Cu and a #10 G between the inverter and the MDP as well as between the inverter and the backup panel. These conduits carry AC and thus may be PVC if the contractor so chooses. Whichever choice is made, it will need to be at least ¾".

7.6 A 45-KW THREE-PHASE PV WITH BESS USING INVERTERS IN TANDEM

7.6.1 Inverter Selection

Some smaller backup systems are limited to 80 A of PV, 40 kWh of storage or 54 kWh of storage, depending on the system. Sometimes, the maximum amount of storage is determined by the local jurisdiction, depending on the occupancy.

When a larger amount of storage is desired, often the problem is solved by combining smaller systems. For example, one larger system can be obtained by connecting 15-kW inverters in tandem to obtain a 45-kW three-phase system. The system is DC coupled and very similar to the basic system described in Section 7.2.2. Each inverter has an individual DC input for a separate array, a DC input for batteries, an AC input for grid connection, an AC input for connection to backup loads and an AC input for connection of an external fossil generator. Optional bypass switches are available to bypass inverters for maintenance purposes without disconnecting the standby loads from the utility. The inverter operates at 48 V DC on either lead-acid or lithium batteries. Figure 7.13 shows a single inverter with inputs, outputs and inverter bypass switch.

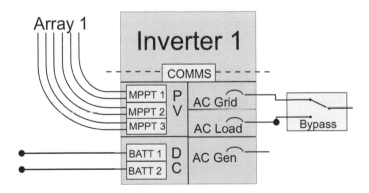

FIGURE 7.13 Basic building block of large ESS.

To minimize connections, string combiner and charge controller functions are built into the inverter. The inverters have three MPPT inputs, each of which can accommodate two strings. RS is achieved by RS devices at the module level to shut down the PV array within the array boundaries. The inverter has built-in GFDI and AFCI features. The communications functions involve communication among inverters, inverters and batteries, and inverters with the rest of the world. In addition to the communication functions via Cat 5 cables, the inverters also track power flows if current transformers are installed in critical locations.

When adding inverters, it is important to recognize how the various inputs and outputs are interconnected. For example, each inverter has its own PV array, limited to a maximum of 6.5 kW per MPPT. Each MPPT must use strings of equal length. For example, if 6 strings are used, 2 per MPPT input, then each string would be limited to 3.25 kW. Strings also have voltage and current limits. String voltages must not exceed 500 V open circuit (V_{OC}) on the coldest day of the year and MPPT inputs cannot exceed 26 A total. If the array wants to produce more than 13 A per string, the inverter limits the total MPPT input current to 26 A. But if the string V_{OC} exceeds 500 V, it can damage the inverter. During operation, the MPPT range for operation is between 125 and 425 V DC.

7.6.2 Module Selection

Optimal module sizing for the inverter can be an involved process, since another important consideration is that the modules need to fit on the roof. Since each roof is presumably different from other roofs, the array layout will presumably use modules with convenient dimensions. To start the process, however, it is convenient to select a possible module that will meet the string power, voltage, and current limits with a reasonable number in the string. For example, if a 405 W module is used, then the array length is limited to 3250 ÷ 405 = 8.02, or 8 modules. But these modules must also meet the voltage and current constraints of the inverter input. The maximum V_{OC} must be < 500 V and the maximum string current is limited to 13 A. This gives enough information to begin a module search.

One 405 W module with a 20.6% efficiency is a 132 half-cell unit that has STC specifications of V_{OC} = 45.09 V, I_{SC} = 11.19 A, V_{mp} = 37.85 V and I_{mp} = 10.70 A. For an eight-module string, this results in V_{OC} = 360.7 V and V_{mp} = 302.8 V. If six strings are used, the total array power will be 19,440 W, which is just under the upper power limit of the inverter.

As a final check, even at elevated temperatures or winter extreme low temperatures, both V_{OC} and V_{mp} remain well within acceptable inverter input limits. Homework Problem 7.12 provides the reader with an opportunity to prove this claim.

Although each inverter has its own array, it is still a good idea to keep the array power ratings as close as possible, since each inverter will ideally provide 1/3 of the total power output to keep the system output balanced. Problem 7.13 offers the reader a chance to explore the phase currents that would result if the inverter outputs are not balanced.

7.6.3 Array Performance

Generally an array of this size will be either ground mounted or mounted on a flat roof of a commercial or industrial building. The modules will be rack mounted at a tilt suitable for local conditions. If on a rooftop, the 144 modules will likely need to dodge some rooftop equipment, such as air conditioning equipment, fans or skylights. Some of these will need specific access space and all will need to be considered for how they might shade the array. In any case, an array of this size will likely be mounted south-facing. For a south-facing array, it is reasonable to assume an annual kWh production of somewhere between 1400 and 1600 kWh/kW. In any case, rather than going through a site-specific performance analysis, it will simply be assumed an annual production of approximately 87,500 kWh, which averages out to 239.7 kWh/day. This provides a number to compare with possible ESS choices.

7.6.4 ESS Selection

The inverters in this example are capable of storage sizes from 50 to 9900 Ah using either lead-acid or lithium storage. The storage is primarily intended for backup if the grid goes down. Since the operating voltage of the system is nominally 48VDC, the allowable storage capacity can be converted to 2.4 kWh up to 475 kWh. If necessary, all or part of this amount can be cycled on a daily basis. This capacity is allowed for anywhere from one to 12 inverters in a system.

Given the daily average kWh production of the array, all that is needed is a criterion or criteria for how this PV production should rate to the ESS capacity. Again, one might argue that without additional load information, preferably on a monthly or daily basis, it is difficult to determine a good estimate of ESS capacity. And, in fact, if one makes this argument, it is an indication that they are well on their way to becoming a good PV system engineer.

Perhaps one of the most important observations is that the maximum available storage is almost twice the average daily PV production and that the average daily PV production is what is available when the utility is down. Thus, depending upon the importance of the backup to the loads, a reasonable choice of ESS size can be made. It is thus interesting to take a look at what sort of increments might be available for 48 V lithium storage.

As with most lithium ESS, large systems are built by combining smaller systems. At this moment in history, it would be foolish to say that there is only one way to do it and this will be the only way for the next 10 years. In fact, by the time the reader reaches this section of the book, the components used in this design may end up as ancient history. Thus, the important issue here is to understand the design process, such that the designer will at least know what to look for.

The first consideration is to look for 48VDC units, keeping in mind that Li ESS units come in both DC and AC configuration, with or without additional DC:DC, DC:AC or AC:DC conversion. One of the reasons for still using 48 V is that it eliminates at least one layer of conversion, which can eliminate 1% or 2% of overall system losses plus the cost of the converter electronics.

One of the compatible ESS units that can be used is a stackable system that can incorporate up to 38.4 kWh in a single package. Up to 15 of these units can be connected in parallel, which adds up to 576 kWh. This means that the largest number that can be used and still stay less than 475 kWh is 12. Thus, anywhere from 1 to 12 units can be selected, with 6 units coming closest to the average daily PV system energy production. From this point, it's between the designer and the customer to arrive at a desirable combination, which will also likely need to take into consideration the actual daily load that is important enough to back up and maybe even the overall system cost.

7.6.5 Wire Sizing

The easy part is the array, since numerous previous experiences have been had with array wiring. Assuming each inverter is connected to six strings, this is 12 #10 (6 positive and 6 negative) plus a #10 G in 1″ metallic conduit. It should be noted that the largest contributor to the "either, or" method will be the combined temperature and conduit fill derate. Using the 75°C rating for #10 Cu (35 A) and applying the two deratings, gives 35×0.5×0.94 = 16.45 A. The 0.94 temperature derating factor is good for ASHRAE 2% high temperatures between 87°F and 95°F, which works for much of the United States. Since I_{SC} = 11.19 A, this is more than adequate for wiring in a single conduit from the roof to the inverter.

The wiring from the ESS to the inverters is the fun part. Relevant numbers include a maximum charging current of 300 A for any one of the ESS units, a maximum of 250 A load on the ESS per inverter when in backup mode and a maximum charging capability for each inverter of 250 A suggests the maximum current on the busbar at any point in time would be 750 A. This explains the 800 A busbar. Remember, even though this is a three-phase system, the busbar is DC and thus none of the 1.732 factors of three-phase apply to it.

As shown in Figure 7.13, each inverter is fed by two sets of battery cables. The maximum input or output current for a single cable set is 160 A. Thus, to allow for a maximum of 250 A, which, incidentally, happens to be the rating of a factory-installed battery circuit breaker in the inverter, two sets of battery cables will be needed for each inverter.

This means each set of cables should be rated for 125 A, since they feed the inverter in parallel. This also means the battery cables for each inverter need to be the same length for all 4 cables.

Since the operating voltage of the ESS is 48 V, %voltage drop is a possible concern. Assuming a two-way voltage drop of 2%, this allows for a total cable resistance for one ESS input of R = 0.02×48V÷125A = 0.00768 Ω. If the distance between the ESS system and the inverter is 12 ft, then, since R = (Ω/kft)×(2L)×1000, solving for (Ω/kft) using R = 0.00768Ω and L = 12 ft shows that the maximum value of (Ω/kft) is 0.32. Thus, two items that determine the battery cable size are 125 A and 0.32 (Ω/kft). If the cables are in conduit, then the size of Cu that meets the (Ω/kft) requirement is any wire larger than #4.

Since the ampacity of #4 for enclosure in a conduit is only 85 A and since the conduit will have four current-carrying wires to each inverter, the ampacity rating will need to be 125 ÷ 0.8 = 156.25 A, which will require 2/0 Cu rated at 75°C. Thus, wiring to each inverter from the ESS will be 4 2/0 plus a #4 EGC in 2″ conduit. If, instead, free air battery cables are used, then the cable sizes can be reduced to #2 Cu, which meets the %VD requirement as well as the ampacity requirement.

On the AC side of each inverter, the maximum grid pass-through current is 200 A. Since this exceeds the maximum inverter output current, the assumption is that this current passes through the inverter from grid input to load and that not all loads can be backed up by the inverter. Otherwise, if the utility connection only needs to handle the backup loads when the utility is available, the inverter current for wire sizing will be 62.5 A. Thus, 125% of 62.5 A is 78.125 A, which will require 3 #4 Cu + 1 #8 Cu in 1¼″ conduit for each inverter. Note that these wires will terminate at the Main Distribution Panel at a 2P80 A circuit breaker. It is important that the three circuit breakers from the inverters are all connected in the MDP such that Inverter 1 is connected across Phase A and Phase B, Inverter 2 is connected across Phase B and Phase C and Inverter 3 is connected across Phase C and Phase A. Since the connection from the inverter to the backup combiner will be the same as between the inverter and MDP, once again, each will terminate at an 80 A circuit breaker in the backup combiner. It should be noted that in the event of a utility failure, it will be the responsibility of one of the three inverters to be the controller, which means it will have remembered the phase sequence of the utility to establish an identical phase relationship for the backup loads if the utility is lost.

Finally, the connection between the backup combiner and the backup panel will be the sum of all the inverter output currents but the three-phase sum version. That is, since each line is fed by two different inverters, the inverter output currents feeding each line will be 120° out of phase. As a result, the line current to the backup panel will be 1.732×62.5 A = 108.25 A. The feeders need to be sized for 125% of this (135.3 A) and need to feed a 3P150A main circuit breaker in the backup panel, since this is the next higher standard circuit breaker size. Since the system is a 120/208 V Y-connected system, the feed to the backup panel needs to include all three lines, a neutral and an equipment grounding conductor. The line and neutral conductors will be 1/0 Cu and the G will be #6 Cu.

Figure 7.14 shows the complete 45 kW 3-phase system with all components. Note that since the inverter outputs are single phase from line-to-line, one of the inverters acts as a controller to be sure all inverters maintain proper phasing if the grid goes down. When the grid is up, phasing is controlled by the grid.

7.7 LARGE BESS DESIGN CONSIDERATIONS

7.7.1 INTRODUCTION

Ultimately, energy storage will most likely end up at the same places where PV is being employed, meaning at the individual building level, the distribution level

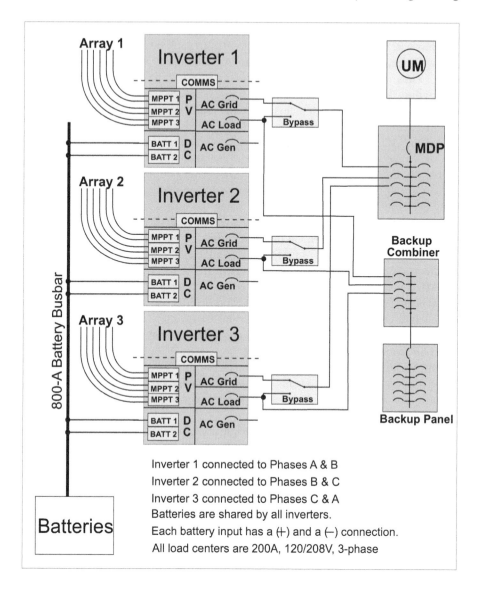

FIGURE 7.14 120/208 V three-phase ESS System with 45-kW AC output rating.

(neighborhood) and the transmission level (system). This scenario is reasonably assured, since it is already happening. One example is a 100MW, 129MWh Li BESS installed in 2017 in South Australia in less than 100 days [6]. Note that this description is in terms of MW first, since one of the important functions of very large storage is grid stability. This particular system can ramp up to full output in less than a second. At the time of this writing, the world's largest BESS is a 7500 MW, 3000 MWh system in Monterey County, California. The second largest BESS (409 MW, 900 MWh) is in Florida [7]. This is one of the systems designed to store daytime PV energy for nighttime use. It is important not to forget that 900 MWh allowed to discharge over a nine-hour period can discharge at a 100 MW rate. The power supplied by the BESS depends upon the needs of the load.

Another recent development has resulted from the observation that when battery packs are no longer satisfactory for vehicular use as their internal resistance rises to the point that acceleration is affected and capacity in miles is reduced, they are still excellent candidates for utility-scale storage for many additional years. They simply need to be operated at slower charging and discharging rates.

Exactly what form the energy storage of the future will take, the system principles will likely remain the same. Storage will be in the form of short term, medium term and long term. Some will be DC-coupled, some will be AC-coupled. Short term (seconds to minutes) will be used for grid stabilization. Medium term (minutes to an hour) will be used for bridging purposes, such as if a cloud covers the sun or if another source needs to be added to the system. Long term is measured in hours and is used for load leveling, firm capacity and deferral of transmission and distribution. Given these classifications, it becomes quite clear why large-scale storage tends to be classified in terms of its power rating first and its energy rating second.

Another difference between large-scale ESS and the systems previously discussed is that all previously discussed systems in this chapter have included a PV source as a part of the system. In some cases, the PV was used as the primary source of charging the ESS. On a larger scale, it is likely that a number of smaller PV and ESS systems will be a part of the overall system, but they need not necessarily be an integral part of larger systems. Perhaps the important issue to remember is that renewable sources tend to be a bit unpredictable and variable. ESS will be needed to iron out the peaks and valleys of this generation. But there will still be incidents where the grid fails for one reason or another but most commonly probably due to weather. For these situations, local storage, including grid isolation, will be essential, since it is difficult to predict in advance exactly where the failure may occur in the grid.

7.7.2 Challenges of Various Storage Options

Maintaining grid stability does not necessarily mean supplying power to the grid when it is needed. It can also require the extraction of power from the grid if needed. For example, the loss of a major load can create a serious grid disturbance when the power that had been supplying the load now needs to either shut down or find another place to go. If shutting down means curtailment of a renewable source, storage becomes a very attractive option. Just as in the case of Lithium BESS, the storage has a single connection that needs to be controlled either to contribute to the grid or to extract from the grid. Since contribution is not an option if the source is discharged, and since extraction is not an option if the source is 100% charged, the ESS will need a sophisticated control system that will be able to determine its optimal state of charge that will accommodate either charging or discharging at the request of the utility. In fact, this may be time-dependent on a daily, weekly or seasonal scale.

Another important consideration is how to connect the various BESS subsystems. Imagine a 1 MW system operating at 48VDC. The current would be 20,833 A! This is certainly not a job for #12 Cu. Clearly, the system will incorporate large DC:DC converters, large transformers or both. At the 7860/13,600 V three-phase level, each MW will require wiring to handle 42.4 A per phase, which is a bit more manageable. The bottom line is that unless large busbars are used for BESS interconnections, the BESS building blocks will need to be sized at reasonable lower voltage sizes such that each building block can tie into one or another form of voltage conversion to enable dealing with reasonable current levels and reasonable wire sizes. This includes the utility-scale systems mentioned earlier that are at the 100 MW level. Usually, the only place for them to be connected to the utility will be at the transmission level if the entire system output is associated with a single connection. For connection at the 120 kV level, the current to or from 100MW of BESS would be 481 A.

7.7.3 System Control and Communication Options

At least part of the control and communications involved in a large PV system is designed to monitor weather conditions, including insolation, such that a calculation can be made automatically as to whether the PV system is performing up to design specifications. For a BESS system, whether or not the sun is shining is not nearly as important as measuring the round-trip losses for the BESS and comparing these values with predicted performance. That said, in fact, the heat from the sun can affect the performance of a BESS, in either a positive or negative way in terms of the operating temperature of the BESS and this must be factored into the performance analysis. In addition to overall performance analysis, cell level management and control need to be constantly monitored by the battery management system for a lithium-based system. For other BESS, such as flow batteries, it should be pretty obvious that the chemical state of the electrodes and electrolytes should be monitored and, perhaps, controlled via a feedback mechanism.

The gateway to the grid is a very important interface, since this is where the charging and discharging of the BESS needs to be monitored for billing purposes. Perhaps even more important is monitoring the state of the grid for any evidence of frequency change, voltage change, power factor status or any indication of instability. For any of these events, the BESS will need to decide how much and how long to either discharge to the grid or charge from the grid. And, of course, all of this information needs to be shared with the utility as a part of a larger-scale utility tracking of the health of the grid. The same rules hold if there is a PV system as a part of the BESS. The overall control mechanism will need to ensure optimal performance of the PV system by either exporting to the grid or sending the PV output to the BESS [8–10].

7.7.4 Utility Interconnection

For large systems, it is not unusual for the utility to run a fiber optic cable from the BESS to either a substation or other utility grid monitoring facility to ensure optimal reliability of the communication link. If the BESS is owned by a third party, it is likely that the utility will want to incorporate manual and automatic disconnects as well as their own production/consumption metering. It is also likely that the utility may want to be able to override the internal system controls in the event of any major utility disturbance.

For systems with utility-scale storage, whether or not BESS, the utility will want to ensure the storage connection cannot create any problems at the distribution level, such as backfeed of too much current to a substation and maybe even the possibility of damage to the transformer taps at the substation. On the other hand, careful control of the storage system output can actually benefit the taps if it can result in fewer transitions back and forth between different voltages.

In any case, it is highly likely that any large ESS system will feed a three-phase distribution-level system or a three-phase transmission-level system. Generally, at distribution and transmission levels, loads are pretty closely balanced, but if for any reason this should not be the case, an algorithm will need to be incorporated into the overall monitoring and control system to do whatever possible to minimize unbalance to minimize line losses.

Discussion of any specific system has been deliberately omitted in this section, simply because the field is changing so rapidly that the specifics of ESS or BESS may very well have changed significantly by the time the reader is able to read this chapter. The important take-away instead should be the understanding of what to expect in terms of control algorithms, wiring and switching equipment. Chapter 9 will explore methods of economic analysis of large ESSs.

Homework Problems

7.1 Design a PV system with battery backup for your home.
  a. Create a table similar to Table 7.1 of the loads at your home that you would like to be able to back up during a utility power outage. If possible, use nameplate ratings of the loads. Otherwise estimate them and be prepared to justify your estimate. Would your list most likely constitute a whole-home backup system or a partial backup system?
  b. Estimate the size of the inverter that would be needed to meet your backup needs as determined in Problem 7.1.
  c. Estimate the mount of storage that would satisfy the backup needs of your system. Express the storage requirements both in terms of kW and kWh.
  d. Using NREL SAM for the location of your backup system, estimate the size of the array that would be needed to meet the backup needs for the month the backup will most likely be needed. Be sure to take into account the most likely orientation of the array based on the orientation of the roof.
  e. Find a suitable module that can be used for building your PV array. Then determine a reasonable choice for a microinverter that will result in a DC:AC ratio of less than 1.5 for the overall array. Be sure that the total output power rating of the microinverters in the array will exceed the inverter size as determined in Problem 7.2. Then simulate again, using the specific modules and microinverters selected to compare estimated system production with the previous simulation.

f. Find an appropriate system controller or hybrid inverter to match the rest of your system. The choice may depend upon whether you need partial backup or whole-home backup, although some controllers can be used for either backup method.

g. Determine the number of strings that will be needed to build your array using the selected modules and microinverters. Note that there may be more than one option for this determination.

7.2 Assume that a three-wire branch circuit is derived from a 120/208 V distribution panel where the voltage on one phase is $120\angle 0°$ and the voltage on the other phase is $120\angle -120°$.

  a. Prove that, for resistive loads on either side of the circuit, the current in the neutral will not exceed the current in either phase.

  b. Prove that when the loads on each phase are equal, the magnitude of the neutral current will equal the magnitude of the current in either phase.

7.3 Show that in the case of a three-phase, four-wire set of branch circuits,

  a. The current in the neutral will not exceed the current in either of the phases if only the fundamental component of the current is present on each branch circuit.

  b. If a balanced, nonlinear load across all 3 phases causes a third harmonic current component to flow in each phase conductor, the neutral current can exceed the individual line current.

7.4 Using your backup load schedule from Problem 7.1, design an optimizer-based ESS PV system to handle the listed backup loads.

7.5 Explain the operation of the inverter bypass switch in the AC component enclosure of the example in Figures 7.13 and 7.14.

7.6 Prove that the eight-module strings used in the three-phase example of Section 7.6 remain well within acceptable inverter input limits during extreme summer or winter temperature conditions. Perform this check assuming the system will be located in Tucson, AZ, and also if it will be located in Roseau, MN.

7.7 Assume inverter 1 of Section 7.6 is connected to an array of 48 modules, inverter 2 is connected to 36 modules and inverter 3 is connected to 24 modules, and each module is generating 70% of its rated power. Assuming a 95% conversion efficiency from DC to AC, calculate the resulting currents in each of the three phases. Assume the system is operating at unity power factor.

7.8 Assume that the system of Section 7.6 is located in Fargo, ND, with all arrays facing south and tilted at 45°.

  a. Using PVWatts with the DC:AC ratio of the system, use NREL SAM to estimate monthly system AC output. Assume overall system losses = 12%.

  b. For the month with the lowest system output, look at the estimates of daily system energy production and determine a three-day period with the lowest overall energy production by the system.

  c. If the estimated daily backup needs are 30 kW and 200 kWh/day, look up available ESS systems that are compatible with this system and specify a suitable backup system to meet the backup needs over this three-day period. If the result exceeds 475 kWh, determine the deficit.

REFERENCES

[1] UL Standard 1741: www.ul.com/news/ul-launches-advanced-inverter-testing-and-certification-program
[2] System Advisor 2022 (SAM 2022.3.14), *National Renewable Energy Laboratory*, Golden, CO, 2022.
[3] *NFPA 70 National Electrical Code*, 2020 ed., National Fire Protection Association, Quincy, MA, 2019.
[4] ASCE Standard 7–16, *Minimum Design Loads for Buildings and Other Structures*, American Society of Civil Engineers, Reston, VA, 2017.
[5] Information on a Variety of Roof Attachments: www.ironridge.com/component/halo-ultragrip/
[6] SouthAustralia 100 MW, 129 MWh Li BESS: https://cleantechnica.com/2017/11/23/
[7] Information on Largest BESS Systems as of 2023: www.saurenergy.com/solar-energy-news/the-top-5-largest-battery-energy-storage-systems-worldwide
[8] Information on Control, Communication and Instrumentation of Large ESS: https://influxdata.com
[9] Information on Control, Communication and Instrumentation of Large ESS: www.emerson.com
[10] Information on Control, Communication and Instrumentation of Large ESS: www.honewell.com

SUGGESTED READING

Following is a short list from among a large number of publications relating to energy storage and energy storage systems.

Barnes, Frank S. and Levine, Jonah G., ed., *Large Energy Storage Systems Handbook*, Routledge, 2011.
Battery Energy Storage Systems by EVLO: https://evloenergy.com/
Fu-Bao, Wu, Yang, Bo and Ye, Ji-Lei, *Grid-Scale Energy Storage Systems and Applications*, 1st ed., Elsevier, 2019.
International Code Council, Energy Storage Systems: Based on the IBC, IFC, IRC and NEC, ISBN-13: 9781957212036

8 Stand-Alone PV Systems

8.1 INTRODUCTION

There are many similarities between stand-alone photovoltaic (PV) systems and grid-connected, battery-backup systems. As previously discussed, when a grid-connected, battery-backup system is disconnected from the grid, it becomes a stand-alone system until grid power is restored.

To some extent, stand-alone systems have more flexibility, since if an inverter is used, it does not necessarily need to comply with UL 1741. Stand-alone systems may have the option of being designed for DC only, for a mix of DC and AC loads, or for AC loads only. Some stand-alone systems will have battery storage and some will not. Grid-connected, battery-backup systems are normally limited to AC loads only.

On the other hand, previous chapter design examples have shown that the tilt of the array for a grid-connected system is generally not critical for annual performance, as long as it is within about 15° of latitude. But stand-alone systems may well need to be designed for seasonal performance, meaning more critical consideration of array tilt angle since no grid is available for backup. In fact, in the far north or the far south, it may be necessary to supplement the energy production of the PV system with another form of generation, such as fossil, geothermal, biomass or wind.

In the grid-connected, battery-backup examples, only one day of battery storage was normally included in the design. In stand-alone systems, it is common to have several days of backup, again, since the PV system is the only source of electrical power year-around. Furthermore, monthly electrical loads may differ. This means a more critical consideration of the loads that must be served by the PV system, often on a monthly basis. For example, for solar street lighting, there is generally abundant sunlight in the summer to charge batteries, but the light is on less because the sun is up longer. The opposite is true for winter. Thus, the array will probably need to be tilted for optimal winter performance with the hope that the winter tilt will provide adequate battery charging for summer use.

This chapter will introduce the design process for several complete self-contained PV systems. The first system will be a simple DC fan powered by a PV module, followed by a DC water pumping system that incorporates a maximum power point tracker (MPPT), a PV-Powered parking lot lighting system, and a PV-Powered remote mountain cabin.

In any stand-alone design, the first task is normally to determine the load. Often there are several choices. For example, a choice might be made between an inexpensive incandescent lamp or an LED lamp. In almost all such cases, the choice of the more efficient load is more cost-effective. In fact, the most efficient load may not necessarily be the most expensive one, especially when the load is a refrigerator.

The load voltage, current, power and daily operating hours are needed for proper sizing of fuses, wires, batteries and other system components. In some cases, not all

Stand-Alone PV Systems

of these quantities are given. For example, the specifications for a refrigerator may only give the voltage and kWh per year. In these cases, it may be necessary to either calculate or estimate the other parameters.

Once the load has been determined, possibly on a monthly or seasonal basis, then the amount of energy storage needs to be determined. Some systems will not need energy storage, some will have minimal storage and some may require sufficient storage to meet critical performance requirements in which the system must operate more than 99% of the time. It will also be the responsibility of the engineer to determine which storage technology to use. In 2024, the choice would likely be lithium iron phosphate (LFP) batteries to achieve a greater energy density per unit weight or per unit volume, but by the time this book is published, another technology may be preferred.

After storage selection, the size of the PV array must be determined. Then the electronic components of the system, such as charge controllers, system controllers and inverters are selected. Finally, the balance of system components are selected, including the array mounts, the wiring, switches, fuses, battery compartments, lightning protection and, perhaps, monitoring instrumentation.

While computer programs may be available for assistance in system design, as in previous chapters, emphasis in this chapter will be placed on sound, common-sense reasoning and the use of good engineering judgment. To keep the design examples as realistic as possible, many of the system components described in the examples of this chapter have been found on an assortment of web sites. It is important for the reader to note that the use of a particular component or web site in an example does not constitute an endorsement of the component or the company. Vendors and component availability change on a daily basis, so when a real system is designed, the designer should carry out an extensive search for the components that best meet the design goals. Hopefully the procedures for component selection outlined in this chapter will prove to be useful when the opportunity to design a real system presents itself.

8.2 THE SIMPLEST CONFIGURATION: MODULE AND FAN

Figure 8.1 shows the simplest of PV systems, a fan motor connected to a PV module. The figure also shows the superimposed performance (I–V) characteristics of the fan and the module. The operation is simple: as the sun shines brighter, the fan turns faster. If the designer has no concern for the exact quantity of air moved, the design becomes nearly trivial. However, if the amount of air moved must meet a code requirement or other constraint, then it will be necessary to consider the design in more detail. Obviously, this system is useful only during daytime hours. Nighttime operation will require adding battery storage to the system.

The system operation point is determined by the intersection of the performance characteristics. Note that as the sun shines brighter, making more PV current and voltage available, the fan consumes more power. It is reasonable to assume that as the fan consumes more power it will move more air.

Perhaps the second observation the reader will make regarding the performance characteristics intersections in Figure 8.1 is that the module is not operating

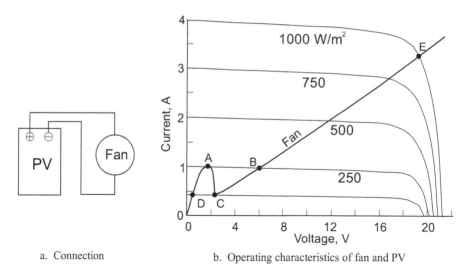

FIGURE 8.1 A simple, PV-powered fan, showing performance characteristics for fan and PV module.

anywhere near maximum power at low light levels. As irradiance approaches 1 kW/m², the fan operating point is close to the module maximum power point. If a module having a higher short circuit current is used, the fan will remain at a more constant speed as the fan current exceeds 3.5 A, but at high illumination levels, the module will be acting more as a voltage source and again will deliver only a fraction of its maximum power capability. Hence, the designer must decide how much air movement is needed at various irradiance levels, and choose the module and fan accordingly. Thus, even in this relatively simple design example, the designer must use discretion. Larger modules will cost more but will deliver more air at lower irradiance levels.

Figure 8.1b also shows the hysteresis effect encountered in starting the fan. Under stalled rotor conditions, the fan motor does not produce a back EMF and thus the fan will draw stalled rotor current until sufficient armature current is present to overcome the starting torque. The irradiance level at point A on the curve is just adequate to provide this current, and the operating point then jumps to point B. As the irradiance level continues to increase, the operating point moves toward point E. When irradiance levels decrease, fan performance follows the fan characteristic to point C, after which the fan stalls and the operating point jumps to point D and eventually approaches the origin as darkness falls.

Another question for the designer to ask is whether it would be better to use a different fan to meet the design requirements. The obvious answer is "maybe." And that is what makes the design of PV systems so much fun. It should be clear from Figure 8.1b that regardless of the choice of fan or module, there will be a significant power mismatch over a relatively wide range of irradiance. Thus, no matter what the choice, there will be some portion of the fan or PV characteristic where maximum power will not be transferred to the fan. If it is desired to optimize fan power for all

Stand-Alone PV Systems

illumination levels, a maximum power point tracker (MPPT) will need to be incorporated into the design.

The MPPT can be particularly useful at irradiance levels between the start and stop irradiance levels of Figure 8.1b, where it will enable the fan to start at a lower irradiance level and to stop at a lower irradiance level, with greater air flow at irradiance levels between these points. Here the interesting part of the trade-off is whether including an MPPT with a smaller module will cost less than using a larger module to obtain comparable system performance.

For example, the PV curve upon which points A and B lie has an approximate power available at the maximum power point of 18.13762 W. But at the intersection of the fan curve and the PV curve, the fan is using only 5.78623 W, which means that an additional 12.35139 W is available, but not used. As a result, the fan is only using 31.9% of the available power. In the ideal case, $P \propto v^3$, where v is the air velocity moved by the fan and P is the power consumed by the fan. Thus, as the power triples, the air velocity increases by a factor of 1.44. Thus, if the simple fan system does not move enough air at low sunlight levels, then an MPPT can make a significant difference in fan performance.

8.3 A PV-POWERED WATER PUMPING SYSTEM

8.3.1 INTRODUCTION

Another common stand-alone PV application is water pumping, especially when the water to be pumped is a long distance from a utility grid. Water pumping applications do not necessarily require battery backup unless the water source will not produce an adequate supply of water to meet the pumping needs during daylight hours. Another reason for using batteries is that the water can be pumped over a longer time at a slower rate with a smaller, less costly pump. Under these circumstances, it may be necessary to charge a battery so the pump can run for an extended period. When the water supply can meet the pumping capacity of the system, then it may be more cost-effective to pump all the water the pump is capable of delivering and storing any excess in a storage tank. In effect, the storage of water replaces the storage of electricity in batteries. It still represents the conversion of kinetic to potential energy.

When designing a water pumping system, it is necessary to determine a number of parameters to properly size the system components. First of all, the daily water needs must be determined. Second, the source must be characterized in terms of available water and the vertical distance over which the water must be pumped. Once these factors are known, along with the number of hours per day available for pumping, the pumping rate can be determined. The pumping rate along with the pumping height equates to the pumping power, once again the product of a pressure quantity with a flow quantity.

Once the size of the pump motor is known, the kWh requirements of the motor can be determined, and, finally, the size of the PV array needed to provide the kWh can be determined. As in the fan example, the inclusion of a LCB or MPPT extends the useful pumping time of the pump motor and enables the use of a smaller motor and a smaller array that is utilized more efficiently.

8.3.2 Selection of System Components

To quantify the pumping problem, it is useful to note that a gallon (3.785 l) of water weighs 8.35 pounds (3.73 kg) and that one horsepower = 550 ft-lb/sec = 746 watts, assuming 100% conversion efficiency. This means that pumping a gallon of water to a height of one foot involves 8.35 ft-lb of work. In the MKS system, converting gallons to liters and feet to meters gives the result that pumping a liter of water to a height of one meter requires 7.23 ft-lb = 9.83 J.

Although this information can be used to determine the horsepower of the pump, it is more common to determine pump size from manufacturers' specifications when daily pumping requirements and pumping height are known. For the design of a pumping system, it is also useful to understand the pressure versus flow characteristics of piping as well as to understand that pumps can be designed with trade-offs between pressure and flow.

Piping friction loss is determined by the type and diameter of pipe used, just as voltage drop is determined by the size and material of the wire used, although the relationship between pressure and flow for a water pipe tends to be somewhat more nonlinear than the I-V relationship for a wire. However, at relatively low flow rates, the flow versus pressure curve for a piping system can be approximated by a linear relationship. Figure 8.2 shows pressure versus flow curves for several sizes of piping [1]. For design purposes, it is common to include a 5% head loss for piping friction loss. Thus, for a 100 ft (30.48 m) pumping height, 105 (32.0 m) ft is used as the design height to compensate for piping friction losses.

Although it is reasonably straightforward to select the horsepower for a pump, it is somewhat more involved to select the pumping system that will perform at maximum efficiency. The reason is that some pumps are designed to deliver higher pressure than others. The pumps that can deliver higher pressure are needed for lifting water to greater heights. Figure 8.3 shows performance curves for two pumps of

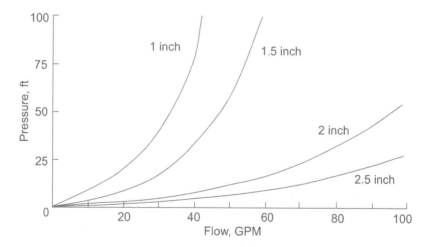

FIGURE 8.2 Pressure versus flow curves for equal lengths of piping of different diameters.

Stand-Alone PV Systems

equal horsepower, one of which is a high-head (pressure) pump and the other is a medium-head pump [2].

Note that the medium-head pump will deliver more volume than the high-head pump at low pressure, but the high-head pump will overcome a greater pumping height. Note also that the performance of a pump depends on the speed at which the pump is operated. If the pump speed decreases, both pressure and flow capacity decrease. It is thus important to select a pump that will be able to overcome the lift requirement under low sun conditions, unless energy storage is included in the design.

8.3.3 Design Approach for Simple Pumping System

To facilitate the selection of system components, the selection will be based upon the design of a pumping system that will deliver 2000 gallons per day from a 200 ft (60.96 m) deep well near Lubbock, Texas. Since Lubbock weather records show annual rainfall to be less than 20 inches (50.8 cm) [3], it is a reasonable assumption that two days of battery storage should be more than adequate if the system is designed with a pump that will need battery storage. During higher-than-average rainfall periods, it is likely that the rain itself will replace at least some of the need for pumped water.

8.3.3.1 Pump Selection

If pumping data are available from a pump manufacturer, then all the designer needs to know is the daily amount of water needed and the overall pumping height. The daily water needs can be converted to gallons or liters per minute over the time the pump will operate and a suitable pump can be selected from manufacturers' tables. Ideally, the desired system will have the lowest cost and the highest reliability. The

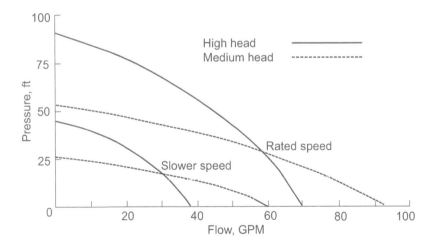

FIGURE 8.3 High-head and medium-head pump performance characteristics at two different speeds.

component choice to achieve this involves deciding on smaller versus larger pumps, DC versus AC pumps, batteries versus no batteries, linear current booster (LCB) versus inverter and maybe even tracker versus fixed mount.

The minimum pump size is determined by assuming continuous pumping, which would require battery storage if the pump is to run during the night. Using the 2000 gal/day (GPD) water requirement means pumping an average of 83.33 gal/hr (GPH) over a 24-hour period. Thus, the pump must be able to deliver 83.33 GPH with a lift of 210 ft. Table 8.1 tabulates the GPH delivered at specific pumping heights when a particular DC submersible pump is operated at 24 V. As expected, the pump current increases as the pumping height increases.

For this pump, the question is whether it will deliver 2000 GPD (7571 l/day) at a pumping height of 210 ft (64.01 m). Linear interpolation between the 200 ft (60.96 m) and 230 ft (70.1 m) data results in GPD (210 ft) = 2112 = 0.52 l/min (64.01.01 m), which is close to the desired pumping rate, since the desired 2000 gal (7571 l) will be delivered in 22.7 hr. Thus, this pump will meet the design criteria, so to evaluate the overall system, the rest of the system can now be designed.

Since the pump will be operating continuously, it will be necessary to provide battery storage. Battery storage generally requires the use of a charge controller to optimize the battery charging process and to prevent overcharging of the batteries. If the pump were an AC pump, it would require an inverter and the inverter would likely have built-in protection for overdischarging the batteries. The DC pump does not have an inverter, so some sort of switching will be needed to protect the batteries from overdischarge. The system will also likely incorporate a water storage tank and may also use a float switch on the tank to prevent the tank from overflowing.

Table 8.1 also shows the daily energy consumption of the pump for various pumping heights. Again, linear interpolation between the 200 ft (61 m) and 230 ft (70.1 m) energy requirements results in a daily energy requirement of 2.25 kWh/day supplied to the pump. Thus, the PV array must produce sufficient energy, such that losses in

TABLE 8.1
Flow versus Lift Showing Operating Conditions for a Specific 24-V DC Submersible Pump

Lift, ft	Flow, gph	Current, A	Power, W	Gal/day	Ah/day	kWh/day
20	117	1.5	36	2808	36	0.864
40	114	1.7	40.8	2736	41	0.979
60	109	2.1	50.4	2616	50	1.210
80	106	2.4	57.6	2544	58	1.382
100	103	2.6	62.4	2472	62	1.498
120	101	2.8	67.2	2424	67	1.613
140	99	3.1	74.4	2376	74	1.786
160	98	3.3	79.2	2352	79	1.901
180	93	3.6	86.4	2232	86	2.074
200	91	3.8	91.2	2184	91	2.189
230	82	4.1	98.4	1968	98	2.362

wiring, charge controller and batteries are taken into account, there will still be 2.25 kWh/day left over for the pump.

8.3.3.2 Battery Selection

Since batteries are often sized in Ah, the first step is to convert the 2.25 kWh/day to Ah/day. This is accomplished by simply converting kWh to Wh and then dividing the result by the battery voltage, yielding a requirement of 93.75 Ah/day at 24 V. Allowing for 90% depth of discharge of the batteries if Lithium batteries are used, a single day of storage will require $93.75 \div 0.9 = 104$ Ah of battery capacity. Since lithium battery packs are available in a wide range of voltages, the next step is to look for 24 V units having 104 Ah ratings. The question is to what discharge rate does the 104 Ah rating apply? The answer is determined by the number of storage days (days of autonomy) are desired. Thus, if two days of storage are desired, then 208 Ah will be needed at the 48 hr discharge rate. If a 24 V LFP (lithium-iron-phosphate) battery rated at 100 Ah at a 24-hr discharge rate is used, two of these batteries in parallel will easily provide 200 Ah. Since two days of storage is a generous allowance, 200 Ah instead of 208 Ah should be a very reasonable selection.

These batteries have built-in battery management systems to ensure uniform charging of all cells and to protect from overcharging. However, for MPPT purposes, an appropriately sized charge controller is still needed. Although the price of the lithium battery packs is higher than comparable lead-acid units, the increase in the number of available cycles, plus the likelihood of using 90% of the available stored energy in any period is very small, the long-term cost of the lithium units is competitive, especially when the minimum need for maintenance is taken into account.

8.3.3.3 Module Selection

Module selection will depend upon the location of the system and the tilt of the array. Since the system is located near Lubbock, TX, then an analysis using the PVWatts model in NREL SAM yields the data of Table 8.2 for an array tilted at 40 degrees [4]. Note that this data is a result of assuming a grid-connected system, which calculates monthly performance data in terms of kWh delivered to the grid, that is, at the inverter output. This number will be close to the charge controller output and will thus be used as a reasonable approximation of the charge controller output to the battery pack. This charge controller output is shown in Table 8.2. Thus, if a battery round-trip efficiency loss of 5% is assumed, the overall efficiency from array output to pump input will be 95% of the charge controller output. Column 3 of Table 8.2 shows this figure on a monthly basis. Column 4 shows the monthly kWh consumption of the pump and Column 5 shows the array size needed to provide the pump monthly kWh consumption. The table shows that for December, the array size needs to be a minimum of 0.52 kW to deliver 69.75 kWh to the pump. For all other months, the array can be smaller, which provides oversizing of the array during all other months if it is sized for December.

If it is felt that this is too close for the winter months, additional PV power can be added to the array or the tilt can be raised. It will then be necessary to check whether increasing the tilt will still provide adequate energy for the summer months. Depending on the selection of charge controller, the array voltage can likely be

TABLE 8.2
Lubbock, TX Monthly kWh Production for PV Array at 40° Tilt

Month	Ch C Out kWh/kW	Batt out kWh/kW	Load kWh/mo	Min Array kW
Jan	160	152	69.8	0.46
Feb	152	144	63.0	0.44
Mar	179	170	69.8	0.41
Apr	167	159	67.5	0.43
May	167	159	69.8	0.44
Jun	158	150	67.5	0.45
Jul	158	150	69.8	0.46
Aug	161	153	69.8	0.46
Sep	160	152	67.5	0.44
Oct	178	169	69.8	0.41
Nov	152	144	67.5	0.47
Dec	140	133	69.8	**0.52**

somewhere between 40 and 150 V, just so the array power rating is at least 520 W. A 520 W array is easily achieved with a single 520-W module. One such module is rated at 520 W with $V_{OC} = 73.5$ V, $V_m = 61.8$ V, $I_{SC} = 8.95$ A and $I_m = 8.41$ A.

8.3.3.4 Charge Controller Selection

If an MPPT charge controller with an input voltage range up to 150 V is used, then, for a 520 W array, the maximum charge controller output current at 24 V will be approximately 21 A, allowing for a 97% efficiency of the charge controller. A number of 40 A MPPT charge controllers are available on the market, so at this point it is a matter of how many features are desired on the unit. For that matter, since the module output will rarely be 520 W, even a 20 A, 24 V charge controller could be considered.

8.3.3.5 BOS and Completion of the System

Figure 8.4 shows a schematic of this system. Now that all the voltages and currents are known, it is left as an exercise for the reader to determine the sizes of wires and switches. Note that the intent of this design is to show one way to design a water-pumping system. Just as in the case of the simple PV fan, water pumping can also be accomplished with an array and a pump and only the wire and piping to deliver the water. When it comes to a serious design, it is recommended that the designer explore more than one option for the system to achieve a system with maximum cost-effectiveness.

8.4 A PV-POWERED PARKING LOT LIGHTING SYSTEM

8.4.1 DETERMINATION OF THE LIGHTING LOAD

It should be evident that if a PV system is to power a parking lot lighting system, a battery storage system will be needed, since if the sun is shining, the lights are not

Stand-Alone PV Systems

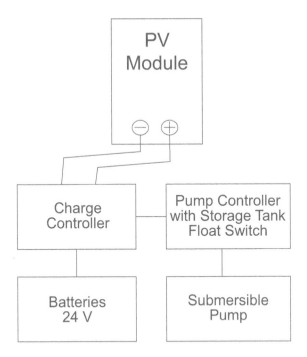

FIGURE 8.4 Electrical schematic diagram of the water pumping system.

needed, and vice-versa. The first step is to determine the lighting load, followed by battery selection and, finally, the number and type of PV modules to use. To determine the lighting load in watts, it is first necessary to determine the amount of light needed and the area of the space to be illuminated. Hence, the design begins with the determination of the necessary illumination level.

While illumination levels could be measured in watts/m², in the United States, illumination levels are most commonly measured in foot candles. A foot-candle is the amount of light received at a distance of one foot from a standard candle. A standard candle is a candle that emits a total amount of light equal to 4π lumens. The lumen is thus the basic quantity of light in the foot-candle system of measurement of light intensity. It compares to the coulomb in the electrostatic realm.

If a closed surface surrounds the standard candle, then all of the 4π lumens of light must ultimately pass through the surface. If the light from the source is emitted uniformly in all directions, and if a sphere of radius 1 ft is centered on the light source, then the light will be uniformly distributed over the surface of the sphere with a density of (4π lumens) ÷ (4π ft²). This light intensity of 1 lumen/ft² defines the foot-candle (f-c).

The Illumination Engineering Society publishes guidelines for illumination levels for various spaces [5]. For example, parking lot lighting should normally be lighted to an average illumination level of approximately 1 f-c, depending on the degree of security desired. A desk for normal work is generally adequately lighted with 50 f-c.

Direct sunlight provides about 10,000 f-c [5]. The *luminous efficacy* of a source is a measure of the efficiency with which the source transforms electrical energy into light energy. It is measured in lumens per watt. Table 8.3 [6–8] shows the luminous efficacies for several light sources.

Determination of the wattage of light needed to accomplish a specific lighting task, then, will depend on the required illumination level and the area to be lighted. It also depends on the luminous efficacy of the source. Other factors include whether the available light can be directed only to the area where the light is needed and whether some of the light will be absorbed by walls or other absorbers, such as the light fixture itself before it reaches the surface to be illuminated. Dust on the fixture and lamp also absorbs useful light output.

In addition to the intensity of light, the color temperature of light is sometimes a factor to be considered. The color temperature of light refers to the equivalent spectral content of radiation from a blackbody at a particular temperature. The color temperatures with which the reader is most familiar are the 5800 K (daylight) temperature of the sun, which produces its characteristic white color, and the various color temperatures of LED lights, some of which are adjustable between 2700 and 5000 K.

Not all light sources can be accurately characterized by a color temperature, since the concept is based on blackbody radiators. Sources with discrete spectral components, such as lasers or gas discharge lamps, can be assigned equivalent color temperatures to indicate the temperature to which the spectrum of the source is most closely matched, but the color temperature is not a precise measure of the color of the source. For example, xenon produces a very white flash, which, on photographic film, appears to be close to the color of daylight. Although the output spectrum of a xenon lamp differs from the AM 1.5 solar spectrum, xenon lamps are commonly used in solar simulators with appropriate correction factors.

Consideration of color temperature can be an important factor in the choice of light sources, but even color temperature is sometimes inadequate for describing a light source. For any non-blackbody radiator, the degree to which the source approximates a blackbody is the color rendition index (CRI). The higher the CRI, up to

TABLE 8.3
Approximate Luminous Efficacy for Several Light Sources [6–8]

Source	Luminous Efficacy, l/w	Lamp Lifetime, hr
25 W incandescent	8.6	2500
100 W incandescent	17.1	750
50 W quartz incandescent	19.0	2000
T-8 fluorescent	75–100	12,000–24,000+
Compact fluorescent	27–80	6,000–10,000
Metal halide	80–115	10,000–20,000
High-pressure sodium	90–140	10,000–24,000+
3.6-W LED array	~130	100,000+
High-bay LED fixture	100–150	100,000+

Stand-Alone PV Systems

100%, the closer the light source spectrum is to a perfect blackbody spectrum, resulting in better color rendition. For PV applications, generally, the most popular and efficient sources are LEDs.

Once a light source and a suitable fixture for the source are chosen, for an outdoor lighting system, the available light from the fixture, expressed in lumens, can be obtained from the formula

$$\text{Lumens} = (FC) \times (\text{area}) \div (CU) \div (MF) \div (RCR), \tag{8.1}$$

where (FC) is the desired illumination level in foot-candles, (area) is the area to be illuminated, measured in ft^2, (CU) is the *coefficient of utilization* of the fixture, (MF) is the *maintenance factor* of the fixture and (RCR) is the *room cavity ratio*. The coefficient of utilization of a fixture is a measure of the fraction of light available from the lamp that is directed to the surface to be illuminated. The maintenance factor of the fixture provides a means of estimating the amount of light from a fixture that can be lost as the fixture and lamp get dirty and the room cavity ratio is a measure of the amount of light from the fixture that will be absorbed by the room and its contents. A good fixture will have a CU of 0.8 or more and a MF of 0.9 or more. Of course, the MF is dependent upon the environment in which the fixture operates and the maintenance interval for the fixture. The RCR depends upon room size, wall color, floor color and room contents. A room full of dark colors will appear darker because it *is* darker. Outdoors, the RCR is generally assumed to be 1.0. Indoors, the RCR is generally < 1.

8.4.2 Lighting Design

8.4.2.1 Introduction

Figure 8.5 shows a parking lot for which it is desired to provide an average illumination of 2 FC. The figure also shows suitable locations for fixtures and an approximate lighting pattern that would be obtained with a well-designed sharp cut-off fixture. The fixture has a CU of 0.8 and an MF of 0.9. The sharp cut-off feature of the fixture helps to prevent spillage of light onto adjacent property, so it will be assumed that any loss of light to adjacent property will be accounted for by the CU. The lot measures 160 ft (48.8 m) × 160 ft, so the total area of the lot is 25,600 ft^2 (2,378 m^2). With the spacing of the fixtures as shown, the distance between fixtures is 113 ft (34.4 m). For parking lots, the spacing of lighting poles is typically set at four pole heights. Thus, in this case, the pole height should be approximately 28 ft (8.5 m).

8.4.2.2 Determination of Lamp Wattage and Daily Load Presented by the Fixture

First, the total lumens needed to illuminate the parking lot must be determined, and then a suitable lamp can be selected. The total lumens can easily be found from (8.1) to be 2 × 160 × 160 ÷ 0.8 ÷ 0.9 = 71,111. If four fixtures are to be used, as shown in Figure 8.5, then each lamp will need to supply 17,778 lumens. Next, a search can be made for LED parking lot fixtures that are rated at a minimum of 17,778 lumens.

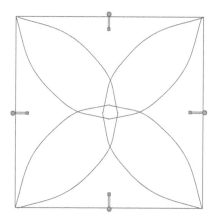

FIGURE 8.5 Parking lot showing locations of fixtures and approximate coverage patterns for each fixture.

One such fixture is a 126-W, 17,724 lumen fixture designed to replace 300 to 525 W high-intensity discharge fixtures. For about $40 additional per fixture, the lumen output can be increased to 22,155 with a 158-watt fixture. The LEDs are rated to last more than 100,000 hours.

If this parking lot is in Miami, FL, and if the lights are to be on from sunset to sunrise, then the maximum amount of time the fixture will need to operate will be 13 hours during the month of December if the lights come on 15 minutes after sunset and shut off 15 minutes before sunrise. During the month of June, the fixture will only need to operate for 9.8 hours [9]. Thus, the winter worst-case daily Wh = 126×13 = 1,638 Wh, and the summer daily energy consumption is 126×9.8 = 1,235 Wh. If the fixtures are 120 V fixtures, then a pure sine inverter will be needed to supply the fixtures. Assuming 95% inverter efficiency, this would increase the energy required from the batteries to 1724 Wh winter and 1300 Wh summer loads. Maximum loads for each month are shown in Table 8.4.

8.4.2.3 Determination of Battery Storage Requirements

The value of stand-alone outdoor lighting systems is enhanced if each fixture can operate independently of others. This saves the effort of running wires from one fixture to another. Thus, all that is needed is to design one system and then build three more identical to it. Since Miami does not normally have many consecutive totally cloudy days, especially in the winter, two days of storage will be used. For a 24 V battery storage system, this means the winter daily *connected* Ah for the load will be 1,724 ÷ 24 = 71.8 Ah, which needs to be supplied by the batteries. Allowing for 2% wiring losses and a 90% depth of discharge, the batteries will need to have a 24-hr discharge rate capacity of 71.8×2 ÷ 0.98 ÷ 0.9 = 163 Ah @ 24 V. Note that the 24-hr discharge rate is used because, over the two-day period, the batteries will be discharged for approximately 12 hours each day. The total discharge will thus take approximately 24 hours. If an inverter with 24-V DC input is used, one way to

TABLE 8.4
Miami NREL SAM PVWatts Simulation for Parking Lot Lighting System with 41° Array Tilt

Month	PV to Batt in kWh/kW	Lights Max Daily on Time	Load kWh/month	Min Array kW
Jan	140	12.8	49.8	0.37
Feb	137	12.1	42.6	0.33
Mar	151	11.3	44.1	0.31
Apr	140	10.5	39.7	0.30
May	129	10.0	39.1	0.32
Jun	110	9.8	36.9	0.35
Jul	122	9.8	38.1	0.33
Aug	130	10.5	41.0	0.33
Sep	128	11.5	43.5	0.36
Oct	139	12.2	47.7	0.36
Nov	138	12.8	48.2	0.37
Dec	133	13.0	50.8	**0.40**

achieve this is to use a 24 V @ 200 Ah Lithium Iron Phosphate battery with a built-in battery management system.

8.4.2.4 Determination of Array Size

The array size can be determined by first calculating the energy that must be supplied to the batteries by the array. Since the connected load on the batteries is 1,724 Wh, and since losses of 5% can be expected for battery charging/discharging and an additional 2% for wiring losses, a total of $1{,}724 \div 0.98 \div 0.95 = 1{,}852$ Wh must be supplied to the batteries by the array each winter day. For summer days, $1300 \div 0.98 \div 0.95 = 1{,}396$ Wh/day will be needed.

Again, the PVWatts version of NREL SAM can be used to determine the array size. Assuming the inverter efficiency of the SAM simulation to be the same as the inverter efficiency of the lighting system, then the resulting AC kWh of the SAM simulation will be essentially the same as the lighting system DC kWh input to the batteries. As a result, an array is needed that will supply 1.852 kWh/day in winter and 1.396 kWh/day in summer. (Note the change from Wh to kWh.) If the array is tilted at latitude + 15° to optimize winter performance, Table 8.4 shows the resulting monthly kWh/kW delivered to the batteries along with the maximum operation time each month, the monthly load kWh and the corresponding array size needed to meet the energy needs of each light fixture. Note that the maximum array size needed is 400 W for December, the month with the least sun and the most nighttime hours, which should come as no surprise.

If a typical 120 half-cell module rated at 400 W, with $V_{OC} = 41.2$ V, $V_m = 34.2$ V, $I_{SC} = 12.28$ A and $I_m = 11.7$ A is selected, a single module may be adequate. Note, however, that if a 48 V system is desired, this module will not have adequate V_m for charging the batteries unless either a DC:DC converter, a 96-cell module or two smaller wattage modules are connected in series are used to increase the array output voltage to an adequate charging voltage. However, this module voltage should be adequate for a charge controller with a 24-V output.

8.4.2.5 Charge Controller and Inverter Selection

This system will use an MPPT charge controller and a pure sine inverter, so the inverter input voltage needs to be consistent with the charge controller output voltage. A wide variety of both devices is available, so if a 24-V inverter is used, the charge controller will need to supply 400W÷24V = 16.7 A. MPPT charge controllers with current outputs in the 10 to 40 A range are readily available. Pure sine inverters that will supply the necessary fixture power are also readily available, but if the inverter will also be required to shut off the light if the batteries discharge too far, then an inverter with this feature will need to be selected. In fact, if it can be verified that the harmonic content of a non-pure-sine inverter will not shorten the life of any of the other system components, it may even be possible to use one of these less expensive devices.

8.4.2.6 Final System Schematic

At this point, all major components have been selected, so the next step is selecting all wire sizes and any switches or controllers that may be needed for proper system operation. The method here is the same as it has been in all previous design examples: calculate all deratings due to temperature and then check voltage drops.

If the module is operated at STC, then V_m = 34.2 V and I_m = 11.7 A for the suggested module choice. For a distance of 25 ft (7.62 m) between the array and the charge controller, #10 Cu wire will have a 2.12% voltage drop at the full module current rating. Since almost all the time the module will be operating below rated output, this should be adequate wire size. The ampacity of the wire must be at least 156% of I_{SC} of the modules, or 1.56×12.28 = 19.2 A. A temperature derating will not be needed for this wire run since it will not be run over a roof space. Thus, in this case, #10 is more than adequate with its 35 A rating for 75°C insulation.

Next, appropriately-sized fuses and switches need to be selected for the input and output of the charge controller and also for the inverter. Figure 8.6 shows a schematic diagram of the system. Note that it includes a photo-control to turn on the light when it is dark outside. Finally, it needs to be noted that the inverter or the charge controller must also be capable of shutting off the light if the battery voltage drops too low. In this circuit, the 20 A fuse is provided at the charge controller input to protect the PV module from excessive backfeed, the only possible origin of which might be the batteries. As a practical matter, both the 20 A fuse and accompanying switch and the 30 A fuse and accompanying switch can be replaced by 20 A circuit breakers, since DC circuit breakers with voltage ratings less than 150 V are relatively inexpensive. The circuit breakers then serve dual roles as overcurrent protection and disconnect. Of course, if circuit breakers are used, a suitable waterproof enclosure will need to be found that will have a means of mounting them.

Stand-Alone PV Systems

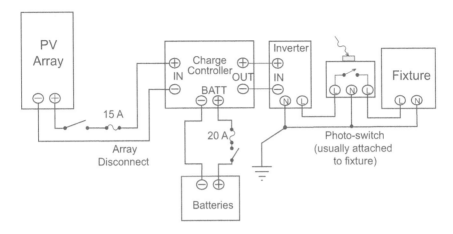

FIGURE 8.6 Schematic diagram of dc parking lot lighting system.

Depending upon the inverter, the array may or may not be required to be grounded. All metal parts, including the mounting pole if it is metal, will also be connected to the ground, making sure that if the array is grounded, the grounded conductor will be connected to the ground at only one point. And, in fact, after all the previous inverter discussions that resulted from the fixture being rated at 120 V AC, it is also possible that a custom fixture could be designed to operate on DC instead of AC, thus eliminating the inverter, since LEDs ultimately operate on about 4 V DC.

8.4.2.7 Structural Comment

Although the optical and electrical design requirements have been met for this system, it is important to note that it will also be necessary to design a sufficiently robust 28-ft (8.53 m) pole to hold all the system components. Since the design wind speed for Miami is 175 mph (282 km/h), this may be a non-trivial exercise. The actual design of such a structure is beyond the scope of this text, and, as such, should be done by a structural engineer.

In the event that the structure should prove to be too expensive, the electrical engineer will be confronted with a redesign of the system. This process may involve exploring the use of more fixtures on shorter poles, shorter lamp operating times, or a compromise on the light level. The point here is that the engineer should not be discouraged if the first try does not work. Often the development of an acceptable solution to a problem will involve a number of redesigns.

8.5 A PV-POWERED MOUNTAIN CABIN

8.5.1 INTRODUCTION

Figure 8.7 shows the floor plan of the cabin, located west of Boulder, CO, to be outfitted with PV-powered electrical loads. It is the same cabin used in previous editions of this book, but two significant additional loads have been added, as will be noted.

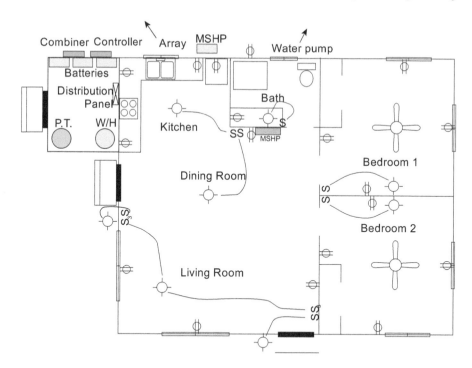

FIGURE 8.7 Floor plan of mountain cabin.

The cabin of previous editions had a kitchen light, a dining room light, a light in each bedroom, a bathroom light, a living room light, a motion sensor light outside each door, a ceiling fan in each bedroom, an entertainment center, a deep-well-water pump, an energy star 16 ft^3 (0.4531 m^3) refrigerator/freezer, a microwave oven plus two additional appliance circuits that will see minimal use unless one day this occupancy is connected to a utility supply. The two new loads are a hybrid electric water heater and a small mini-split heat pump for heating and air conditioning.

The PV array will be located approximately 10 ft (3 m) from the back of the cabin as shown. The 200-ft-deep (61 m) well will be located 30 ft (9.14 m) from the cabin in the same relative direction. The pump will be a submersible unit, so the total distance to the pump will be 230 ft (70.1 m). Most loads will operate at 120 VAC, but since 240 V microinverters will be used in this design, a neutral will need to be derived to operate the 120 V loads. Using 2024 technology, a matching controller will be needed to provide the necessary control of microinverters and batteries to be sure the microinverters operate when the sun is shining and the batteries do not become either overcharged or overdischarged on a daily basis.

The system loads, with the exception of the refrigerator and the heat pump, will operate three days per week to accommodate weekend use over long weekends. The refrigerator will be left on continuously to prevent inadvertent odor build-up in the event it gets too warm inside the unit. At least this is the stated reason. It is highly suspected that the real reason is to have cold beer ready as soon as someone arrives at the cabin.

Stand-Alone PV Systems 267

The heat pump will be left on for seven days per week for several reasons. First of all, it will prevent anything inside from freezing in the winter or getting too hot in the summer. Second, since it and the refrigerator will be the only operating loads every day, the power and energy produced by the PV array those days may otherwise go to waste.

Since the *NEC* requires that a building be wired as though it were to be connected to the grid, the cabin will be wired for all 120 or 240 V loads. The PV array will be rack mounted at ground level, so the feasibility of seasonal adjustment of array tilt will be explored as a means of optimizing system performance. Also, maintenance-free, lithium energy storage units will be used as a part of keeping system maintenance to an absolute minimum.

8.5.2 Load Determination

At this point, it should be abundantly clear that the only reasonable choice for the refrigerator and other loads will be high-efficiency units. The refrigerator that will be used has separate freezer and refrigerator compartments and is rated at 306 kWh/yr, which equates to 0.84 kWh/day. Assuming six hours of compressor operation per day equates to a power consumption of 140 W.

For the water pump, it is first necessary to determine the water needs for the cabin. This determination can be quite a wild guess unless the uses of the water are reasonably well defined. The water will be heated by a water heating heat pump (WHHP), which is also known as a hybrid water heater, but it is still reasonable to assume that showers will not be too lengthy. A reasonable estimate for water usage might be 30 gal (113 l) per day per person, with an anticipated occupancy of four persons, assuming no lawn to water and no cars to wash. This amounts to a total water requirement of 120 gal (364 l) per day. If more people use the cabin, it may be necessary to do some water rationing or maybe just backing off on some other electrical load. In the worst case, where drinking is the only use of the water, 120 people would still have a gallon (3.785 l) each per day for drinking over the three-day weekend. It is assumed that less water is used in winter, spring and fall than the 120 gal per day of summer use.

The next step is to go to the manufacturers' catalogs or to the internet or to both. For example [10], lists a number of different water pumps. For deep wells, as noted in the earlier example, normally a submersible pump is the best choice. The first check to make on a pump to be used in a system with battery storage is whether it can pump the needed water in either a day or a week. If the needed water can be pumped in a day, then a relatively small pressure tank may be needed. The pumping power will depend on the flow rate and the lift distance. If the top of the storage tank is 6 ft (1.82 m) above ground and the tank is pressurized to 30 psi (69 ft of head, 207 kPa), and, allowing for 5% piping losses, the equivalent lift for the pump will be (200+10+69)×1.05 = 293 ft (89 m).

In Section 8.3 it was observed that it is generally less costly to use a small pump that pumps over a longer period of time than a larger pump that pumps the needed water in a short time. Of the many pumps on the market, the AC pumps tend to have higher horsepower than the DC pumps. To pump 120 gal/day to a height of 293 ft with a 0.5 hp pump operating at 40% wire to water efficiency, it can be shown (Problem 8.11) that it will take 43.6 minutes per day and the power consumption of 240-V submersible pump will be 380 W. Thus, the pump current will be 1.98 A, assuming a power factor of 0.8, and the daily energy consumption when the cabin is occupied will be 0.276 kWh. A 20 gal (75 l) pressure tank is a reasonable choice to protect the pump from short cycling. When the pressure in the tank reaches a predetermined value, the pump shuts off and does not start again until the pressure drops below another predetermined value, thus creating a hysteresis effect delay between the pump start and stop times.

Next in line for energy use estimation is the hybrid water heater. By using a reverse refrigeration cycle to extract heat from the air, similar to an air-to-air or water-to-air heat pump for air conditioning and heating, up to three additional units of heat can be added to the water for each unit of electrical heat added. Thus, one unit of electrical energy will yield four units of water heating energy. Since a conventional electric water heater adds 3413 BTUs/hr/kW, the WHHP can add up to 13,652 BTUs/h/kW. Figure 8.8 illustrates the heat pump concept.

For one brand of hybrid water heaters, the 40-gallon unit operating in heat pump mode will add 4800 BTUs/hr to the water [11]. Since 1 gallon of water weighs 8.3 lb (3.765 kg), 40 gallons will weigh 332 lb (150.6 kg). Thus, the water will be heated at the rate of (4800 BTUs/hr) ÷ 332 lb = 14.46 BTUs/lb/hr. Thus, since 1 BTU will heat 1 lb of water 1°F, 14.46 BTU/lb/hr will raise the temperature of 1 lb of water to 14.46°F (8.033°C). In other words, the heat pump heats the 40 gal at a rate of 14.46°F per hour. If the incoming water is at 55°F, then adding 75°F to raise the temperature of the 40 gal to 130°F, will take 75 ÷ 14.46 = 5.2 hr. For each gallon (3.79 l) of hot water used, it will take 5.2 ÷ 40 = 0.13 hr = 7.8 min to replace it.

While the hot water may be the concern of the cabin occupants, the concern of the design engineer is how to determine the electrical power required to operate the heat pump. For this heat pump, the ratio of heat out to electricity is 3.75. Thus, for an output of 4800 BTUs/hr of heat, it will take an electrical input of 4800 ÷ 3.75 = 1280

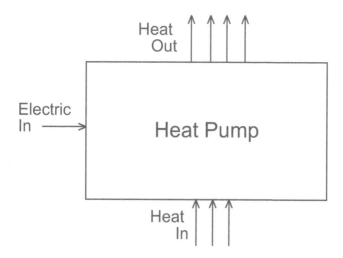

FIGURE 8.8 Heat pump energy flow diagram.

Stand-Alone PV Systems

BTUs/hr of electrical equivalent. Thus, since 1000 W equates to 3413 BTUs/hr, 375 W will deliver the required 4800 BTUs/hr.

As an interesting coincidence, for four people, a 40-gal hybrid water heater is recommended. Even though in the heat pump mode, only 375 W is required from the electrical supply, the unit plugs into a 30 A, 240 V outlet because it has other operation modes, including 5500 W (18,772 BTU/hr) of straight electric resistance heating. In the event that the occupants prefer faster heating of the water at a particular time, it can be heated at 26.7% of the heat pump mode efficiency, by sacrificing some kWh that might otherwise have been used for another load. On a daily basis, assuming each occupant uses 5 gal of hot water, this equates to 12 hours of operation of the heat pump. Thus, $0.375 \times 12 = 4.5$ kWh/day, provided the occupants are willing to wait for their hot water.

A relatively small (9000 BTUs/hr, SEER 23) mini-split, ductless heat pump will be used for space heating and cooling [12]. Noting that the SEER of a unit is defined to be the BTU/watt-hr ratio of the unit, for an SEER of 23 and 9000 BTUs/hr, the power consumption will be $9000 \div 23 = 391$ A. Assuming a duty cycle of 50% on a hot or cold day yields a daily energy consumption of 0.391 kW × 12 hr = 4.69 kWh/day. Note that it is expected that this level of energy consumption will apply to less than 40% of the time the unit is occupied, but perhaps for longer daily times when the unit is unoccupied if it is set to prevent the indoor temperature from either freezing or getting too hot. Of course, when the unit is not occupied, most of the other loads will not be in use, but there will still be daily sunshine to keep the PV system operating, which will presumably provide enough energy to operate the refrigerator and heat pump. It is also possible that in the future, the cabin will be equipped with a heat recovery ventilation unit for air exchange, but it is also anticipated that if the HRV is installed, the cabin energy consumption will decrease slightly.

The cabin loads are listed in Table 8.5. Allowing for an inverter efficiency of 96% and other losses of 4%, the actual loads are divided by 0.92 to determine the load on the batteries. Notice that in this example, the monthly variations in loads are accounted for. For example, the ceiling fans will be used more in the summer and the lights will be used more in the winter. The monthly variations in loads are important to use when determining the array tilt.

Note that the 120 V branch circuit loads are computed at the values required by *NEC* 690.10, even though the total power and energy used will be limited by the array size, the inverter size and the battery size but, most importantly, by the habits of the occupants and the loads of interest, such as computers, cell phone chargers, etc. The idea is that one day when the grid is everywhere, some future owner may elect to connect to the grid. For the moment, however, *NEC* 710.15(A) and (B) will apply while the system is a stand-alone system. In other words, the inverter(s) will not be required to simultaneously power all loads. Per *NEC*, each duplex outlet, except in the kitchen or bathroom, must be rated at 180 VA, so the total load assigned to the duplex outlets in the two bedrooms and the living/dining room area is 2160 W. The load assigned to the kitchen circuit is 1500 W as is the load assigned to the bathroom (hair dryers). Even though all the lighting outlets will have no more than a 20 W LED lamp, they still must be rated at 90 W each. But to calculate the load on the inverters and the storage, the expected off-grid loads

TABLE 8.5
Summary of Monthly Variation in Daily kWh Loads for Mountain Cabin

Load Description	P watts	Days/week	Nov–Feb		Mar	
			Hr/week	kWh/week	Hr/week	kWh/week
Kit Light	20	3	12.0	0.24	10.5	0.21
BR 1 Light	10	3	6.0	0.06	4.5	0.05
BR 2 Light	10	3	3.0	0.03	3.0	0.03
LR Light	10	3	15.0	0.15	12.0	0.12
Outdoor Lights	10	3	3.0	0.03	1.5	0.02
DR Light	15	3	12.0	0.18	9.0	0.14
Bath Light	10	3	6.0	0.06	6.0	0.06
Refrigerator	152	7	42.0	6.38	42.0	6.38
Water Pump	380	3	2.2	0.84	2.2	0.84
BR 1 Fan	32	3	0	0.00	0.0	0.00
BR 2 Fan	32	3	0	0.00	0.0	0.00
Receptacles	2500	3	2.0	5.00	1.8	4.38
Hybrid W/H	375	3	36.0	13.50	36.0	13.50
Heat Pump	391	7	84.0	32.84	56.0	21.90
Totals	3947			59.3		47.6
Corrected Loads				64.5		51.7

Load Description	Apr, Oct		May, Sep		Jun–Aug	
	Hr/week	kWh/week	Hr/week	kWh/week	Hr/week	kWh/week
Kit Light	10.5	0.21	9.0	0.18	7.5	0.15
BR 1 Light	4.5	0.05	3.0	0.03	3.0	0.03
BR 2 Light	3.0	0.03	3.0	0.03	3.0	0.03
LR Light	9.0	0.09	6.0	0.06	6.0	0.06
Outdoor Lights	1.5	0.02	1.5	0.02	1.5	0.02
DR Light	7.5	0.11	6.0	0.09	4.5	0.07
Bath Light	5.0	0.05	4.0	0.04	3.0	0.03
Refrigerator	42.0	6.38	42.0	6.38	42.0	6.38
Water Pump	2.2	0.84	2.2	0.84	2.2	0.84
BR 1 Fan	3.0	0.10	8.0	0.26	24.0	0.77
BR 2 Fan	3.0	0.10	8.0	0.26	24.0	0.77
Receptacles	1.5	3.75	1.3	3.13	1.0	2.50
Hybrid W/H	36.0	13.50	36.0	13.50	36.0	13.50
Heat Pump	7.0	2.74	21.0	8.21	84.0	32.84
Totals		28.0		33.0		58.0
Corrected Loads		30.4		35.9		63.0

on these circuits are used in Table 8.5, with the understanding that the circuits are wired adequately to handle the loads as required by *NEC* 210 and 220.

For each load, the average kWh/week if the load operates for n hours per day and d days per week is determined by

Stand-Alone PV Systems 271

$$\text{kWh/week} = Pnd, \quad (8.2)$$

where P is the load power in kW. All loads operate at 120 VAC except for the hybrid water heater and the well pump, which operate at 240 VAC. In fact, the total power in column 1 suggests that the total rated output power of the microinverters used should be at least 3935 W. Note that the power assigned to the receptacles is unusually high, but the hourly use of the receptacles is estimated to be comparatively low to correspond to a few minutes of microwave or hair dryer time and longer times for other, relatively low power plug-in devices, such as a TV or a computer or a small LED reading lamp. The entries in Table 8.5 are based on three days per week of usage for all loads except the refrigerator, which remains on for seven days per week, the water pump, which will pump as needed, to keep the pressure tank pressurized and the heat pump that will remain on to prevent freezing conditions inside the cabin. The pump is controlled by a pressure switch in the storage tank.

It is important to note that the average kWh/week represents the weekly average, taking into account that the water pump, the mini-split heat pump and the refrigerator may run seven days per week, but the other loads are only operational for three days per week while the cabin is occupied. While the weekly average provides an interesting set of numbers that relate to actual usage, the daily average each month is what is needed for the calculation of array size and necessary storage. Daily averages can be determined by simply dividing the weekly kWh consumption of any of the loads by the number of days per week of operation of the load.

The next step is to compute the system-corrected loads, which will be the loads on the batteries. As noted earlier in this section, the corrected kWh loads for each day, week or month are obtained by dividing the load kWh by 0.92. The corrected weekly loads for each month are also shown in Table 8.5.

8.5.3 Array Sizing, Array Tilt and Microinverter Selection

Array sizing involves determining the optimum system design current by computing the design kWh for each month of the year at each of three tilt angles, then choosing the tilt angle that yields the lowest design kWh for each month. Since the array size should be capable of meeting the daily loads when the cabin is occupied to avoid additional expensive storage, it is first advisable to calculate the anticipated daily loads and then multiply them by the number of days in each month to determine the monthly loads. At this point, the best array size and tilt can be determined for each month. Thus, it is useful to develop a table of actual daily loads for occupied days and then calculate array and storage sizes. Table 8.6 shows the conversion of weekly loads to daily loads of the cabin while it is occupied.

The next step is to assume the daily loads to apply to every day of the month and use this information to calculate the array size necessary to produce the monthly kWh consumption of the loads for each month of the year.

Table 8.7 shows the corrected load in kWh for each month of the year based on the monthly Corrected load information that is shown in Table 8.6.

TABLE 8.6
Conversion of Weekly Loads to Daily Loads During Times of Occupancy

		Nov–Feb		Mar	
Load Description	P watts	Daily hr	Daily kWh	Daily hr	Daily kWh
Kit Light	20	4	0.08	3.50	0.07
BR 1 Light	10	2	0.02	1.50	0.02
BR 2 Light	10	1	0.01	1.00	0.01
LR Light	10	5	0.05	4.00	0.04
Outdoor Lights	10	1	0.01	0.50	0.01
DR Light	15	4	0.06	3.00	0.05
Bath Light	10	2	0.02	2.00	0.02
Refrigerator	152	6	0.91	6.00	0.91
Water Pump	380	0.73	0.28	0.73	0.28
BR 1 Fan	32	0	0.00	0.00	0.00
BR 2 Fan	32	0	0.00	0.00	0.00
Receptacles	2500	0.7	1.67	0.58	1.46
Hybrid W/H	375	12	4.50	12.00	4.50
Heat Pump	391	12	4.69	8.00	3.13
TOTALS	3692		12.3		10.5
Corrected Loads			13.4		11.4

	Apr, Oct		May, Sep		Jun-Aug	
Load Description	Daily hr	Daily kWh	Daily hr	Daily kWh	Daily hr	Daily kWh
Kit Light	3.50	0.07	3.00	0.06	2.50	0.05
BR 1 Light	1.50	0.02	1.00	0.01	1.00	0.01
BR 2 Light	1.00	0.01	1.00	0.01	1.00	0.01
LR Light	3.00	0.03	2.00	0.02	2.00	0.02
Outdoor Lights	0.50	0.01	0.50	0.01	0.50	0.01
DR Light	2.50	0.04	2.00	0.03	1.50	0.02
Bath Light	1.67	0.02	1.33	0.01	1.00	0.01
Refrigerator	6.00	0.91	6.00	0.91	6.00	0.91
Water Pump	0.73	0.28	0.73	0.28	0.73	0.28
BR 1 Fan	1.00	0.03	2.67	0.09	8.00	0.26
BR 2 Fan	1.00	0.03	2.67	0.09	8.00	0.26
Receptacles	0.50	1.25	0.42	1.04	0.33	0.83
Hybrid W/H	12.00	4.50	12.00	4.50	12.00	4.50
Heat Pump	1.00	0.39	3.00	1.17	12.00	4.69
Totals		7.58		8.22		11.86
Corrected Loads		8.2		8.9		12.9

The monthly kWh/kW information for the three different array tilts is the result of simulating the performance of a 1.0 kW array using the PV Watts model of SAM with a DC/AC ratio of 1.38, an inverter efficiency of 97% and overall system losses of 10%. Since the monthly corrected load requirements are available, the necessary array size to achieve the necessary monthly system kWh production is found by

Stand-Alone PV Systems

dividing the monthly corrected load by the system kWh/kW for each tilt and for each month of the year. The array size and tilt that will meet the needs of the system for each month are then found by comparing the results for each month at each of the three tilt angles and selecting the tilt that results in the smallest array size, as shown in Table 8.7. In this table, the optimal tilt angle for the array requires the minimum array size.

The table shows that for the months of April through August, the latitude − 15° tilt (25°) gives optimum performance. For example, in July, to meet the system's daily load requirements, a 1.71 kW array is needed. A tilt of latitude (40°) yields the best array performance for March, September, and October and a tilt of latitude + 15° gives the best performance for November through February. Thus, to ensure that the needs of all months can be met by the array, one simply selects the worst case of these best cases, that is, the month that requires the maximum array size. In this case, the month is December, which requires a 2.25 kW array. Since the simulation numbers are a result of an "average" year, increasing the array size to 2.4 kW gives some allowance for the possible less-than-average year. This can easily be achieved with six 400 W modules and accompanying microinverters. Since the DC/AC ratios of the simulations were based upon the use of 290 W microinverters and 400 W modules, it is now a simple matter to adopt the 290 W microinverter for the system. The constraint on the microinverter is simply its compatibility with the modules, the system combiner, the system controller and the system energy storage. In terms of the modules, all that is needed is to use 60, 66, or 72-cell modules or 120, 132, or 144 half-cell modules with rated power somewhere between 320 and 440 W to match the module outputs. The 400 W module fits nicely within this range.

The inverters used in the simulation are rated at 290 W, with an output voltage of 240 V and an output current rating of 1.21 A. The maximum DC input voltage is 60 V, and the maximum DC input current is 15 A. The modules have V_{OC} = 47.2

TABLE 8.7
Determination of Necessary Array kWh Production and Array Tilt Angle

Month	CorrLoad kWh/mo	Latitude−15° kWh/ kW	kW	Latitude kWh/ kW	kW	Latitude+15° kWh/ kW	kW
Jan	283	112	2.53	123	2.30	128	**2.21**
Feb	256	120	2.13	127	2.01	129	**1.98**
Mar	227	150	1.51	153	**1.48**	149	1.52
Apr	128	150	**0.85**	146	0.88	137	0.93
May	156	157	**1.00**	145	1.08	128	1.22
Jun	268	162	**1.65**	147	1.82	124	2.16
Jul	277	162	**1.71**	149	1.86	126	2.20
Aug	277	155	**1.79**	149	1.86	133	2.08
Sep	151	147	1.03	147	**1.03**	141	1.07
Oct	132	133	0.99	140	**0.94**	139	0.95
Nov	274	122	2.25	133	2.06	137	2.00
Dec	283	110	2.57	121	2.34	126	**2.25**

V, I_{SC} = 10.9 A, V_{mp} = 38.7 V and I_{mp} = 10.34 A. Thus, even if it gets cold enough to increase V_{OC} by 20%, V_{OC} will be 56.6 V, which is well within the allowed input voltage limits.

Before moving on to energy storage selection, it can be noted that although the six modules will need only a single string, since the maximum allowed string current is 16 A and 6 × 1.21 A = 7.26 A, it is still convenient to use a compatible combiner box with communications that are compatible with both the microinverters and the system controller. The system controller will need to be compatible with the microinverters, the combiner and the energy storage units. Since the system controller will be the same as what might be used for a grid-connected system operating as a battery backup system when the grid is down, the selection of system controller will have only a small number of choices.

In backup mode, the system controller will provide synchronization for the grid-forming microinverters, an autotransformer to provide a neutral for the system and regulation of the storage system output to prevent overcharging or overdischarging of the energy storage system. Communication with the microinverters will be an essential part of the limiting process. One day if a grid suddenly appears, if it is connected to the system controller grid connections, the system controller does its duty as it did in previous system designs in Chapter 7.

At this point, the system designer has all the information needed to determine the energy storage needs.

8.5.4 Energy Storage Selection

In this case, the "batteries" will have their own internal electronics to convert the AC from the array to DC, to control charging and discharging rates, to control charging levels, to ensure uniform cell charging and to convert battery DC to AC with grid-forming inverter(s) to provide a stable voltage and frequency for the output utility-grade sine wave to simulate the utility for the array microinverters. In addition, the energy storage electronics will provide the communications and control necessary to manage the battery charging by increasing or decreasing the array output as may be necessary. The batteries will provide a 240 V, 60 Hz output voltage. What still needs to be determined is the total kWh of storage needed and the maximum required rate of discharge in terms of maximum storage output power and current.

Table 8.8 shows the calculation of required storage capacity on a daily basis. It is assumed that a Lithium Iron Phosphate unit will be chosen with a maximum depth of discharge of 90%. To make selection simpler, the manufacturer simply specifies the useable amount of storage. Once the total storage capacity is determined, a selection of a unit or units that will have enough storage to meet the system requirements can be made. In this case, an LFP system with built-in battery management is a reasonable choice, as indicated by the ratio of total capacity to battery capacity. If a 10-kWh unit is selected, then three of them would be needed to achieve the 26.6 kWh shown in Table 8.8 for two days of storage during the winter months. However, before settling on this value, it might be interesting and useful to estimate exactly WHEN during the day will all these kWh be used.

Stand-Alone PV Systems

Lights will be used when the sun is down, but the daily lighting energy never exceeds 0.3 kWh over the year. And, of course, this energy needs to come from the storage system at night. The largest load is the heat pump, with 6.26 kWh/day in the winter and 6.26 kWh/day in the summer. In the summer, a significant fraction of these kWh are used during the day, given that the days have more daylight time and the temperatures are hotter during daylight hours. It is not unreasonable to estimate that 70% of these 4.69 kWh (Table 8.6) will be used during the day, thus removing 3.28 kWh from the nighttime load.

In addition, if the hybrid water heater is controlled by a load controller, it can be limited to daytime use, removing another 4.5 kWh/day from the nighttime load. Thus, limiting these two loads to mostly daytime use can reduce the summer daily storage needs by nearly 7.8 kWh, resulting in a revised storage estimate of 5.0 kWh per day for the summer months. If two days of storage is chosen to take into account cloudy days, the total summer storage increases to 10.0 kWh overall.

On the other hand, winter estimates are a bit different, since it can get pretty cold at night. In this case, it is probably reasonable to estimate nighttime energy use for the heat pump as 70% of the daily total, which means the winter reduction in daily storage requirements for the heat pump will be 1.4 kWh. However, if similar use limits are imposed for the water heater, then the storage load can be decreased by 4.5 kWh for the water heater, resulting in a total reduction of 5.9 kWh for the two loads. The remaining winter storage load is thus reduced to 7.4 kWh. For two days of storage, this adds up to 14.8 kWh. A look at the other months of the year shows that with similar adjustments for nighttime loads, 14.8 kWh will be adequate for two days of storage.

Another item that deserves a closer look is that if the array tilt is adjustable, the summer daily array production will exceed the winter daily energy needs by 3.3 kWh. In fact, even if the tilt of the array is not adjusted for summer, the array output will still exceed the daily needs by 2.0 kWh. By increasing daily array production for all seasons, more energy will be available for charging the energy storage system, even on cloudy days.

Finally, all that needs to be done is to find the best way to come close to 14.8 kWh of storage with compatible storage components. Most small storage systems can be built with increments of 3, 5, 10 or 13 kWh. In terms of using the same manufacturer to increase the likelihood that all system components will be compatible, one choice

TABLE 8.8
Determination of System Battery Capacity Requirements

Month	Jan	Feb	Mar	Apr	May	Jun
kWh/day	13.3	13.3	11.3	8.2	8.9	12.8
Two days' storage	26.6	26.6	22.6	16.4	17.8	25.6
Month	Jul	Aug	Sep	Oct	Nov	Dec
kWh/day	12.8	12.8	8.9	8.2	13.3	13.3
Two days' storage	25.6	25.6	17.8	16.4	26.6	26.6

would be to use three 5 kWh units in parallel, each of which is capable of delivering 16 A @ 240 V, or 3.84 kVA. This would result in a 15 kWh storage system with the capability of simultaneously delivering 48 A. Of course, it is unlikely that the storage would supply 48 A other than instantaneously, since this would use up all the stored energy in 1.3 hr instead of two days. In any case, 11.5 kW could be delivered for a very short time, if needed.

8.5.5 Excess Electrical Production

It is interesting to look at the amount of excess kWh produced by the system on a month-by-month basis, since the array has been selected to produce adequate energy during the worst month. By multiplying the monthly kWh/kW by the array size in kW (using the 2400 W array) the available load kWh/mo can be obtained. Then simply subtract the corrected monthly kWh from the available monthly kWh to obtain the excess monthly kWh. The results are shown in Table 8.9 as excess system daily kWh. Note that technically these numbers apply to the three days a week the cabin is expected to be occupied, when, in fact, for the remaining four days, only the heat pump and refrigerator will be operating. Neither the water pump nor the water heater will be turned off, but if nobody is in the cabin, the water heater will use minimal energy just to overcome heat loss from a very well-insulated tank and the well pump might occasionally turn on if the pressure in the pressure tank slowly decreases. As a result, during these four days, even more excess electricity will be available, since on a daily basis, a maximum of just over 5 kWh will be used.

The next logical challenge for the designer is to figure out what to do with this extra electricity. Comparison with the kWh requirements of the cabin loads shows that the spring and fall excess is approximately equal to the weekly receptacle kWh. One idea might be to use a vacuum cleaner in spring and fall and a broom in winter and summer. Then there is the possibility that more hot coffee, tea or other beverages might be needed during cold winter months via an electric teapot or coffee pot for heating the water. Another possibility would be to pump extra water. It is left as an exercise for the reader to determine the additional water that could be pumped with the leftover electricity (Problem 8.10).

Even the heat pump will likely use less energy during unoccupied times, since there will be no need to heat the space to much more than 50°F or to cool below 78°F. However, estimating the heat gain or loss of the cabin on hot or cold days is a bit beyond the scope of this design exercise. It can be noted that several computer programs are available from either the Florida Solar Energy Center or the National Renewable Energy Laboratories that will facilitate these building energy use calculations [13, 14].

TABLE 8.9
Average Daily Excess kWh Produced by the Selected Array for the Cabin

Mo	Jan	Feb	Mar	Apr	May	Jun
kWh	0.8	1.9	4.5	7.7	7.1	4.0
Mo	Jul	Aug	Sep	Oct	Nov	Dec
kWh	3.6	3.1	6.7	6.6	1.8	0.6

Stand-Alone PV Systems 277

It is up to the engineer to decide which use, if any, will make economic sense for this extra electricity. The determination will ultimately be dependent upon the specific preferences of the cabin owner. Who knows, since it is turning out to be such a nice cabin, maybe the owner will want to spend four or more days per week at the cabin during the months with the greatest excess production.

8.5.6 Balance of System Component Selection

8.5.6.1 Wire, Circuit Breaker and Switch Selection

The nice thing about an AC cabin is that no extraordinary effort is required for sizing the wiring to most of the loads, as long as they are no more than 100 ft (30.5 m) from the distribution panel. In this case, everything is close with the exception of the well pump, for which a voltage drop calculation will be needed. Table 8.10 summarizes the maximum system component currents as well as one possible selection of circuits along with wire sizing and sizing of fuses. In Table 8.10, the *NEC* currents incorporate the appropriate 125% multipliers for continuous duty. It turns out that all the calculations in this table lead to wire sizing based on necessary ampacity. Note that the cabin lights and outlets are wired with #14 (2.08 mm^2) wire as would be done in a grid-connected residence. The kitchen circuits are 20 A circuits that are wired with #12 (3.31 mm^2). The 2020 *NEC* requires arc-fault circuit breakers on all the 15 A circuits except the well pump and ground-fault circuit interrupters on the kitchen, bathroom and outside receptacles.

Switches should have the same current rating as the fuses or circuit breakers along with an adequate voltage rating for the purpose. For lighting, the switches are relatively inexpensive at most home improvement stores. Fans come with fan controllers rated for the purpose. The pump will need to be controlled with a pressure switch in the pressure tank and possibly with a time switch to limit pumping to certain hours. These switches need to be adequately rated to carry the pump motor current, which is a comparatively small amount. Because the pump current is only 1.6 A, even though the distance from the pump to the distribution panel is 240 ft (73.15 m), the percent voltage drop in the pump wiring ends up at less than 1%.

8.5.6.2 Other Items

Other items that need to be included are an array mount, distribution panel, lightning arrestors, ground rod(s) and miscellaneous connectors and junction boxes. Most of these components, except perhaps those related to the storage system, will be used in the electrical system regardless of the source of electricity. Once wire length and size are known, wire cost can be calculated. For the cabin, a distribution panel with AC circuit breakers is used to supply individual circuits to the various loads, with some doubling up of loads on a single circuit as indicated in Table 8.10.

The array mount may take many shapes and forms, from commercially available units to homemade wooden frames. For the cabin, a commercial ground mount has been selected.

Figure 8.9 shows the cabin system electrical schematic diagram. Note that a separate set of 3 #12 (L_1, L_2, N) is included with the PV to controller wiring. These wires supply power from a 20 A CB in the controller to the communication module in the

TABLE 8.10
Summary of Circuits, Wiring and Overcurrent Protection for the Cabin

Wire location	Max A	NEC A[a]	Voltage	Length, ft	Wire Size	C.B. Size, A
Array to combiner	7.26	9.08	240	30	#12	20
Combiner to controller	7.26	9.08	240	6	#12	20
Storage to storage combiner	16	20	240	6	#12	20
Storage comb to controller	48	60	240	25	#6	60
Contr to Distribution Panel	100[b]	100	240	6	#4	100
Refrigerator	2.0	20	120	<100	#12	20
Water pump	1.6	2.0	240	230	#14	15
Kitchen receptacles	16	20	120	<100	#12	20
LR receptacles	12	15	120	<100	#14	15
BR 1 receptacles	12	15	120	<100	#14	15
BR 2 receptacles	12	15	120	<100	#14	15
Bath and outside receptacles	12	15	120	<100	#14	15
Lights and fans	12	15	120	<100	#14	15

[a] Required for ampacity calculations. All wire sizes are based on ampacity requirements.
[b] Minimum allowed NEC service size.

combiner. Note also that the occupancy is required to meet certain NEC requirements for grid-connected residences, including a minimum of 100 A main distribution panels as noted in Table 8.10. In effect, the cabin is grid-ready, since all that is needed is to install a utility meter and connect to the line side of the system controller, possibly with an outside main disconnect, depending upon local utility requirements.

8.6 SUMMARY OF DESIGN PROCEDURES

At this point, a relatively wide range of stand-alone PV systems have been designed in this chapter. An effort has been made to present alternate, common-sense design procedures to illustrate that no single methodology is necessarily the best one for any given problem. However, there are many commonalities among the designs, including the procedure for determining loads, the procedure for sizing energy storage and the procedure for sizing arrays. These procedures are summarized in the following listings. The designer should remember, however, that these procedures are intended to be used as guidelines. Specific design requirements may result in the need to modify these procedures. For example, the guidelines include both AC and DC loads. In most cases, loads will be only AC or only DC as in the previous examples.

Stand-Alone PV Systems

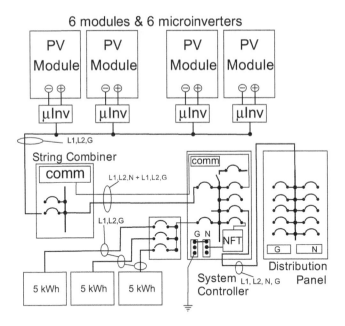

FIGURE 8.9 Mountain cabin electrical riser diagram.

Determination of Average Daily PV System Load

1. Identify all loads to be connected to the PV system.
2. For each load, determine its voltage, current, power and daily operating hours. For some loads, the operation may vary on a daily, monthly or seasonal basis. If so, this must be accounted for in calculating daily averages.
3. Separate AC loads from DC loads.
4. Determine the average daily kWh for each load from current and operating hours data. If operating hours differ from day to day during the week, the daily average over the week should be calculated. If average daily operating hours vary from month to month, then the load calculation may need to be determined for each month.
5. Add up the Ah for the DC loads, being sure all are at the same voltage.
6. If some DC loads are at a different voltage, which will require a DC-to-DC converter, then the converter input kWh for these loads needs to account for the conversion efficiency of the converter.
7. For AC loads, the DC input current to the inverter must be determined and the DC kWh is then determined from the DC input current. The DC input current is determined by equating the AC load power to the DC input power and then dividing it by the efficiency of the inverter.
8. Add the kWh for the DC loads to the kWh for the AC loads, and then divide by the wire efficiency factor and the battery efficiency factor to obtain the corrected average daily kWh for the total load.

9. The total AC load power will determine the required size of the inverter. Individual load powers will be needed to determine wire sizing to the loads. The total load current will be compared with the total array current when sizing the wire from the battery to the controller. The power lost in the wiring is determined by the product of the current and the voltage drop. Depending on the nature of the load, a drop in voltage may result in either a drop or a rise in current. Normally the current will drop slightly, but, in some motors with constant loads, the current will rise. However, for a voltage drop of only 2%, it is reasonable to assume the load current will remain essentially constant. With this assumption, the power loss can be approximated by the product of load current at nominal load voltage and the voltage drop. Hence, a 2% voltage drop implies a 2% power loss, or, in terms of Ah, a 2% Ah loss. This loss must be added to the PV array power output requirements. Thus, a 2% voltage drop implies a wiring efficiency of 98%, and the load Ah per day must be increased by dividing by 0.98 to account for the wiring loss.

Battery Selection Procedure

1. Determine the number of days of storage required, depending on whether the load will be noncritical or critical. The designer may use discretion for special cases such as weekend occupancy or seasonal variations. It is possible that different seasons may have different numbers of storage days.
2. Determine the amount of storage required in kWh. This is the product of the corrected kWh per day and the number of days of storage required. This amount may vary with season. If so, list all values.
3. Determine the allowable level of discharge. Divide the required kWh by the level of allowed discharge, expressed as a fraction. For example, using 80% of the total charge requires dividing by 0.8. This result is the total corrected kWh required for storage. Note that many of the newer batteries simply list the kWh available from the battery without mentioning the percent of charge or discharge.
4. Check to see whether an additional correction for the discharge rate will be needed. If so, apply this correction to the result obtained in (3). Again, since most modern storage elements are currently limited, this correction will not be needed.
5. Check to see whether a temperature correction factor is required. If so, apply this to the result of (3) or (4). Again, modern lithium systems have built-in temperature control.
6. Divide the final corrected battery capacity by the capacity of the chosen battery. The result may be rounded up or down, depending on the judgment of the system designer. Often the result can be rounded down, provided that the system receives at least some diffuse sunlight over prolonged overcast periods.

Stand-Alone PV Systems

Array Sizing and Tilt Procedure

1. For each tilt angle of interest, simulate the array performance for the system geographic location using the PVWatts version of the NREL SAM simulation program and a 1-kW array. Determine the required design array power in kW for each month of the year by dividing the corrected kWh load for the system each month by the monthly average kWh supplied by the 1-kW array at each array tilt angle.
2. Determine the worst-case (highest monthly) design array power for each tilt angle.
3. For a fixed mount, select the tilt angle that results in the lowest worst-case design array power.
4. If tracking mounts are considered, then determine the design array size in kW for the chosen trackers. Note that the one-axis tracker tracks from east to west and needs to have its design power checked for the three angles of fixed arrays.
5. Select a module that has rated output current and voltage at maximum power consistent with system needs.
6. Determine the number of modules in the array by dividing the required array power by the rated power of the selected module. Alternatively, pick a reasonable integer and divide the required array power by this integer to determine the necessary module power rating. Keep in mind that the number of modules in the series will be limited by the maximum value of V_{OC} of the series combination as evaluated at the lowest expected array operating temperature. In addition, the minimum number of modules in the series will be limited by the value of V_{mp} of the series combination as evaluated at the highest expected array operating temperature.

Homework Problems

8.1 For the fan example, indicate applications for which it would be desirable to have an oversized PV module and indicate when it would be satisfactory to have a smaller module that would produce a significantly lower fan speed at lower light levels. Can you envision an application for which the use of an MPPT/LCB would be advantageous?

8.2 How much power could be recovered at low light levels by using an MPPT on the fan of the first example in the text, assuming the MPPT is 95% efficient? Estimate the additional maximum power output that would be required of a PV array that would produce the same low-light-level power as the system with the MPPT. If the additional PV costs $1.50/watt (installed), how much could you spend on the MPPT to produce the same effect?

8.3 It is desired to pump 50 gpm from a depth of 100 ft, using a pump that is claimed to have a wire-to-water efficiency of 35%.
   a. Determine the horsepower the pump will need to develop to achieve this pumping rate.
   b. Determine the daily kWh needed to pump 3000 gal.
8.4 Using the module and charge controller of Section 8.4, determine the wire size between the array and the charge controller if the distance between the two is 25 ft and the voltage drop is to be limited to 2%.
8.5 Recalculate the battery size and the array size for the parking lot fixture example if the 158-W fixture is chosen.
8.6 Determine the lamp wattage required to obtain an illumination level of 50 f-c over a 100 ft^2 area if a fixture is used with a CU of 0.75 and 80% of the available light reaches the work surface, the rest being absorbed by walls and other items in the space. Assume a luminous efficacy of 120 lumens/watt.
8.7 Assume 10 days of operation are desired for a 20 kWh battery system, with a maximum depth of discharge of 90%. Then assume 12 consecutive days occur during which array output averages only 20% of the predicted amount generated on an average day. Estimate the state of charge of the battery system after the end of the 12th day if the load on the batteries is 2 kWh/day and the array is designed to generate 2 kWh/day on an average day.
8.8 Determine the wire-to-water efficiency of the well pump used for the cabin.
8.9 Research 20 gal pressure tanks and list on and off pressures for what you find. Are these pressures consistent with the choice of pump?
8.10 Determine the additional water that can be pumped with excess summer electricity produced by the cabin array for unoccupied periods in June, July and August.
8.11 Find a 9000 Btu/hr heat pump with a SEER > 30 and calculate the effect on the daily kW and weekly kWh load of the cabin.
8.12 Write a program or develop a spreadsheet that will size a pump and wire to the pump for a PV-powered pumping system if the pumping height and daily volume are known along with the distance from the array to the pump.
8.13 Design your own little off-grid hideaway. Specify your own loads, occupancy, peak sun and storage requirements, and determine the number of batteries and the number of modules you would need to implement the system. The system may be AC or DC or both and may use a string inverter, microinverters or no inverter at all. Then specify BOS components.
8.14 Explore the effect of adding additional storage to the size of the array needed for the mountain cabin.
8.15 Explore the effect of adding additional array power to the amount of storage needed for the mountain cabin.

REFERENCES

[1] Table of Head Loss vs Flow for Various Pipe Sizes: www.engineeringtoolbox.com/pressure-loss-plastic-pipes-d_404.html
[2] Comparison Curves for High Head and Medium Head Water Pumps: www.inyopools.com/blog/which-is-best-high-head-or-medium-head-pumps/
[3] National Weather Service Lubbock, TX Local Climate Data: www.weather.gov/lub/climate-klbb-pcpn
[4] System Advisor 2022 (SAM 2022.3.14), National Renewable Energy Laboratory, Golden, CO, 2022.
[5] *IES Lighting Handbook*, 10th ed., Illuminating Engineering Society of North America, New York, 2011.
[6] *9200 Lamp Catalog*, 22nd ed., GE Lighting, General Electric Co., 1995.
[7] *Lamp Specification and Application Guide*, Philips Lighting Co., Somerset, NJ, 1999. https://www.usa.lighting.philips.com/support/purchase/literature
[8] Information on LED Performance, April 2023: www.lumistrips.com/lumistrips-blog/led_efficacy_efficencty_explained/
[9] U.S. Naval Observatory Table of Sunrise/Sunset Moonrise . . .: http://aa.usno.navy.mil/data/docs/RS_OneYear.php
[10] Listing of 141 Different Well Pumps: www.waterpumpsdirect.com/pumps/240v-deep-well-pumps.html
[11] Specifications for Hybrid Water Heater: www.rheem.com/heatpumpwaterheater/
[12] Specifications for 9000 BTUs/hr Mini-Split Heat Pump: www.pioneerminisplit.com/products/
[13] Computer Software for Building Energy Analysis: Florida Solar Energy Center, Energy Gauge USA (2 versions).
[14] More Software for Building Energy Analysis: National Renewable Energy Laboratories, BEOpt.

SUGGESTED READING

Anderson, K., Slope-Aware Backtracking for Single-Axis Trackers: www.nrel.gov>docs
For Additional Information on Array Mounts: www.ironridge.com
NFPA 70 National Electrical Code, 2020 ed., National Fire Protection Association, Quincy, MA, 2019.
www.batteries4everything.com for battery information.
www.unirac.com for information on array mounts.

9 Economic and Environmental Considerations

9.1 INTRODUCTION

The costs of a PV system include acquisition costs, operating costs, maintenance costs and replacement costs. At the end of the life of a system, the system may have a salvage value or it may have a decommissioning cost. This chapter introduces the method of life cycle costing, which accounts for all costs associated with a system over its lifetime, taking into account the time value of money. Life cycle costing is used in the design of the PV system that will cost the least amount over its lifetime. Life cycle costing, in general, constitutes a sensible means for evaluating any purchase options when a decision has been made to purchase. Sometimes a potential PV system owner wants to know how long it will take for the system to pay for itself, and sometimes a potential PV system owner simply wants to know what can be expected from the system, independent of cost. For larger systems, to be able to compare the costs of electricity from different sources, the levelized cost of electricity (LCOE) method has been introduced. If the system includes storage, or if the system is only storage, the levelized cost of storage (LCOS) has been introduced. This chapter explores some of the economic considerations that enter the decision process when someone expresses an interest in owning a PV system.

If it is necessary to borrow money to purchase an item, the cost of the loan may also need to be incorporated into the total cost of a system. At this point in time, there are a number of lenders that will establish a payment system that allows a fraction of the savings in utility bills to be used as payment for the PV installation loan. This creates a system that immediately begins saving the owner the monthly difference between electrical savings and loan costs. If incentive programs are available in any of a number of forms, they may significantly affect the life cycle cost (LCC) or payback of a system. Thus, some attention will also be paid to the payback question as well as various forms of incentive programs currently in place in certain areas.

Another economic option is third-party ownership, in which a for-profit company offers to install and own a PV system on a non-profit property and then sell electricity from the system to the property owner at a preferred rate. As the cost of PV systems continues to fall and as tax incentives continue to be adopted, this option is becoming more attractive, since it involves minimal risk for the property owner or for the system owner, as long as the sun rises and as long as the equipment functions properly with high reliability. It is also possible that the contract with the third party might include a performance clause that would link the payments to the third party to the performance of the system.

Economic and Environmental Considerations   285

This chapter also introduces the concept of externalities. Externalities are costs that are not normally directly associated with an item. A few decades ago, acid rain, caused by sulfur emissions from smokestacks, attracted a lot of attention because it was causing damage to buildings and lakes. Today the culprits are CO_2 and other greenhouse gases that cause global climate change. Yet, the increasing costs to society from global climate change still are generally not paid for directly by the entity that generates the emissions. Externalities will be considered in more detail in the final section of this chapter.

9.2 LIFE CYCLE COSTING

9.2.1 THE TIME VALUE OF MONEY

The LCC of an item consists of the total cost of owning and operating an item over its lifetime. Some costs involved in the owning and operating of an item are incurred at the time of acquisition, and other costs are incurred at later times. To compare two similar items, which may have different costs at different times, it is convenient to refer all costs to the time of acquisition. For example, one air conditioner may be initially less expensive than another, but it may require more electrical energy and more repairs over its lifetime. The additional costs of electrical energy and repairs may more than offset the lower acquisition cost.

Two phenomena affect the value of money over time. The *inflation rate* is a measure of the decline in the value of money. For example, if the inflation rate is 3% per year, then an item will cost 3% more next year. Since it takes more money to purchase the same thing, the value of the unit of currency, in effect, is decreased. Note that the inflation rate for any item need not necessarily follow the general inflation rate. Recently health care costs have exceeded the general inflation rate in the United States, while the cost of most electronic goods has fallen far below the general inflation rate.

The *discount rate*, d, relates to the amount of interest that can be earned on principal that is saved. If money is invested in an account that has a positive interest rate, the principal will increase from year to year. The real challenge, then, in investing money, is to invest at a discount rate that is greater than the inflation rate.

As an example, assume an initial amount of money is invested at a rate of 100d% per year, where d is the percentage rate expressed as a fraction. After n years, the value of the investment will be

$$N(n) = N_0 (1+d)^n. \tag{9.1}$$

However, in terms of the purchasing power of this investment, N(n) dollars may not purchase the same amount as this amount of money would have been purchased at the time the investment was made. To account for inflation (or deflation), note that if the cost of an item at the time the investment was made is C_0, then the cost of the item after n years if the inflation rate is 100i% per year, will be

$$C(n) = C_0 (1+i)^n. \tag{9.2}$$

One might argue that if the cost of an item increases at a rate that exceeds the rate at which the value of saved money increases, the item should be purchased right away. Similarly, if the cost of the item increases more slowly, or, perhaps, actually decreases over time, then one should wait before making the purchase since the relative cost will be less at a later time. The disadvantage of this purchasing algorithm, of course, is that the item to be purchased will not be available for use until it is purchased. Hence, the new computer that becomes less and less expensive while the invested money continues to increase in value, is not available for computing when it should be purchased. In other words, economics may not be the only consideration in making a purchase. Sometimes people buy things simply because they want them.

It is important to remember that choosing values for d and i is tantamount to predicting the future, since d and i fluctuate over time. Depending upon the saving mechanism, the rate of return may be fixed or may be variable. Inflation is, at best, unpredictable. The worldwide recession of 2009 surprised many observers and supports the observation that the future will be different from the past. The important point is to be able to distinguish between predictable events, such as death, taxes, sunrise and loan repayment amounts and unpredictable events, such as the value of the stock market, the inflation rate and available interest rates on savings plans.

The consumer price index (CPI) is often used as a measure of inflation, since it tracks increases or decreases in the cost of various commodities over time. Since 1985, the CPI has risen relatively steadily from just over 100 to about 300 in 2023 [1]. Thus, on average, one could expect that the cost of items has risen by a factor of 3 over this period. However, if electronics and a number of items for which the prices have actually decreased over this period are subtracted out of the total mix, this means that the cost of many other items has risen faster. Assuming everyone needs pretty much everything on the list, one could argue that the cost of living is three times higher now than in 1985. However, in all fairness, one could also argue that this is NOT the case.

The Dow-Jones Industrial Average (DJIA) is often used as a measure of the discount rate. Since 1985, the DJIA has risen from below about 3800 to 38,873 on March 18, 2024 [2]. This suggested that a dollar invested in 1985 would be worth about $10 today if it had been invested at the average rate. However, because of the inflation rate, the invested dollar would only purchase about three times as much instead of 10 times as much.

When one looks at where the dollars that are lent out come from, although some come from individual savings, the bulk of available cash comes from borrowing from the government. Thus, the government has a means of presumably either slowing the economy by increasing the "effective funds rate" or stimulating the economy by decreasing the effective funds rate. Thus, during a recession, typically the government lending rate goes way down and during an inflation period, the government lending rate goes up. Figure 9.1 [3] shows how the U.S effective federal funds rate, as a measure of the absolute lowest minimum borrowing interest rate, has varied over the period between 1985 and 2023.

Economic and Environmental Considerations

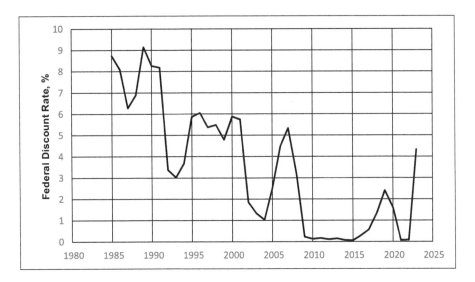

FIGURE 9.1 U.S. effective funds rate, 1985–2023.

(Courtesy Federal Reserve Bank of St Louis https://fred.stlouisfed.org/series/DFF)

It is interesting to note how the effective federal funds rate is adjusted with the intent of either controlling inflation by discouraging borrowing or stimulating the economy by encouraging borrowing. The high effective federal funds rate in the early 1980s reflects the attempt to control high inflation in the late 1970s resulting from significant increases in energy prices during this period. The low rates in 2002, 2008 and 2021 reflect the attempt to stimulate the economy and bring about a recovery of the stock market. In 2023, the Effective Funds Rate rose sharply in response to high inflation. The corresponding changes in the CPI and the Dow Jones industrial average suggest there just may be a correlation here.

One additional factor sometimes involved in the stimulation of the economy, or, perhaps, certain sectors of the economy, is Federal incentive programs. The Inflation Reduction Act of 2022 [4] involved a number of different incentives, from keeping the tax credit on PV systems at 30%, providing incentives for nonprofits to install renewable energy systems, to providing incentives for bringing certain types of manufacturing, such as solar modules, back to the United States as well as providing support for development of new technology for energy storage to incentives for purchase of electric vehicles and installation of EV charging stations.

9.2.2 Present Worth Factors and Present Worth

If $C_o = N_o$, the ratio of C(n) to N(n) becomes Pr, which represents the *present worth factor* of an item that will be purchased n years later, and is given by

$$\text{Pr} = \left(\frac{1+i}{1+d}\right)^n. \tag{9.3}$$

The present worth of an item is defined as the amount of money that would need to be invested at the present time with a return of 100d% to be able to purchase the item at a future time, assuming an inflation rate of 100i%. Hence, for the item to be purchased n years later, the present worth is given by

$$PW = (Pr)C_o. \qquad (9.4)$$

Sometimes it is necessary to determine the present worth of a recurring expense, such as fuel cost. Since recurring expenses can be broken down into a series of individual expenses at later times, it is possible to determine the present worth of a recurring expense by simply summing up the present worth of each of the series. For example, suppose a commodity such as diesel fuel is to be used over the lifetime of a diesel generator. It is desired to determine how much money must be invested at present at an annual interest rate of 100d%, under conditions of 100i% annual inflation to purchase fuel for n years. If the first year's supply of fuel is purchased at the time the system is put into operation, and each successive year's fuel supply is purchased at the beginning of the year, the present worth of the fuel acquisitions will be

$$PW = C_o + C_o\left(\frac{1+i}{1+d}\right) + C_o\left(\frac{1+i}{1+d}\right)^2 + C_o\left(\frac{1+i}{1+d}\right)^3 + \ldots + C_o\left(\frac{1+i}{1+d}\right)^{n-1}. \qquad (9.5)$$

Letting $x = \left(\frac{1+i}{1+d}\right)$, (9.5) becomes

$$PW = C_o(1 + x + x^2 + \ldots + x^{n-1}). \qquad (9.5a)$$

This expression can be simplified by observing that

$$\frac{1}{1-x} = 1 + x + x^2 + x^3 + \ldots = \sum_{i=0}^{\infty} x^i. \qquad (9.6)$$

Now, the *cumulative present worth factor* can be defined as

$$Pa = PW/C_o = \frac{1}{1-x} - \sum_{i=n}^{\infty} x^i = \frac{1}{1-x} - x^n \sum_{i=0}^{\infty} x^i,$$

or, finally,

$$Pa = \frac{1-x^n}{1-x}. \qquad (9.7)$$

It is important to recognize that (9.7) is based on the assumption that the first year's supply is purchased at the beginning of the year at a time when the fuel is at its present value. The fuel is then purchased annually with the last purchase occurring one year before the system lifetime has expired. In other words, there are n purchases of fuel, each at the beginning of the nth year.

Economic and Environmental Considerations

If the recurring purchase does not begin until the end of the first year, and if the last purchase occurs at the end of the useful life of the system, there will still be n purchases, but, using x again, the cumulative present worth factor becomes

$$Pa_1 = x + x^2 + x^3 + \ldots + x^n$$
$$= x(1 + x + x^2 + \ldots + x^{n-1}) = xPa. \qquad (9.8)$$

Since x will typically be in the range $0.95 < x < 1.05$, and since determination of i and d are at best, good guesses, and since it would be unusual to purchase an entire year's supply of many things all at once at the beginning of the year or at the end of the year, either (9.7) or (9.8) will generally provide a good estimate of the present worth of a cumulative expenditure. Often the values for Pr and for Pa are tabulated. Since most engineers have programmable calculators and computers, there is little point in repeating tabulated values for these expressions. Once the values for the variables have been decided upon, the present worth of quantities can be calculated. For that matter, if an engineer wanted to assume different values for d and i for different years, the same methodology could be used to determine the present worth of a quantity.

9.2.3 LIFE CYCLE COST

Once the PW is known for all cost categories relating to the purchase, maintenance and operation of an item, the LCC is defined as the sum of the PWs of all the components. The LCC may contain elements pertaining to the original purchase price, replacement prices of components, maintenance costs, fuel and/or operation costs and salvage costs or salvage revenues. Calculating the LCC of an item provides important information for use in the process of deciding which choice is the most economical. The following example demonstrates the use of LCC.

Example 9.1. Refrigerator A costs $1200 and uses 50 kWh of electricity per month. It is designed to last 10 years with no repairs. Refrigerator B costs $1400 and uses 35 kWh of electricity per month. It is also designed to last 10 years with no repairs. Assuming all the other features of the two refrigerators are the same, which is the better buy if the cost of electricity is $0.17/kWh? What if the cost of electricity is $0.22/kWh? Assume a discount rate of 4% and assume an inflation rate of 6% for the electrical costs. Also, note that this example is being written out in early 2023, a time when estimating either i or d is a bit speculative.

Solution: The solution, of course, is to perform LCC analyses on each refrigerator. First note that $x = 1.06 \div 1.04 = 1.0192$. For a 10-year period, since electricity purchase begins at the time of purchase of the refrigerator, using (9.7) gives Pa = 10.91.

Then note that for refrigerator A, the electrical cost for the first year will be (12 mo) × (50 kWh/mo) × ($0.17/kWh) = $102, and for refrigerator B, the electrical cost for the first year will be $71. Multiplying the first year cost by Pa yields the PW of the electrical cost. A simple table may be constructed to compare the two refrigerators. Note that separate columns are used for the PW associated with electricity at $0.17/kWh and electricity at $0.22/kWh.

From Table 9.1 it is evident that the $1400 refrigerator has a lower LCC than the $1200 refrigerator. It is also evident that as the price of electricity is increased, the LCC of the more expensive first-cost refrigerator becomes more and more attractive. Federal law requires that certain appliances, including refrigerators, have labels that disclose the energy consumption. *It is not necessarily the case that the more expensive units use less electricity.* One should check the labels carefully when contemplating a purchase.

Example 9.2. Compare the LCC of a highway construction warning sign that is PV powered versus using a gasoline generator only and then a gasoline generator with storage batteries to power the same sign. The system is to be capable of 24-hour-per-day operation with minimal down time. Assume the load to be 2 kWh per day with a 20-year lifetime.

To power this load with a PV system, it will take a 500-watt array of PV modules, plus array mount, at a cost of $1.25 per watt, $1000 worth of storage batteries, which need to be replaced every eight years, and a $250 charge controller. Assume a system maintenance cost of $150 per year and a 25-year lifetime for the PV array and mount.

Although the average power requirement is only 83 watts, it is unlikely that an 83-watt generator will be used. For the purposes of this example, assume that a 500-watt gasoline generator can be purchased for $250. Since it is running at about 20% of the rated load, a generous efficiency estimate is 1 kWh per gallon, and will thus use about 730 gallons of gasoline per year at an initial cost of $4/gal, and will require frequent maintenance with an annual cost of about $1500 for oil changes, tune-ups and engine rebuilds. Because of the heavy use, after five years the generator must be replaced. Assume an inflation rate of 3% and a discount rate of 5%.

For the generator with a battery system, assume only one day of storage will be needed. The idea here is to run the generator closer to its most efficient operating level for a shorter time each day to charge the batteries, thus prolonging the life and reducing maintenance on the generator. However, since the generator will now run intermittently, depending on the state of charge of the batteries, it will need an auto-start feature, which adds $100 to its cost. The smaller battery pack, priced at $500, is acceptable since the PV system battery pack needs to have several days of storage capacity to ensure system availability. In addition to the batteries, a $100 battery charger will also be needed to convert the generator AC output to DC. The generator will be delivering 400 W to the batteries at the rate of 4 kWh/gal, and will thus take 5 hours and 0.5 gal of fuel to generate 2 kWh each day. For this system, the generator will need replacement after ten years and is estimated to have annual maintenance

TABLE 9.1
Life Cycle Cost Analysis for Two Refrigerators at $0.17/kWh and $0.22/kWh

	Refrigerator A			Refrigerator B		
	First Year	PW	PW	First Year	PW	PW
Purchase price	$1200	$1452	$1452	$1400	$1694	$1694
Electrical cost @ $.17/kWh	$102	$1113	N/A	$71	$775	N/A
Electrical cost @ $.22/kWh	$132	N/A	$1440	$92	N/A	$1004
LCC		$2565	$2892		$2469	$2698

Economic and Environmental Considerations

costs of approximately $500. Battery life is expected to be eight years. Assume an annual inflation rate of 3% and a discount rate of 5%.

Solution: For the PV system, Pr is needed for 6 years, 12 years and 18 years, using (9.3). For the generator, Pr is needed for the generator replacement after 5 years, 10 years and 15 years. For the PV system, Pa_1, using (9.8) is needed for maintenance costs and Pa using (9.7) is needed for generator fuel and maintenance costs. For the given inflation and discount figures, $x = 0.9809$, $Pa = 16.763$ and $Pa_1 = 16.443$. Table 9.2 can now be completed.

Hence, even though the initial cost of the PV system is significantly higher, its LCC is significantly lower. Could this possibly explain the rapid deployment of these signs?

9.2.4 Annualized LCC

It is sometimes useful to compare the LCC of a system on an annualized basis. Dividing the system LCC by the expected lifetime of the system may appear to be the way to arrive at an annual cost. This would assume the cost per year to be the same for every year of operation of the system. But this does not take into account

TABLE 9.2
Comparison of LCCs for PV, Generator, and Generator with Battery System for Highway Sign

PV System				Generator System			
Component	Initial Cost	PW		Component	Initial Cost	Annual Cost	PW
Array + mount	$625	$625		Generator	$250		$250
Batteries	$1000	$1000		Fuel		$2920	$48,948
Batt 8 yr	$1000	$891		Gen 5 yr	$250		$227
Batt 16 yr	$1000	$794		Gen 10 yr	$250		$206
				Gen 15 yr	$250		$188
Annual Maintenance	$150	$2474		Annual Maintenance		$1500	$24,735
LCC		$6742		LCC			$74,554

Generator System with Batteries			
Component	Initial Cost	Annual Cost	PW
Generator	$350		$350
Battery Charger	$100		$100
Batteries	$500		$500
Fuel		$730	$12,237
Batt 8 yr	$500		$429
Batt 16 yr	$500		$368
Gen 10 yr	$350		$289
Annual Maintenance		$500	$8381
LCC			$22,654

the time value of money. Hence, to find the annualized LCC (ALCC) in present-day dollars, it is necessary to divide the LCC by the value of Pa or Pa_1 used in the PW analyses for the system components.

For the refrigerators in Example 9.1, this means dividing the LCC by 10.91 for each LCC evaluated. For the PV system of Example 9.2, the ALCC = $6742 ÷ 10.91 = $618, for the gasoline generator system of Example 9.2, the ALCC = $74,554 ÷ 10.91 = $6834 and for the generator plus battery system, ALCC = 22,654 ÷ 10.91 = $2076.

9.2.5 Unit Electrical Cost

One especially valuable use of the ALCC is to determine the unit cost of electricity produced by an electrical generating system. Obviously if the electricity is to be sold, it is necessary to know the price for which it should be sold to either earn a profit or to at least know how much will be lost in the process. Once the ALCC is known, the unit electrical cost is simply the ALCC divided by the annual electrical production. If the annual electrical production is measured in kWh, then the unit electrical cost will be measured in $/kWh.

For the PV system of Example 9.2, the unit electrical cost is ALCC/kWh = $618 ÷ 730 = $0.847/kWh. For the gasoline generator system, the unit electrical cost is $6834 ÷ 730 = $9.36/kWh and for the gasoline generator plus battery system, the unit electrical cost is $2076 ÷ 730 = $2.84/kWh. Clearly, all three of these costs far exceed the cost of utility-generated electricity, but since this is a portable application, where the sign is moved around from day to day, the cost of hooking up and disconnecting the system from utility power is impractical, at best.

9.2.6 LCOE Analysis

The concept of a method to evaluate the relative cost of all energy sources was introduced by Walter Short and his colleagues at NREL in March 1995 [5]. The simplicity of the method is in the determination of the cost of a unit of energy by dividing the net present value of all costs and expenses of running a power generator through its expected life, by the total energy generated over the same period.

The method has been adopted by most planners around the world as a standard for energy unit costs whether the generation is fossil or nuclear based or renewable. It has also been modified and adapted to account for depreciation, tax credits, degradation and efficiency losses, and governmental or utility policies. The basic LCOE formula can be represented by:

$$\text{LCOE} = \frac{(\text{Project Cost}) \frac{\text{RES}}{(1+\text{DR})^{(pl)}} + \sum_{n=1}^{N} \frac{\text{OM}}{(1+\text{DR})^n}}{\sum_{n=1}^{N} \frac{(\text{Initial kWh}) \times (1-\text{System degradation rate})^n}{(1+\text{DR})^n}} \quad (9.9)$$

where OM is the operations and maintenance cost for each year n,
RES is the end-of-life residual value of the generating plant in year n = (pl), and
DR is the assumed discount rate.

Economic and Environmental Considerations

The Department of Energy Sunshot goal for LCOE of PV-generated electricity was $0.06 U.S. per kWh by 2020 [6]. It should be noted that if decommissioning of a generating plant involves environmental mitigation, disposal fees or other costs, the minus (−) in the formula may become a plus (+), thus increasing the LCOE. LCOE can also be modified to include the contribution of carbon credits which would of course lower the unit cost. A simple spreadsheet exercise to determine the LCOE of a solar PV plant has been left as a homework problem in this chapter.

Example 9.3: The following LCOE Excel exercise is designed to show the utility of the method and the ease with which it can be applied to virtually any electricity generation scenario. The model is built around a 1 MW plant in the Phoenix, AZ, area that has several possible implementations, beginning with a fixed crystalline Si array, followed by a single-axis tracking crystalline Si array with 20% additional annual kWh production. Other options considered include dual-axis tracking crystalline Si with 39% more annual kWh production than the fixed array, a concentrating PV (CPV) system with 18% higher annual generation, a diesel generator operating 8 hr/day that uses 120 gal of diesel fuel per day with a first-year fuel cost of $3.00/gal and a 1 MW gas turbine generator operating 8 hr/day that uses 102,960 ft^3/day (2894.4 m^3/day) at a cost of $5.00/MCF ($0.177/m^3). Note that the cost of land is not included in this analysis. If the design criterion is to maximize the kWh/area ratio, the results could be different.

TABLE 9.3A
List of Assumptions Relating to First-Year Costs of Several 1 MW Electrical Generation Options, Including Resulting LCOE After 25 Years of Operation

System	Disc Rate (%)	kWh Annual Inflation Rate (%)	Initial Cost U.S. $	Initial Annual Gen MWh	Annual Degrad (%)	Annual O & M	Annual O & M Inflation (%)	LCOE U.S.$/kWh
1 MW Si PV Fixed array	3	3	1.0M	1778	0.5	$10,000	4	0.042
1 MW Si PV one-axis tracking	3	3	1.15M	2112	0.5	$20,000	4	0.048
1 MW Si PV two-axis tracking	3	3	1.25M	2473	0.5	$32,000	4	0.051
1 MW CPV	3	3	2.0M	2100	0.5	$36,000	4	0.084
1 MW diesel	3	3	0.26M	8000	0.0	$695,200	4	0.138
1 MW gas turbine	3	3	0.53M	8000	0.5	$325,600	4	0.08

The same basic assumptions for discount rate and inflation are used for each system. To give the model more flexibility, the inflation rate for the cost of electricity is decoupled from the general inflation rate. This is because, in recent years, the volatility of oil and gas prices has had more of an impact on the cost of electricity than the general inflation rate both on the upside and the downside. The tracking solar PV and CPV plants are also assumed to be operating in regions where tracking or concentrating solar is competitive. These conditions have been discussed in other chapters of this book and as a reminder, they are in areas with high direct beam normal incidence.

The costs and the generation characteristics shown in this example are based on the general information currently available to the authors, but they need to be researched before any definitive analysis is conducted. For example, the assumption that a 1 MW PV system generates 1.78 MWh after all the losses and de-rating factors in the first year of operation is based on numerous factors and is purely for the purpose of this exercise. A serious LCOE analysis should consider all the information provided in other chapters to develop a reliable kWh generation profile based on the solar conditions, regional data, module inclination, the soiling and washing cycles, etc. The assumption that a single tracking system generates 20% more electricity than non-tracking may be true in a few regions but is most likely not realistic for Florida, or regions with higher diffuse to direct radiation solar components. The analyst should rely on accepted national and regional data to conduct her analysis. Also, the assumptions regarding the cost and 40% higher generation of CPV are highly variable depending on the specific tracking platform and the architecture of the cell assemblies. The degradation rates for CPV are also based on relatively limited data points compared with other PV systems and may be statistically insignificant. The daily run time for the diesel and gas turbine generators has been selected to achieve annual energy generation times comparable to the PV options. *Solution:* The assumptions for each system option are shown in Table 9.3a along with the LCOE for each system as determined using (8.9) and the assumptions. The results of the analysis of the fixed array system using (9.9) are tabulated in Table 9.3b to show how the results in Table 9.3a are obtained. Note that less run time results in higher LCOE, since the initial cost now must be made up for by charging more per unit for a smaller amount of total energy production. Thus, gas turbines used as peaking generators may only have a 5–10% capacity factor on an annual basis, whereas the unit in the example has a 33% annual capacity factor. The important issue with LCOE is that it is important to check the assumptions, since different sources using different assumptions will likely report different values.

9.2.7 LCOS Analysis

In the "good old days," when storage was used simply to provide electric power when no utility service was available, the main concern was providing power when it was needed, such as to keep the refrigerator and a few lights running during a utility outage, as has been discussed in previous chapters. In essence, the storage served as

TABLE 9.3B
Example of LCOE Analysis for 1-MW Fixed Crystalline Si Array

Year	Discount Rate	Initial Cost (×10³)	OM (×10³)	NPV OM (×10³)	Residual (×10³)	Total kWh (×10⁶)	NPV kWh (×10⁶)
0		1000					
1	0.03		10.00	10	0	1.78	1.78
2	0.03		10.40	10.10	0	1.77	1.72
3	0.03		10.82	10.20	0	1.76	1.66
4	0.03		11.25	10.30	0	1.75	1.60
5	0.03		11.70	10.40	0	1.74	1.55
6	0.03		12.17	10.50	0	1.73	1.50
7	0.03		12.65	10.60	0	1.725	1.44
8	0.03		13.16	10.70	0	1.72	1.40
9	0.03		13.69	10.80	0	1.71	1.35
10	0.03		14.23	10.91	0	1.70	1.30
11	0.03		14.80	11.01	0	1.69	1.26
12	0.03		15.40	11.12	0	1.68	1.22
13	0.03		16.01	11.23	0	1.67	1.17
14	0.03		16.65	11.34	0	1.665	1.13
15	0.03		17.32	11.45	0	1.66	1.10
16	0.03		18.01	11.56	0	1.65	1.06
17	0.03		18.73	11.67	0	1.64	1.02
18	0.03		19.48	11.79	0	1.63	0.988
19	0.03		20.26	11.90	0	1.62	0.954
20	0.03		21.07	12.02	0	1.616	0.922
21	0.03		21.91	12.13	0	1.61	0.891
22	0.05		22.79	12.25	0	1.60	0.860
23	0.05		23.70	12.37	0	1.59	0.831
24	0.05		24.65	12.49	0	1.58	0.803
25	0.05		25.63	12.61	0	1.576	0.776
Totals		1000		281.42	0	41.88	30.28
LCOE	0.042						

an insurance policy. On the other hand, by 2018, it had become evident that storage could serve perhaps an even more interesting and important purpose—backup of the utility. If the utility had excess electricity, rather than curtail the supply, the excess could be stored and then made available to the utility during peak time or failure of other generation sources, including PV at night. Under these circumstances, the storage becomes a part of a larger system and to provide investors with an incentive, a method of evaluating the value of the stored energy is needed. A number of different factors enter the picture when trying to assign a value to stored electricity in terms of value over time and how to balance storage with generation in the most efficient manner. This implies a comparison of both power and energy values for generation and storage.

Some of the factors that affect the value of an investment in storage include:

- The initial cost of the storage mechanism
- The energy capacity of the storage mechanism
- The difference in cost between the stored energy and power and the delivered energy and power
- The number of charge/discharge cycles or, alternatively, the total number of kWh that can be discharged over the lifetime of the storage system
- The number of charge/discharge cycles or, alternatively, the total number of kWh that can be discharged by the system in one day
- The round-trip efficiency of the storage mechanism
- The maximum discharge rate and the maximum storage rate of the system
- The seasonal system hourly load profile
- The daily and hourly variation in the value of energy and power on the grid
- The effect of storage on the value of energy and power on the grid
- Operation and maintenance costs
- Degradation over time of charging and discharging efficiencies
- Disposal costs

Clearly, with all these variables that change with time on an hourly, daily, monthly and yearly basis, playing the energy storage for profit game looks a lot like trying to optimize return on investment in the stock market. A comprehensive treatment of all these variables is far beyond the scope of this book, but that does not mean a brief sample analysis should not be undertaken to illustrate the nature of the challenge.

As a very simple example, suppose a BESS is rated at 1 MWh (0 to 100%) and can operate between 10% and 100% of its capacity. Suppose its initial maximum charge or discharge rate is C/2 and its round-trip efficiency is 90% AC to AC. This means that for every kWh delivered, it will take 1.11111 kWh of charging to replace it. It also means the maximum discharge or charge rate will be 500 kW. Thus, over a three-hour period, if the storage begins at 100% capacity, the highest rate of discharge will be 500 kW for 1.8 hr to discharge the system to 10% of its capacity. At this point, it will either need to rest or charge up again. If it begins to recharge at the maximum rate, it will have 1.2 hr left in the period for charging at 500 kW, which will result in an increase of 600 kWh to bring the state of charge up to 700 kWh (70%) at the end of the third hour. This will allow the system to continue charging or to discharge, if required, at the beginning of the 3rd hour at a rate that is useful to the grid as well as to the BESS unit.

Then suppose it is connected to a utility under the conditions that the average purchase rate per kWh for charging from the grid will be $0.05 (when the grid has excess capacity) and the average discharging price paid by the utility will be $0.20 per kWh. For the purposes of this example, no charges or payments will be made for demand (power). This might be the next step in adding to the complexity of the problem. It is important to recognize that neither charging nor discharging will necessarily occur over any single time period in a day. Furthermore, discharging or charging will not necessarily take place at a constant rate during the cycle. A simple analysis based only on purchase price, selling price, round-trip efficiency and an estimate of total purchases and sales over three-hour periods over a 24-hour period is shown in

Economic and Environmental Considerations

Table 9.4. The assumption is made that the system starts at 50% capacity at midnight and ends at 50% 24 hours later and that most of the purchased electricity comes from renewable sources, most during off-peak hours but not ruling out the possible purchase of some during on-peak hours.

Clearly, many of the factors listed previously are missing in this table, since it makes a lot of assumptions about the first year of operation. For example, based on the assumed daily purchase price and the assumed daily resale price, the daily profit would be $112, resulting in an annual profit of $40,880. Assuming a system cost of $250,000, the resulting profit in the first year would be 16%, which would be reduced by many of the remaining factors but would be increased by optimizing charge/discharge cycles to deliver a larger amount of energy in a day. In addition, had the storage system been needed to replace the utility during an outage, a totally different value system would have been introduced, such as the value of the food in the freezers and refrigerators that did not spoil or the value of gasoline that might otherwise have been used in fossil backup generators.

Other important items to add would be the time value of money, perhaps a depreciation factor, the expected lifetime, total energy delivery, a value for available instantaneous power and any degradation over time as storage internal resistance increases, or administrative costs. On the other hand, the guesses were merely a means of demonstrating a simple analysis method. A more refined method is emerging to represent the LCOS. One of the formulas looks similar to the analysis of (9.9). In essence, fuel costs have been replaced by the cost of purchased electricity for storage and use of the discharged electricity rather than generated electricity, as used for traditional renewable and non-renewable power sources. As more operation experience is gathered, the analysis will no doubt be refined in a manner similar to the refinements typically made to the most important metrics to incorporate more of the previously mentioned variables. Meanwhile, the following equation can at least be used to get the process started. Note that the round-trip efficiency (RTE) has been incorporated in the denominator.

$$LCOS = \frac{(\text{Project Cost})\frac{\text{RES}}{(1+DR)^{(pl)}} + \sum_{n=1}^{N}\frac{\text{OM}}{(1+DR)^n}}{\sum_{n=1}^{N}\frac{(\text{Initial kWh})(\text{RTE})(1-\text{System degradation rate})^n}{(1+DR)^n}} \quad (9\text{--}10)$$

TABLE 9.4
Sample Daily Charge/Discharge Cycling for 1 MWh with 90% RTE

Time	12M-3a	3-6	6-9	9-N	N-3	3-6	6-9	9-M	Totals	Value
KWh in	0	0	100	260	375	100	25	0	860	$43
Net kWh in	0	0	90	234	338	90	23	0	774	
kWh out	150	150	50	0	0	150	150	125	775	$155
End SOC	350 kWh	200	240	474	812	752	624	499		
End SOC %	35%	20%	24%	47%	81%	75%	62%	50%		

9.3 BORROWING MONEY

9.3.1 INTRODUCTION

Sometimes the desire to own something causes the potential owner to realize that money does not grow on trees. While some money comes from paychecks, usually larger sums of money come from borrowing from banks or other lending institutions. The question at this point for the engineer-turned-economist is whether borrowed money is any different from paycheck money or money from a savings account, a mattress or other form of liquid asset. For example, if the money to purchase one of the refrigerators of Example 9.1 had to be borrowed rather than taken from a billfold, would that affect the LCC of the refrigerator?

Once again, tables are readily available for looking up the annual payments on a loan of C_o dollars taken out at $100i\%$ annual interest over a period of n years. For that matter, it is possible to purchase a calculator that will automatically yield the answer when it has been given the terms of the loan or to look it up on the internet. Engineers, of course, will want to know how the numbers are obtained and do their own calculations.

9.3.2 DETERMINATION OF ANNUAL PAYMENTS ON BORROWED MONEY

To satisfy this curiosity, consider Table 9.5, representing the principal, interest, total payments and principal balance at the end of the kth year for repayment of an n-year loan at $100i\%$ annual interest, where C_o represents the amount borrowed. Note that during any of the years, it is not yet known how much will be paid on the principal to repay all of the principal in n years. The challenge is to arrive at an appropriate formula for equal total annual payments over the period of the loan.

In Table 9.5, A_k represents the amount paid on the principal after the kth year. To pay the principal fully in n years requires that the sum of the annual payments on principal must add up to the loan amount, C_o.

Setting the total payments of each year to be equal, yields a solution for A_1. For example,

$$A_1 + iC_o = A_2 + iC_o - iA_1,$$

TABLE 9.5
Breakdown of Portions of Loan Payment Allocated to Principal and Interest

Yr	Payment on Principal	Interest Payment	Total Payment	Balance of Principal
1	A_1	iC_o	$A_1 + iC_o$	$C_o - A_1$
2	A_2	$i(C_o - A_1)$	$A_2 + i(C_o - A_1)$	$C_o - A_1 - A_2$
3	A_3	$i(C_o - A_1 - A_2)$	$A_3 + i(C_o - A_1 - A_2)$	$C_o - A_1 - A_2 - A_3$
n	A_n	$i(C_o - \ldots - A_{n-1})$	$A_n + i(C_o - \ldots - A_{n-1})$	0

Economic and Environmental Considerations

which yields

$$A_2 = A_1(1+i).$$

In general,

$$A_n = A_{n-1}(1+i) = A_1(1+i)^{n-1}.$$

Next, letting $x = (1 + i)$ and summing all the payments toward principal, yields the amount borrowed.

$$A_1 + A_1x + A_1x^2 + \ldots + A_1x^{n-1} = A_1(1+x+\ldots+x^{n-1}) = C_o.$$

But (9.5a) and (9.7) have shown that

$$1+x+x^2+\ldots+x^{n-1} = \frac{1-x^n}{1-x},$$

which yields

$$A_1 = \frac{C_o(1-x)}{(1-x^n)}. \tag{9.11}$$

Now all that remains is to add the interest payment for the first year to the principal payment, given by (9.11), to get the total payment for the first year, which will be equal to the total payment for each succeeding year. Proceeding yields

$$\text{ANN PMT} = \frac{C_o(1-x)}{(1-x^n)} + C_o i = C_o\left(\frac{i}{(1+i)^n - 1} + i\right).$$

Simplifying this result yields, finally,

$$\text{ANN PMT} = C_o i \left(\frac{(1+i)^n}{(1+i)^n - 1}\right). \tag{9.12}$$

Usually, payments are made monthly rather than annually. In this case, one need only note that rather than n payments, there will be 12n payments, and the monthly interest rate will simply be the annual rate divided by 12. Doing so converts (9.12) into an equation for monthly payments. Obviously, if one wants to split hairs even more finely, (9.12) could be modified for weekly, daily or hourly payments. Equation (9.12) is also an equation that might be conveniently tabulated for various values of i and n, but again it is left to the reader to use either a computer or a programmable calculator to generate the numbers that apply to the problem at hand.

9.3.3 THE EFFECT OF BORROWING ON LCC

Depending on the nature of a purchase, it is interesting to compare whether the cost of borrowing money will render the purchase undesirable. For example, if money is borrowed to purchase something that will provide a return on the investment, it may make economic sense to borrow the money. This is the standard criterion for commercial loans. The better the return, the better the reason to borrow the money. But what if it is necessary to borrow the money for the initial cost of a system that does not have an obvious return on investment? Does it make sense to borrow for something having a greater first cost if it results in higher loan repayment costs? Although there are several ways to evaluate the worthiness of a purchase, once again, people sometimes borrow money to purchase things that do not have a measurable monetary return on investment, such as automobiles, simply because they want them.

As a simple example, suppose it is possible to spend $2000 on maintenance-free systems that will last for 25 years and will reduce an electric bill by a certain amount per year. The exact system might be insulation, good windows, a hybrid electric water heater, an energy recovery ventilator or any number of other energy efficiency measures. Suppose also that money is borrowed to purchase the system and the period of the loan is 25 years. This would be the case if the system is purchased for a dwelling at the time of construction and the item is included in the 25-year mortgage. If the mortgage rate is 4%, then (9.12) shows the additional monthly mortgage payments would be $11, or $131 annually. If the annual savings on the electric bill exceed $131 then it is worth borrowing the money.

One should note, however, that the additional $131 mortgage payment will be constant over the life of the loan. In an inflationary environment, this means that even though the electric bill savings may not be $131 the first year, after a certain number of years, the annual savings may well exceed $131, making the investment worth considering.

What if the money is not borrowed? What if the $2000 is at hand and available for investing? What would be the return on investment if it is spent on one of the items of the previous example? The simple answer would be that the return is equal to the resulting savings less the resulting operating or maintenance costs. This works fine for the first year, but for successive years to yield a more precise estimate, the time value of money must be taken into account.

Assuming the value of the quantity saved increases at the inflation rate, i, and the discount rate is d, the present worth of the savings accrued over n years will be given by (9.5). As a result, the PW of the annual savings over n years is given by (9.7). Hence, (9.7) can be applicable in either an expenditure mode or a savings mode. By converting costs and savings to LCC, it is easy to compare savings with costs to determine whether to make the purchase.

Assuming that a system must be acquired to do something, and assuming that the system with the least LCC has been identified, it is simply a matter of deciding whether to borrow money or to use money on hand, if, indeed, the money is on hand. If the money is not on hand, then the only option is to borrow. If the money is on hand, then the criterion is whether the lending rate is less than or more than the discount rate. If money can be borrowed at a rate less than what it can earn, then it

Economic and Environmental Considerations 301

makes sense to borrow for the acquisition and to invest the money that might have been used for the purchase. This is a choice often made by the buyer of a new automobile. Should the savings account be used or should the money be borrowed? The answer depends on the relative interest on savings versus the interest on the loan. Once again, some would prefer to just pay the bill up front and then forget about it.

9.4 PAYBACK ANALYSIS

As mentioned in the introduction to the chapter, some people are not interested in a renewable energy system unless they know how long it will take the system to pay for itself. Often these same people are owners of expensive automobiles who did not even think about how long it would take the expensive automobile to pay for itself. In other cases, the potential owners are business people who are interested in "greening" their business, but only if the "greening" has a reasonable payback period.

Once again, determining a payback period depends on predicting the future, since the idea is to install a system that will generate enough electricity in the first few years to offset the net cost of the installation. The question is what will be the value of the electricity? The part of the future that is relatively predictable is the annual available sunlight and the system's performance. The hard part is to figure out what will happen with electric rates and discount rates over the life of the system. Thus, the process involves some guessing.

One way to take some of the uncertainty out of the guessing is to establish lower and upper limits of assumed utility cost escalation rates and then analyze the resulting payback periods. Rather than attempting to take into account the time value of money in the analysis, which would add even more uncertainty, the time value of money will not be included in this analysis.

Example 9.4. Assume that the installed cost of a 50 kW straight grid-connect PV system is $2.50/watt. Also assume that this system will qualify for a 30% investment tax credit and that the cost of the system can be depreciated at the rate of 20%/yr for 5 years. For system performance, assume NREL SAM PVWatts predicts an annual system production of 64,000 kWh to the grid. Assume the money is borrowed at 4% over 15 years to pay for the system and that the overall utility cost/kWh is $0.15. A more sophisticated analysis would separate out the energy and the demand charges of the electric bill, but this analysis simply divides the total bill by the total kWh. Perform a payback analysis on the system. To simplify the analysis a bit, assume the discount rate equals the inflation rate. This analysis will not assume any increase or decrease in utility rates.

Solution: First determine the annual loan repayment from (9.12). Also note that the interest on the loan will be tax deductible for a business, but this will need to be handed over to the accountant at the end of the year. It will not be considered in this analysis. Using (9.12), the annual loan repayment amount is found to be $11,243.

Next, the annual system kWh production is determined to be 64,000, which has a first-year value of $9,600. At this rate, it might appear that the system will never pay for itself, but it is still useful to construct a table that looks at income versus expenses on an annual basis over the lifetime of the system, which will be assumed to be 30 years. Note that this exercise is perfectly suited to a spreadsheet analysis, where many variables can be included as suits the fancy of the engineer and/or the accountant, including such items as tax refunds on interest paid, replacement system components, such as inverter and system maintenance. Table 9.6 shows the simplified results. In the table, note that the annual depreciation is considered an operating expense, so the assumption is that at the end of the year after all expenses are included, the owner will still have tax liability.

An examination of Table 9.6 shows that over the first five years, the combination of a 30% investment tax credit plus rapid depreciation more than offset the loan repayment value so the fact that the value of the system electricity is less than the loan repayment is offset by the investment tax credit and rapid depreciation. By the end of year 5, after the system has been fully depreciated, the cumulative value of savings is about $154,000 depending upon the utility escalation rate and the system degradation rates, which have not been considered. Then, beginning in year 6, with no further depreciation or tax

TABLE 9.6
Payback Analysis for Example 9.4

Year	1	2	3	4	5	6
Costs						
Loan pmt	11,243	11,243	11,243	11,243	11,243	11,243
Credits						
Inv Tax Cr	37,500					
Ann Depr	25,000	25,000	25,000	25,000	25,000	
Elect value	9,600	9,600	9,600	9,600	9,600	9,600
Net/yr	60,857	23,357	23,357	23,357	23,357	−1,643
Cum gain	60,857	84,214	107,571	130,928	154,285	152,642
Year	7	8	9	10	11	12
Costs						
Loan payment	11,243	11,243	11,243	11,243	11,243	11,243
Credits						
Inv Tax Cr						
Ann Depr						
Elect value	9600	9600	9600	9600	9600	9600
Net/yr	−1643	−1643	−1643	−1643	−1643	−1643
Cum gain	150,999	149,356	147,713	146,070	144,427	142,784
Year	13	14	15	16	17	18
Costs						
Loan pmt	11,243	11,243	11,243	0	0	0
Credits						
Inv Tax Cr						
Ann Depr						
Elect value	9600	9600	9600	9600	9600	9600
Net/yr	−1643	−1643	−1643	9600	9600	9600
Cum gain	141,141	139,498	137855	147,455	157,055	166,655

Economic and Environmental Considerations 303

credits, the system begins to lose about $1600 each year for a while. Thus, the cumulative value of savings begins to fall but never goes below zero. After year 15, there are no further loan payments, so the value of the electricity is now the only contributing factor to the system cumulative savings. Even though this analysis has been considerably simplified, it should at least convince the reader that it is worth going through the exercise, even including additional assumptions (wild guesses?), if so desired.

9.5 EXTERNALITIES

9.5.1 INTRODUCTION

What happens if something owned and operated by one party causes damage to something owned by another party? A quick response would be that the entity causing the damage would be liable and that the cost of repair of the damaged property should be the responsibility of the party doing the damage. The problem is complicated in many instances, however, when more than one party is responsible for creating the cause of the damage, and when more than one party suffers damage as a result of the cause. And things get even more complicated when debate ensues over whether the alleged cause is really the cause.

Externalities are factors resulting from energy production or other manufacturing that are not incorporated directly into the selling price of a product. That is, once the smoke is out of the smokestack, it is no longer the fiscal responsibility of the smoke producer. In this section, externalities will be explored in greater detail. In particular, environmental effects, health and safety issues and subsidies will be discussed in the context of comparing energy produced by photovoltaic means with energy produced by other means. In general, the externalities discussed are not associated with costs that appear in a LCC analysis of an energy source. The reader can expect that in the future, greater attention will be paid to those issues currently considered externalities, as means for attaching monetary value to these issues are developed and adopted. Some readers may find the research necessary to affix monetary value to externalities of sufficient interest and challenge that they may choose to pursue this interesting and important area to a greater extent.

In control engineering terms, where stability is determined by the locations of poles and zeroes of the system in the complex plane, the feedback processes can be characterized in control terms by observing the locations of the poles of the processes. If the poles are in the left half of the complex plane, then the process is stable. If the poles are in the right half plane, then the process is unstable. Stable implies that the process will tend to correct itself. Unstable implies an unbounded increase.

While many processes appear to be stable, examples of unstable processes have also been noted. In particular, increased CO_2, the most commonly known greenhouse gas, from nonpolar regions causes warming in the polar regions, resulting in the thawing of permafrost regions, in which large quantities of CH_4, another greenhouse gas, are trapped. Liberation of the CH_4 causes additional warming, resulting in additional liberation of CH_4 and so forth. The question is whether another natural process exists to counteract this cycle. The process itself is unstable.

The accumulated impact of human activity is at a point where serious questions are now being asked about the ability of the earth to recover from the results of this activity. One might argue that as long as the processes remain stable, the pollution is at acceptable levels for the planet to sustain life. The challenge becomes one of defining acceptable limits on pollution and the resulting changes in natural cycles caused by human activity.

Another factor that needs consideration in establishing cost is the effect of various forms of subsidies. Some subsidies may enter the picture as direct costs, while others may appear as externalities.

9.5.2 Subsidies

When performing an LCC, sometimes the cost of a component, a fuel or the operation of a system may be affected by a subsidy or subsidies. For example, the cost of military presence in a region to ensure the steady flow of fuel from the region is never included in the selling price of the product. Mineral depletion allowances, however, can be factored into the selling price of a product. In other cases, governments have been known to offer price support to ensure competitive prices in a world market. Tariffs, in effect, are a form of subsidy, since they ensure that domestic production will be sold at a profit, thus not competing directly with less costly products from outside a country.

Green pricing is a form of subsidy for the acquisition of clean energy sources. An example is when the customers of a utility express a willingness to pay extra every month to ensure that a part of their energy mix comes from renewable sources, such as photovoltaics. Subsidies in the form of tax breaks tend to come and go on a year-by-year basis. Some states have initiated grant programs to encourage homeowners to install solar systems and some utilities established rebate programs for part of the cost of installing various energy conservation measures, such as more efficient air conditioning or attic insulation.

The argument is often set forth that all competing interests must compete on a *level playing field*. The meaning is simply that with so many subsidies, some of which are obvious and others of which are hidden, it is difficult for two competing interests to engage in fair competition. This is true in many industries and particularly in the energy industry. The reader is thus reminded that no economic analysis or comparison is complete until all forms of subsidy have been considered.

9.5.3 Renewables and the Environment

9.5.3.1 Introduction

The cost of an electrical generation source often excludes externalities. Subsequently, if a source is cleaner from an environmental viewpoint but has a higher LCC based on the parameters considered, that source may not be chosen. A proper treatment of externalities includes not only the operating externalities but the externalities associated with the construction and salvage or decommissioning of the facility. In both categories, photovoltaics shows significant advantages over nonrenewable sources.

Economic and Environmental Considerations 305

9.5.3.2 The Cost of Carbon Dioxide

The cost of CO_2 can be considered in two ways: the cost of controlling CO_2 and the benefit of controlling CO_2. This method is applicable to other pollutants as well, but only CO_2 will be discussed in this section.

According to the World Resources Institute [7], holding CO_2 at 1990 levels will cost approximately 1 to 2% of the Gross Domestic Product for developed countries over the long term, with reductions below 1990 levels cost increasing to 3%. An alternate analysis, however, suggests that by reducing CO_2 levels significantly by the incorporation of energy conservation and renewable energy measures, every dollar spent on these measures may very well save more than a dollar in energy costs.

If action is not taken to stop the increase of CO_2 atmospheric concentration, it has been estimated that a doubling of atmospheric CO_2 will lead to an average warming of 2.5°C [7]. According to current projections, this doubling will occur during the 21st century if no curtailment measures are taken. In this case, it is estimated that the damage due to the warming will reach approximately 1% to 1.5% of GDP per year in developed countries, with substantially more in island nations. If CO_2 levels continue to rise beyond the double point, the damage may come closer to 6% of GDP. At the Paris climate conference of Nov–Dec 2015, it was pointed out that an increase of 1°C has already occurred and a strong push has been made to agree to limit future increases to less than 1.5°C [8].

Hence, it appears certain that damage will occur if nothing is done, while it is debatable whether there will be cost or cost savings if a program to reduce CO_2 levels is pursued. Considering that CO_2 production is also generally accompanied by the production of NO_2, SO_2 and particulates, if the reduction of CO_2 is accompanied by the reduction of other pollutants, then the benefits become subject to a multiplier effect. What is also being supported here is that it may be a better strategy to focus on reducing the production of CO_2 rather than sequestering CO_2 already produced. Perhaps a carefully planned combination of these two strategies may produce the most cost-effective results.

9.5.3.3 Sequestering CO_2 With Trees

To quantify the cost of externalities, it is necessary to arrive at a methodology for determining the value of a ton of SO_2 or NO_2 or CO_2 or other pollutants. Since trees, other green plants and oceans have maintained the balance between O_2 and CO_2 over the millennia, it would seem that a tree might be valued in terms of the number of tons of CO_2 it will remove from the atmosphere in its lifetime. Since SO_2 and NO_2 generally are removed from the atmosphere by rain, which dissolves them as weak acids, presumably a price per ton can be assigned in terms of the monetary value of any damage that may result to structures or the environment from acid rain. For example, if a building requires repair more often as a result of acid rain, this is a measurable cost, provided that the increased repair costs can be documented. If the fish die in a lake as a result of acid rain and, as a result, cause a decrease in tourism to the area, this also may constitute a documented cost if the source or sources can be identified.

The cost of CO_2 and other pollutants can be measured in several ways, including the cost of repairing the damage, the cost of controlling the damage by reducing the emission of pollutants and the cost of mitigating the damage, such as by absorbing the additional CO_2 by trees. Hodas (1990) [9] reported costs of $240 per ton of carbon for the removal of CO_2 from exhaust gases, with significant technical difficulty. The scrubbed CO_2 would then be liquefied and pumped to a depth of about 500 feet in the ocean, where it would then dissolve and become available for plankton and ocean vegetation. The cost of such an operation would likely be close to $1000/kW for scrubbing and disposal.

There would also be a 25% energy penalty, since the energy for the scrubbing and disposal would not be available for other end users, as well as a 22% capacity penalty and additional annual maintenance costs.

Supply, and, demand, side efficiency improvements are estimated to cost in the range of $20 per ton of carbon avoided. Automobile efficiency increases also save 20 pounds of CO_2 per gallon not burned, with an estimated cost of $0.53 per gallon for the effort required to increase automobile efficiency to 44 miles per gallon, assuming continued use of gasoline as a fuel [9]. Depending upon the source of electricity, electric vehicles use significantly less CO_2 per mile traveled than gasoline vehicles. This fact, along with performance, maintenance and other price advantages, results in rapid acceptance of these vehicles.

It has been estimated that the United States would require 1.5 billion hectares of forest to absorb its annual CO_2 emissions [9]. The problem is that the total land area of the United States is only 913 million hectares, and forest in Death Valley, CA, and quite a few other areas appears as an unlikely prospect. Thus, it will be essential to incorporate additional means of reducing the production of CO_2 and other greenhouse gases.

The idea of planting trees to mitigate the effects of CO_2 generation was first attempted in a project proposed by the Applied Energy Service. This project involved planting approximately 50 million trees in Costa Rica to mitigate the CO_2 generated by a fossil fuel electric generator in Connecticut. The project also involved a fire protection program to save 2400 hectares of forest from fires, with an overall estimated sequestering of 387,000 metric tons of carbon per year [9].

When trees are used for CO_2 mitigation, it is necessary to consider what will be done with the trees after they mature. If they are used for firewood, then the sequestered carbon is returned to its CO_2 form. If they are used for construction, then the carbon remains sequestered. However, even if they are burned, on the assumption that something will be burned, the burning of these trees displaces the burning of other trees. If the trees spared from fire are genetically diverse, then an additional value can be assigned to the preservation of diversity in the biosphere.

Perhaps an even more valuable location for trees is the urban environment, particularly in tropic or near-tropic regions. Hodas [9] observes that by planting trees near buildings, the microclimate of the buildings is cooled by several degrees, thus reducing air conditioning needs and the corresponding energy required for cooling. If the energy for cooling comes from fossil-fueled generation, then the trees reduce atmospheric CO_2 directly as a result of their photosynthesis as well as indirectly by reducing the output demands and corresponding fuel requirements of the fossil-fueled generator. On the other hand, if the trees shade a PV site, then the situation deserves further analysis.

9.5.3.4 Attainment Levels as Commodities

Another interesting method of assigning value to pollutants comes as a consequence of the Clean Air Act [10] and regulated attainment levels. Companies are granted a certain allowable number of pollution units for their facilities. For example, an electrical generating utility may be allowed a certain number of tons of SO_2 emissions per year. If the utility does not generate as many tons as allowed, it may transfer the remainder of its allowed emissions to another entity. In a free enterprise society, this is not normally done as a simple, friendly gesture. However, the U.S. Environmental Protection Agency publishes the Green Book, which provides detailed information about area designations, classifications and nonattainment status [11]. Pollutants include ozone, particulate matter, sulfur dioxide, lead, carbon monoxide, nitrogen dioxide, national and county area multi-pollutants as well as other resources such as data, maps, GIS, frequent questions and related air quality and designation websites.

Other related actions have also been offered by energy producers in trade for permission to operate, such as improving navigation channels to reduce the likelihood of accidental fuel spills. These added costs represent the attachment of a price to certain externalities, and, as a result, end up incorporating the costs of the externalities into the cost of the product.

For some products, such as gasoline, CH_4 or coal, the amount of CO_2 produced when a specified quantity is burned is well-defined. Bill Nye, the Science Guy, along with other knowledgeable members of the scientific community, has suggested a carbon fee be imposed on the purchase of these fuels. The fee would then be applied to a trust fund, similar to the highway trust fund, to be used specifically for research on reducing CO_2 emissions [12].

9.5.4 HEALTH AND SAFETY AS EXTERNALITIES

Litigation against tobacco companies, resulting in the award of billions of dollars in damages to states to help reimburse the costs of tobacco-related illnesses, has shown that the cost of public health has become a recognized externality, with a significant price tag. Other lawsuits have also been successfully prosecuted against a variety of forms of pollution resulting from careless or indiscriminate disposal of many forms of toxic waste. These cases confirm the linkage between public health and its economic cost.

Public safety is also an issue when energy sources are considered. The Chernobyl, Three Mile Island and Fukushima Daiichi nuclear accidents brought to the forefront public concern over nuclear safety. The location of large fuel tanks in highly populated areas also is of concern to public advocacy groups. Transport of various fuels is also problematic, especially when large tanker trucks and railroad tanker cars are involved in accidents that result in the spilling of their contents and subsequent major efforts to clean up the damage before aquifers are affected. Such an accident occurred on June 4, 2016, along the Washington-Oregon border near Mosier, OR, which resulted in the derailment of 16 of 96 tank cars, the burning of four, and an oil slick on the Columbia River [13].

Even in the case of renewable energy sources, concern has been expressed about the possibility of construction accidents, since installation of renewable sources tends to be more labor intensive per installed kW than conventional sources. Manufacture of the materials used in conventional generation facilities carries with it both energy costs and materials costs, along with certain levels of exposure to hazardous or toxic materials. Similar costs are associated with the manufacture of photovoltaic cells and system components. Although these costs tend to be incorporated into the production cost of the materials, when toxic waste byproducts result, they are often treated as externalities.

The balance of this chapter explores the externalities associated with the production, deployment, operation and decommissioning of PV power systems. Particular attention is paid to the environmental concerns associated with each of the phases, recognizing that to produce PV systems, initially other energy sources must be exploited.

9.5.5 Externalities Associated With PV Systems

9.5.5.1 Environmental Effects of PV System Implementation

The implementation phase of each PV technology can be described in terms of a production cycle, which includes potential pollutants or hazardous waste associated with each production step. Fthenakis [21–26] has done extensive studies of the areas of environmental or health concern for current PV technologies and for future technologies that may have promising possibilities for large-scale production. Hodas [9] has explored the relative significance of many of these concerns, including mining, cleaning solvents, fabrication materials, glass, CO_2, CH_4, various dusts and toxic compounds such as silicon dust, B_2H_6, AsH_3, H_2Se, Cd, Li (fire) and Te. From an environmental perspective, he gives cleaning solvents the highest significance, followed by the production of steel, aluminum and concrete. Fortunately, he has also found that all but the cleaning solvents have relatively low cost to control and control strategies are in place for everything on his list.

The associated CO_2 for each technology represents the CO_2 resulting from the conversion of the primary energy used to produce the PV systems, so, in effect, it represents an energy cost of production. In fact, in 2023, the CO_2/kWh of energy for PV production differs by a factor of 3 between PV module producers in countries with higher CO_2/kWh ratios in their electricity generation and producers with lower CO_2/kWh. As the CO_2/kWh for the production of electricity continues to decrease, as a result of PV or other renewable sources, CO_2 and other pollutants associated with the production of PV cells can be expected to continue to decrease. This concept is feasible since current PV systems, including cells, module materials and balance of system will produce, over their useful lifetime, between 8 and 32 times the energy used in their fabrication [14]. It is expected that new technologies and processing techniques will lead to energy returns on investment ranging as high as 45.

Aside from the energy cost, each technology has its own specific toxic waste areas of concern. Some concerns are common to all technologies, such as the production of the materials for support structures and encapsulants for the modules. Perhaps the most important common denominator associated with all technologies is the material used for cleaning the cells, as previously mentioned. In all cases, high levels of cleanliness are required to maximize performance, and in some cases, the solvents

Economic and Environmental Considerations 309

used are highly toxic. Fortunately, these solvents have been in common use in the semiconductor industry and careful means of control are well understood within the industry. The common means of dealing with these solvents is to confine them and then recycle them.

In the case of Si PV cells, the most significant concerns are the proper handling of the dopants for n-type and p-type. Since the semiconductor industry has been handling these components effectively, again the processes are well defined. Once the impurities are in the Si, the amounts are so insignificant that the doped Si is considered to be benign.

For CIS cells and CdTe cells, Cd is the primary item of concern, but since the films in these devices are generally less than 2 μm thick, the total amount of Cd and Te in these cells is very small. Another concern for CIS is the use of H_2Se for the deposit of Se into the cell, but again, the use of this highly toxic substance can be kept well under control with minimal risk of having it escape into the environment.

9.5.5.2 Environmental Effects of PV System Deployment and Operation

The deployment of PV systems has associated environmental and health costs similar to the deployment of other energy technologies [15]. Steel, aluminum and concrete can be expected to be part of the structures upon which the PV arrays and associated BOS components will be mounted. Hence, the production costs and associated environmental costs, such as CO production in the reduction of iron oxides, become associated with PV deployment.

Just as construction accidents occur in any construction project, it is anticipated that PV project construction will also result in construction accidents. For the most part, materials used in the construction of PV facilities are nontoxic. It is thus the responsibility of the installation contractor to ensure that the workplace and work practices follow Occupational Safety and Health Administration (OSHA) rules.

Once the PV system is installed, it quietly generates pollution-free electricity any time the sun is shining. The main environmental concern associated with the operation of PV systems is whether a system containing Cd or Se or Te should be exposed to fire. Analysis has shown, however, that since these materials appear in such minute quantities in thin-film PVs anyone approaching close enough to encounter significant exposure to any of the toxins would face significantly more danger from the fire itself. For example, only 400 g of Cd are present in a 1 MW CIS PV system, whereas 5 g/m^2 of Cd are present in CdTe systems, or about 25 times the amount in CIS systems.

Other dangers associated with the installed system include, of course, the possibility of electrical shock or fire. Systems installed in accordance with the *National Electrical Code* [16] and local, regional and/or national fire safety guidelines as required by local code enforcement jurisdictions will reduce any shock or fire hazard to a minimum, both for the general public and for maintenance personnel. The fact that PV systems are inherently current limiting tends to reduce their potential to produce high-current arcing if they are inadvertently shorted, except for very large systems as discussed in Chapter 4. The incorporation of ground fault and arc fault protection in systems adds further protection.

Since most PV components and materials are benign, the transportation of modules and other system components poses minimal environmental risk compared to the transportation of oil and gas.

A final safety consideration for installed PV systems relates to their ability to withstand high winds. Inadequately mounted PV modules in high winds may tear loose from their support structure and become projectiles. Installations compliant with local building code wind-loading requirements, such as ASCE 7 [17], present minimal risk as long as the PV modules are listed to UL 1703 or IEC61730.

9.5.5.3 Environmental Impact of Large-Scale PV Installations

Of all solar technologies, PV installations appear to have the least serious environmental impact. As a start, conventional PV requires the least amount of water for washing and dust control. Since PV does not require a working fluid or cooling tower (in contrast to CSP technologies such as parabolic troughs, tracking dishes or power towers), the impact on local water resources is negligible. Once a PV installation is completed, it is basically an autonomous operation with no need for round-the-clock operational staff, except perhaps for a small security staff. This minimizes the effect of vehicular traffic on pollution, noise and lighting interference with wildlife habitat and movements. The following mitigation strategies should be considered when designing utility-scale PV installations:

1. PV installations use relatively less water per MW than all other solar technologies. The average annual water for washing and dust control is 24,700 m^3/400 MW, or 20 acre-ft/400 MW [18]. This impact must be mitigated. There are no chemicals used for the PV panel washing except mild soap. However, if any chemicals are used for the control of vegetation, a mitigation plan should be prepared.
2. PV installations have a potential impact on topography that could result in changes to the drainage and water runoff patterns. A well-designed PV installation can minimize this impact. The EPA requires that developments greater than 20 acres (8 hectares) must comply with turbidity concentration guidelines (EPA-2007c). Soil erosion resulting from the construction activity should also be mitigated through standard erosion control practices.
3. The impact on vegetation from trimming to avoid shading of panels is difficult to mitigate. This is one of the more important decisions that must be made in siting a PV plant. The Bureau of Land Management has developed a map [19] showing the regions that would be least impacted by solar installations. To reduce operation and maintenance costs and to minimize the impact on local vegetation and the wildlife that depends on tall grass or vegetation, it is best to study these maps and consult with local land use experts.
4. There is a negligible site lighting impact as PV plants do not operate at night. Lighting is limited to security.
5. PV installations should be planned to avoid crucial life stages of local wildlife such as breeding or nesting seasons. If the site selected is on a path of

Economic and Environmental Considerations 311

endangered species, an expert familiar with the Endangered Species Act [20] and its guidelines should be consulted.
6. Habitat fragmentation that results from the installation of fences or uninterrupted linear components such as rows of solar panels or transmission lines must be considered carefully. Fencing serves the dual purpose of security and also protects wildlife from the collision with solar panels, especially at night. But it can also result in injury or confusion for some of the species that are hardwired to follow a certain path. Fences must be designed to allow reptiles, tortoises and other small animals to get through but also be designed to minimize impact on all local wildlife, and migratory routes.
7. There are some positives, however. Fthenakis [21–24] and colleagues who have written extensively on this subject, have also found that in certain regions, solar PV shading may actually improve the habitat for certain species.
8. In recent years, significant research has been done on agrivoltaics—as was done for the large array example in Chapter 4. A number of installations have demonstrated that by careful planning, PV and agriculture can share the same space. In fact, they can benefit each other by providing shade and feed for animals and by allowing for food crops between rows.

The design of a utility-scale PV installation should start by cataloging a complete list of all potential impacts and developing a mitigation plan to address the more critical factors.

9.5.5.4 Environmental Impact of PV System Decommissioning

Certain regulatory requirements apply to the decommissioning of PV systems. Assuming that the system will be disassembled at the end of its useful lifetime, it is then necessary to consider the destination of the disassembled components. Some of the components will require disposal according to toxic waste regulations unless they can be recycled, and other components will not be subject to quite as stringent requirements.

The Resource and Conservation Recovery Act constitutes the primary set of rules governing wastes containing Cd, Se, Pb, Cu or Ag provided that these wastes are considered to be discarded material and are not included in any specific exclusions [25]. The EPA defines the Toxicity Characteristic Leaching Procedure (TCLP) for 39 materials to determine whether waste products containing them may be classified as hazardous. Any waste product containing any of the 39 materials that yield a soluble concentration in excess of its TCLP limits is considered hazardous.

The California Hazardous Waste Control Law introduces an additional test for toxicity characterization, with two additional indicators. The Waste Extraction Test (WET) is used to identify non-RCRA toxic wastes. The WET threshold limits are defined as the soluble threshold limit concentration (STLC) and the total threshold limit concentration (TTLC), that is the total concentration of listed materials, in any form.

In terms of STLC and TTLC, it appears that CdTe modules may be characterized as hazardous in terms of TTLC and STLC for Cd. CIS modules may be subject to TTLC for Se and STLC for Se and Cd. Polycrystalline Si modules may have a problem with TTLC and STLC for Ag or Pb. In some cases, Pb and Cu may be a problem.

However, a-Si:H modules do not seem to have any toxicity problems [25, 26]. If CdTe and CIS modules are recycled, they are exempt from the disposal regulations. Studies have projected recycling costs of approximately $0.01 to $0.04 per watt for both materials, using relatively standard separation techniques.

Homework Problems

9.1 Obtain the data for CPI, Dow Jones industrial average and effective federal funds rate for the past 20 years from references 1–3. Plot the data and attempt to fit curves to the data to show trends. For example, use Excel graphs with trendlines, equations and R^2 values added to show the "goodness of fit." Compare the R^2 values for linear, exponential and polynomial fits to your curves.

9.2 Determine the present electricity cost for which the $1200 refrigerator of Example 9.1 will have the same LCC as the $1400 unit, assuming all other variables to be the same.

9.3 Next time you are in an appliance store, record the first cost and annual operating costs of several refrigerators that are comparable. Then, making reasonable assumptions about discount rates, inflation rates and appliance lifetimes, compare the LCCs of the units. You might want to ask a salesperson for information on the expected lifetime and repair costs of the units, including annual maintenance contracts.

9.4 Rework Example 9.2 for a system that uses 4 kWh per day. This will require double the PV array and double the batteries, as well as double the fuel for each generator scenario. However, the cost of the generator will remain the same and the cost of the maintenance will remain the same for each case. Assume the cost of the charge controller or battery charger will not increase.

9.5 Consider the LCOE analysis of the 1-MW non-tracking PV shown in the Excel example. The analysis assumes that all the equipment is rated and guaranteed 25 year-life. However current inverter technology is guaranteed for ten years only with the expectation that inverters may need to be replaced every ten years.
   a. Assuming the total cost of labor and equipment change out is fixed at $80/kW, determine how the LCOE would change for inverter replacements at year 10 and at year 20.
   b. Assuming an initial cost of $1.00/W, with all other parameters the same except the Discount Rate, calculate the discount rate that would give an LCOE of $0.035/kWh

9.6 Using the spreadsheet from Problem 9.5 that was based on (9.9), determine a realistic set of values for project cost, O&M, project lifetime, DR, RES, initial kWh/yr and annual degradation rate for a 15 kW residential system that will result in an LCOE of $0.07/kWh. You may want to consider rebates and tax incentives as a part of the project cost.

9.7 a. Construct a spreadsheet similar to the one in Table 9.4 that applies to the storage system from which Table 9.4 was developed. Then

experiment with different charge/discharge rates over the 24-hour period to maximize the profit from the system, assuming the same purchase and sell-back rates that were used in the example.
b. Keep everything the same in the previous example except assume the system has a maximum charge/discharge rate of C/1.0. Use the spreadsheet to explore how to maximize profit again. For both parts a and b, be sure that the charge does not drop below 10% SOC or exceed 100%. Also be sure that charging and discharging are not happening simultaneously.

9.8 Use (9.11) to compare 12 equal monthly payments to one annual payment by modifying the equation to account for monthly interest rate versus annual rate and monthly payments versus annual payments. How does the sum of 12 monthly payments compare with a single annual payment?

9.9 CPV is more sensitive to tracking as the modules need to be within 0.5-degree alignment for proper focus through lenses or reflectors. The lenses or reflectors should also be clean for optimum optical conditions. The cost of OM is therefore higher since modules need to be clean of dust, cell cooling module fins need to be free of sand and debris, and continuous recalibration of the tracker is essential. In fact, based on national data, CPV needs 10 times more water for system washing than PV.

Considering that water availability may be an issue in the coming decades, perform a sensitivity analysis comparing the LCOE shown previously for CPV with a case where water cost escalates. Assuming that the cost of water comprises 90% of the OM annual cost, draw a graph to show the impact of various escalation scenarios on the LCOE using the following:
Case 1: The base case of example 9.4
Case 2: 2% inflation rate for the first 10 years, and 5% for the last 15 years
Case 3: 6% inflation rate for the first 10 years and 4% for the last 15 years
Case 4: 4% inflation rate for the first 10 years and 10% for the last 15 years

9.10 Calculate a set of reasonable conditions on interest rate, term of loan and cost per installed kW, for a PV system that generates an average of 50 kWh per day so the annual loan repayment can be recovered if the value of the electricity generated is $0.20/kWh.

9.11 Net metering is when a utility pays the same rate for electricity sold back to the utility as is charged to the consumer for purchased electricity. Assume a straight grid-connected 10 kW PV system can be installed for $2/watt after all incentive payments are accounted for. Also assume the system is installed in an area where it will produce 1500 kWh/kW/yr at the inverter output. Then assume that the present utility electric rate is $0.15/kWh and that the rate appears to be increasing at 4%/year. If money can be borrowed at 6%/yr,
a. Calculate the annual loan repayment if the loan is for 20 years.
b. Calculate the value of this electricity at the present utility rate.

c. Construct a bar graph that shows the difference between the annual value of the electricity generated and the annual loan repayment rate for a 30-year period. Keep in mind that the loan payments cease after 20 years.

d. Construct the equivalent of the integral of the bar graph of part c) that shows the cumulative return on investment for the system.

9.12 Assume that the system of Problem 9.10 is installed using a single-axis tracking array to increase the annual kWh production by 25%. The additional cost of the tracking array mount increases the cost after incentives to $2.50/watt.

a. If all the other conditions of Problem 9.10 are the same for this case, evaluate parts a–d for the tracking system.

b. Which system appears to be the best choice in terms of payback?

9.13 A battery energy storage system is to be designed to provide a storage capacity of somewhere between 20 and 27 kWh @ 240 V at a C/2 discharge rate. Four battery storage systems are under consideration:

Battery	Capacity	Type	Lifetime	Max P_{out}	Weight	Cost Each
A	5 kWh	LFP	15 yr	3.84 kW	175 lb	$4000
B	10 kWh	LFP	10 yr	3.84 kW	315 lb	$6000
C	13.5 kWh	LFP	10 yr	5 kW	340 lb	$9200
D	25 kWh	Ni-Cd	25 yr	20 kW	1200 lb	$13,000

Assume $i = 2\%$, $d = 5\%$ and a 24-year system lifetime. Perform an LCC for the four battery types and discuss other considerations that may influence the choice of batteries. If the site were a homeowner, what would you recommend? Why? If the site were a remote communication system, what would you recommend? Why?

9.14 a. List five examples of how humans have upset the balance of various natural processes, such as photosynthesis and water purification.

b. List five examples of how humans either have or are currently working to restore the natural balance.

9.15 Look up the U.S. nonattainment regions, if any, on the U.S. Environmental Protection Agency web site for the six toxic air pollutants defined in the Clean Air Act.

9.16 If recycling of CdTe PV modules will cost approximately $0.04/watt and the lifetime of a CdTe module is 20 years, what is the present value of the recycling cost for a 10 kW array? Make reasonable assumptions.

9.17 If burning a gallon of gasoline produces 19.6 pounds of CO_2, estimate the CO_2 that your vehicle emits in a year. Compare this with the amount that would have been generated had you been driving an electric vehicle charged from a PV array or a hybrid auto or a fuel cell–powered auto. Take into account the origin of the H_2 used in the fuel cell.

9.18 In the CO_2 scrubbing and liquefaction scheme reported by Hodas, estimate the increase in electrical cost that would be necessary to offset the cost of the process.

9.19 Compare Article 690 in the 2020 or 2023 *National Electrical Code* with Article 690 in an earlier edition, preferably prior to 2008, in terms of additional safety measures required by the most recent edition. If your local library does not have a copy of an earlier edition, a friendly local electrical contractor or electrical engineer who has been in business for a long time may have one.

9.20 Referring to the water use given in Section 9.4.3 for utility-scale PV installations, assuming the modules in the 400 MW system to be 400 W modules,
   a. Express the water use in terms of gallons per year per module.
   b. Assuming the array will produce 2000 kWh/kW annually, express the water use in terms of gallons per kWh.

REFERENCES

[1] Consumer Price Index Data from 1913 to 2023: www.usinflationcalculator.com/inflation/consumer-price-index-and-annual-percent-changes
[2] Dow Jones Indexes: www.macrotrends.net/1319/dow-jones-100-year-historical-chart
[3] Historical Changes of the Target Federal Funds and Discount Rates: https://fred.stlouisfed.org/series/DFF
[4] Inflation Reduction Act: https://home.treasury.gov/policy-issues/inflation-reduction-act
[5] Short, W., Packey, D. and Holt, T., *A Manual for the Economic Evaluation of Energy Efficiency and Renewable Energy Technologies*, National Renewable Energy Laboratory, Golden, CO, March 1995.
[6] U.S. Department of Energy Goal for 2020 PV LCOE: http://energy.gov/eere/sunshot/sunshot-vision-study
[7] World Resources Institute, Insights: WRI's Blog: www.wri.org/blog/2015/12/4-things-you-need-know-about-current-trends
[8] 21st Conference of the Parties to the U .N. Framework Convention on Climate Change, Paris, France, December 2015.
[9] Hodas, D. R., *External Environmental Costs of Electric Power*, Springer-Verlag, Berlin, 1990, 59–78.
[10] 42 U. S. C. § 7401 et. seq. (Clean Air Act).
[11] Information on a Variety of Pollutants in Areas Within the U.S.: www.epa.gov/green-book.
[12] Nye, Bill, *Unstoppable: Harnessing Science to Change the World*, St. Martin's Press, New York, 2015.
[13] "Protest and Oil Sheen on Columbia River Follow Oregon Train Derailment," *The Guardian*, June 4, 2016, www.theguardian.com/us-news/2016/jun/04/protest-oil-sheen-columbia-river-oregon-train-derailment
[14] National Renewable Energy Laboratories, Information on Energy Return on Investment for Current and New PV Technologies: www.nrel.gov/docs/fy04osti/35489.pdf
[15] Markvart, T., ed., *Solar Electricity*, John Wiley & Sons, Chichester, UK, 1994.
[16] *NFPA 70 National Electrical Code*, 2020 ed., National Fire Protection Association, Quincy, MA, 2019.
[17] *Minimum Design Loads for Buildings and Other Structures, ASCE 7-22*, American Society of Civil Engineers, Reston, VA, 2022.
[18] Frisvold, G. B. and Marquez, T., "Water Requirements for Large Scale Solar Energy Projects in the West," *Journal of Contemporary Water Research and Education*, vol. 151, no. 1, pp. 106–116, August 2013.

[19] U. S. Bureau of Land Management: www.blm.gov/programs/energy-and-minerals/renewable-energy/solar-energy
[20] 16 U. S. C. § 1531 et. seq. (Endangered Species Act).
[21] Fthenakis, V. M. and Kim, H. C., "Greenhouse Gas Emissions from Solar Electric and Nuclear Power: A Life Cycle Study," *Energy Policy*, vol. 35, pp. 2549–2557, 2007.
[22] Fthenakis, V. M., "End of Life Management and Recycling of PV Modules," *Energy Policy*, vol. 28, pp. 105–108, 2000.
[23] Fthenakis, V. M., Fuhrmann, M., et al., "Emissions and Encapsulation of Cadmium in CdTe PV Modules During Fires," *Progress in Photovoltaics: Research and Application*, vol. 13, pp. 713–723, 2005.
[24] Fthenakis, V. M., Green, T., et al., "Large Photovoltaic Power Plants: Wildlife Impacts and Benefits," *Processing of 37th IEEE Photovoltaic Specialists Conference*, Seattle, WA, June 2011.
[25] Fthenakis, V. M. and Moskowitz, P. D., "Emerging Photovoltaic Technologies: Environmental and Health Issues Update," *NREL/SNL Photovoltaics Program Review*, AIP Press, New York, 1997.
[26] Fthenakis, Vasilis, et al., "Emissions from Photovoltaic Life Cycles," *Environmental Science & Technology*, vol. 42, no. 6, p. 2168, March 2008.

SUGGESTED READING

42 U. S. C. § 7401 et. seq. (Clean Air Act).

Markvart, T., ed., *Solar Electricity*, John Wiley & Sons, Chichester, UK, 1994.

"National Appliance Energy Conservation Act of 1987" (PL 100–12, March 17, 1987, 101 § 103).

10 The Physics of Photovoltaic Cells

10.1 INTRODUCTION

By this point, it is possible that the reader, now highly skilled at photovoltaic (PV) system design, might be interested in what goes on inside the PV cell during the process of converting light energy into electrical energy. This chapter is presented with the purpose of enabling the reader to become familiar with the challenges facing those engineers and physicists who spend their lives working on processes and materials aimed at reducing the cost and increasing the efficiency of PV cells. Initially, the basics of PV energy conversion are presented, followed by a discussion of present limitations of cell production and some of the ideas that have emerged toward overcoming these limitations. Most readers will have already had at least an exposure to the theory of operation of the pn junction and the semiconductor diode in an electronics course, so it will be assumed that the reader will at least be familiar with the basic diode equation.

10.2 OPTICAL ABSORPTION

10.2.1 Introduction

When light shines on a material, it is reflected, transmitted or absorbed. Absorption of light is simply the conversion of the energy contained in the incident photon to some other form of energy, typically heat. Some materials, however, happen to have just the right properties needed to convert the energy in the incident photons to electrical energy.

When a photon is absorbed, it interacts with an atom in the absorbing material by giving off its energy to an electron in the material. This energy transfer is governed by the rules of conservation of momentum and conservation of energy. Since the zero mass photon has very small momentum compared to the electrons and holes, the transfer of energy from a photon to a material occurs with inconsequential momentum transfer. Depending on the energy of the photon, an electron may be raised to a higher energy state within an atom or it may be liberated from the atom. Liberated electrons are then capable of moving through the crystal in accordance with whatever phenomena may be present that could cause the electron to move, such as temperature, diffusion or an electric field.

10.2.2 Semiconductor Materials

Semiconductor materials are characterized as being perfect insulators at absolute zero temperature, with charge carriers being made available for conduction as the temperature of the material is increased. This phenomenon can be explained on the basis of quantum theory, by noting that semiconductor materials have an energy band gap between the valence band and the conduction band. The valence band represents the allowable energies of valence electrons that are bound to host atoms. The conduction band represents the allowable energies of electrons that have received energy from some mechanism and are now no longer bound to specific host atoms.

At T = 0 K, all allowable energy states in the valence band of a semiconductor are occupied by electrons, and no allowable energy states in the conduction band are occupied. Since the conduction process requires that charge carriers move from one state to another within an energy band, no conduction can take place when all states are occupied or when all states are empty. This is illustrated in Figure 10.1a.

As the temperature of a semiconductor sample is increased, sufficient energy is imparted to a small fraction of the electrons in the valence band for them to move to the conduction band. In effect, these electrons leave covalent bonds in the semiconductor host material. When an electron leaves the valence band, an opening is left which may now be occupied by another electron, provided that the other electron moves to the opening. If this happens, of course, the electron that moves in the valence band to the opening leaves behind an opening in the location from which it moved. If one engages in an elegant quantum mechanical explanation of this phenomenon, it must be concluded that the electron moving in the valence band must have either a negative effective mass and a negative charge or, alternatively, a positive effective mass and a positive charge. The latter has been the popular description, and, as a result, the electron motion in the valence band is called hole motion, where "holes" is the name chosen for the positive charges, since they relate to the moving holes that the electrons have left in the valence band.

What is important to note about these conduction electrons and valence holes is that they have occurred in pairs. Hence, when an electron is moved from the valence band to the conduction band in a semiconductor by whatever means, it constitutes the creation of an electron–hole pair (EHP). Both charge carriers are then free to become a part of the conduction process in the material.

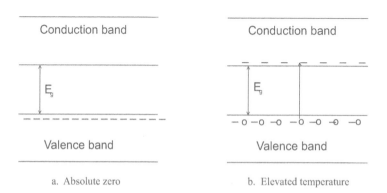

FIGURE 10.1 Illustration of availability of states in the valence band and conduction band for the semiconductor material.

10.2.3 Generation of EHP by Photon Absorption

The energy in a photon is given by the familiar equation,

$$E = h\nu = \frac{hc}{\lambda}, \tag{10.1}$$

where h is Planck's constant (h = 6.63×10^{-34} joule-sec), c is the speed of light (c = 2.998×10^8 m/sec), ν is the frequency of the photon in Hz and λ is the wavelength of the photon in meters. Since energies at the atomic level are typically expressed in electron volts (1 eV = 1.6×10^{-19} joules) and wavelengths are typically expressed in micrometers (μm), it is possible to express hc in appropriate units so that if λ is expressed in μm, then E will be expressed in eV. The conversion yields

$$E = \frac{1.24}{\lambda} (eV). \tag{10.1a}$$

The energy in a photon must exceed the semiconductor bandgap energy, E_g, to be absorbed. Photons with energies at and just above E_g are most readily absorbed because they most closely match bandgap energy and momentum considerations. If a photon has energy greater than the bandgap, it still can produce only a single EHP. The remainder of the photon energy is lost to the cell as heat. It is thus desirable that the semiconductor used for photoabsorption have a bandgap energy such that a maximum percentage of the solar spectrum will be efficiently absorbed.

Now, referring back to the Planck formula for blackbody radiation in Chapter 2 (2.1), note that the solar spectrum peaks at λ ≈ 0.5 μm. Equation (10.1a) shows that a bandgap energy of approximately 2.5 eV corresponds to the peak in the solar spectrum. In fact, since the peak of the solar spectrum is relatively broad, bandgap energies down to 1.0 eV can still be relatively efficient absorbers, and in certain special cell configurations to be discussed later, even smaller bandgap materials are appropriate.

The nature of the bandgap also affects the efficiency of absorption in a material. A more complete representation of semiconductor bandgaps must show the relationship between bandgap energy as well as bandgap momentum. As electrons make transitions between the conduction band and valence band, both energy and momentum transfer normally take place, and both must be properly balanced in accordance with conservation of energy and conservation of momentum laws.

Some semiconducting materials are classified as direct bandgap materials, while others are classified as indirect bandgap materials. Figure 10.2 shows the bandgap diagrams for two materials considering momentum as well as energy. Note that, for silicon, the bottom of the conduction band is displaced in the momentum direction from the peak of the valence band. This is an indirect bandgap, while the GaAs diagram shows a direct bandgap, where the bottom of the conduction band is aligned with the top of the valence band.

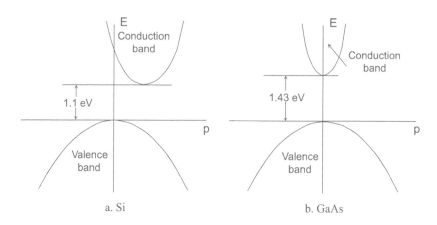

FIGURE 10.2 The energy and momentum diagram for valence and conduction bands in Si and GaAs.

What these diagrams show is that the allowed energies of a particle in the valence band or the conduction band depend on the particle momentum in these bands. An electron transition from a point in the valence band to a point in the conduction band must involve conservation of momentum as well as energy. For example, in Si, even though the separation of the bottom of the conduction band and the top of the valence band is 1.1 eV, it is difficult for a 1.1 eV photon to excite a valence electron to the conduction band because the transition needs to be accompanied with sufficient momentum to cause displacement along the momentum axis, and photons carry little momentum. The valence electron must thus simultaneously gain momentum from another source as it absorbs energy from the incident photon. Since such simultaneous events are unlikely, absorption of photons at the Si bandgap energy is several orders of magnitude less likely than absorption of higher energy photons.

Since photons have so little momentum, it turns out that the direct bandgap materials, such as gallium arsenide (GaAs), cadmium telluride (CdTe), copper indium diselenide (CIS), Perovskites and amorphous silicon (a-Si:H) absorb photons with energy near the material bandgap energy much more readily than do the indirect materials, such as crystalline silicon. As a result, the direct-bandgap absorbing material can be several orders of magnitude thinner than indirect bandgap materials and still absorb a significant part of the incident radiation.

The absorption process is similar to many other physical processes, in that the change in intensity with position is proportional to the initial intensity. As an equation, this becomes

$$\frac{dI}{dx} = -\alpha I, \tag{10.2}$$

with the solution

$$I = I_0 e^{-\alpha x}, \tag{10.3}$$

The Physics of Photovoltaic Cells

where I is the intensity of the light at a depth x in the material, I_o is the intensity at the surface and α is the absorption constant. The absorption constant depends on the material and on the wavelength [1]. Equation (10.3) shows that the thickness of material needed for significant absorption needs to be several times the reciprocal of the absorption constant. This is important information for the designer of a PV cell, since the cell must be sufficiently thick to absorb the incident light. In some cases, the path length is increased by causing the incident light to reflect from the front and back surfaces while inside the material until it ultimately generates an EHP. For all materials, no photon absorption takes place at wavelengths corresponding to photon energies less than the bandgap.

Figure 10.3 [2] shows the maximum theoretical photon collection efficiency versus bandgap energy. It is interesting to note that even though smaller bandgap energy suggests collection of more photons, while this may be true, what happens to the photons after collection is what determines the value of the material for creation of EHPs or other means that can contribute to the conversion of photon energy to electrical energy. It is important to remember that photons collected at energies greater than the bandgap energy lose excess energy to heat. Furthermore, as the photon energy moves away from the peak of the solar spectrum, fewer photons are available at that energy for conversion to any form of energy. In any case, when the photon is absorbed, it generates an EHP. The question, then, is what happens to the EHP?

10.2.4 Photoconductors

Once an EHP is generated, it becomes a question of how long the EHP lasts before the conduction electron returns to the valence band. Remember that the creation of an EHP does not imply that the electron will remain in the conduction band

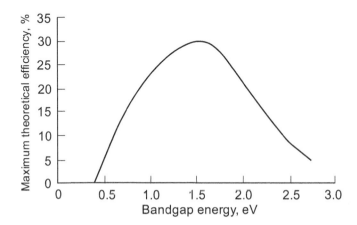

FIGURE 10.3 Maximum theoretical photon collection efficiency versus bandgap energy. (Adapted from Granata, J., Sites, J. R. and Tuttle, J. R. in Proc 14th NREL PV Program Review, AIP Conference Proceedings 394, Lakewood, CO, 1996, 621–630.)

and the hole will remain in the valence band. Thermal equilibrium in a semiconductor comprises a constant generation and recombination of EHPs, so that, on average, the population of electrons and the population of holes remain constant. In fact, the product of the concentration of holes and the concentration of electrons in thermal equilibrium is a constant, which depends on temperature and the bandgap energy of the semiconductor, along with a few other parameters unique to each semiconductor. If n_o represents the thermal equilibrium concentration of electrons per cm³ and if p_o represents the thermal equilibrium concentration of holes per cm³, then, in thermal equilibrium, with n_o and p_o at any single location in the material,

$$n_o p_o = n_i^2(T) = KT^3 e^{-\frac{E_g}{kT}} \qquad (10.4)$$

where n_i represents the intrinsic carrier concentration, K is semiconductor-material dependent, as will be discussed later, E_g is the bandgap energy, k is Boltzmann's constant and T is the temperature in K [3]. Note that materials with smaller bandgap energies have higher intrinsic carrier concentrations. In addition, intrinsic carrier concentrations have significant temperature dependence, with a rapid increase as temperature increases.

Electrons and holes generated as a result of optical absorption bring the material into a state of nonthermal equilibrium [4]. The photon-generated excess electrons and holes remain, on average, for a time, τ, which is defined as the excess carrier lifetime. The recombination of EHPs is a statistical process in which some EHPs recombine in times shorter than the carrier lifetime and some take longer to recombine. EHPs, then, are subject to a generation rate, measured in number per cm³ per sec, which is proportional to the incident photon flux, and are subject to a recombination rate that is proportional to the departure of np from its equilibrium value. Under steady-state conditions, these two rates are equal.

The conductivity of a material is proportional to the density of free charge carriers and is given by

$$\sigma = q(\mu_n n + \mu_p p) \qquad (10.5)$$

where μ_n and μ_p represent, respectively, the mobility of electrons and the mobility of holes in the material. Mobility is simply a measure of how easily the particles can move around in the material when an electric field is present.

Equation (10.5) clearly indicates that, if shining a light on a piece of material can create excess charge carriers, then the conductivity of the material will increase, causing a corresponding decrease in the resistance of the material. The material becomes a light-sensitive resistor. A common material used for such devices is cadmium sulfide (CdS), which is used in many of the light sensors that turn lights on after dark. The most sensitive photoconductors are materials that have long lifetimes of excess EHPs. However, if the problem is to detect short bursts of light occurring at a high repetition rate, it is necessary to have any excess population of electrons and holes quickly die out as soon as the light source is removed. The trade-off between speed and sensitivity is similar to the trade-off between bandwidth and gain for an amplifier as described by the familiar gain-bandwidth product.

Note that even though EHPs are generated in the host material, the material remains passive because the generated EHPs have random thermal velocities. This means that no net current flow results from their creation and, since no separation of charges occurs, no voltage is produced. With no resulting voltage or current, the only effect of the creation of the additional charge carriers is the reduction in resistance of the host material. The next step, then, is to figure out a way to get work out of these photon-generated charges.

10.3 EXTRINSIC SEMICONDUCTORS AND THE PN JUNCTION

10.3.1 Extrinsic Semiconductors

Up to this point, the semiconductors discussed have been intrinsic semiconductors, meaning that the populations of holes and electrons have been equal. A somewhat more formal definition of an intrinsic semiconductor takes into account differences in electron and hole mobilities and defines intrinsic semiconductors as materials for which the Fermi level energy is at the center of the bandgap. Since Fermi levels have not been discussed and since electron and hole mobilities are generally close enough to keep the Fermi level quite close to the center of the bandgap, the equal carrier definition will suffice for the following discussion. The reader is referred to a text on semiconductor device physics for more details on Fermi levels [1, 3].

At $T = 0$ K, intrinsic semiconductors have all covalent bonds completed with no leftover electrons or holes. If certain impurities are introduced into intrinsic semiconductors, there can be leftover electrons or holes at $T = 0$ K. For example, consider silicon, which is a group IV element, which covalently bonds with four nearest neighbor atoms to complete the outer electron shells of all the atoms. At $T = 0$ K, all the covalently bonded electrons are in place, whereas at room temperature, about one in 10^{12} of these covalent bonds will break, forming an EHP, resulting in minimal charge carriers for current flow.

If, however, phosphorous, a group V element, is introduced into the silicon in small quantities, such as one part in 10^6, four of the valence electrons of the phosphorous atoms will covalently bond to the neighboring silicon atoms, while the fifth valence electron will have no electrons with which to covalently bond. This fifth electron remains weakly coupled to the phosphorous atom, readily dislodged by temperature, since it requires only 0.04 eV to excite the electron from the atom to the conduction band. At room temperature, sufficient thermal energy is available to dislodge essentially all of these extra electrons from the phosphorous impurities. These electrons thus enter the conduction band under thermal equilibrium conditions, and the concentration of electrons in the conduction band becomes nearly equal to the concentration of phosphorous atoms, since the impurity concentration is normally on the order of 10^8 times larger than the intrinsic carrier concentration. Since the phosphorous atoms donate electrons to the material, they are called *donor* atoms and are represented by the concentration, N_D. Note that the phosphorous, or other group V impurities, *do not add holes* to the material. They only add electrons. They are thus designated as n-type impurities.

On the other hand, if group III atoms such as boron are added to the intrinsic silicon, they have only three valence electrons to covalently bond with nearest silicon neighbors. The missing covalent bond appears the same as a hole, which can be released into the material with a small amount of thermal energy. Again, at room temperature, nearly all of the available holes from the group III impurity are donated to the conduction process in the host material. Since the concentration of impurities will normally be much larger than the intrinsic carrier concentration, the concentration of holes in the material will be approximately equal to the concentration of impurities.

Historically, group III impurities in silicon have been viewed as electron acceptors, which, in effect, donate holes to the material. Rather than being termed hole donors, however, they have been called *acceptors*. Thus, acceptor impurities donate holes, but no electrons, to the material, and the resulting hole density is approximately equal to the density of acceptors, which is represented as N_A. Figure 10.4 shows the effects of donor and acceptor impurities on the intrinsic material along with the positions of the energy levels of the impurities in the bandgap of the material.

Equation (10.5) shows that adding a mere one part in a million of a donor or acceptor impurity can increase the conductivity of the material by a factor of 10^8 for silicon. Equation (10.4) shows that in thermal equilibrium, if either an n-type or a p-type impurity is added to the host material, the concentration of the other charge carrier will decrease dramatically, since it is still necessary to satisfy (10.4). In extrinsic semiconductors, the charge carrier with the highest concentration is called the *majority carrier* and the charge carrier with the lowest concentration is called the *minority carrier*. Hence, electrons are the majority carriers in n-type material, and holes are the minority carriers in n-type material. The opposite is true for p-type material.

If both n-type and p-type impurities are added to a material, then whichever impurity has the higher concentration will become the dominant impurity. However, it is then necessary to acknowledge a net impurity concentration that is given by the difference between the donor and acceptor concentrations. If, for example, $N_D > N_A$, then the net impurity concentration is defined as $N_d = N_D - N_A$. Similarly, if $N_A > N_D$, then $N_a = N_A - N_D$.

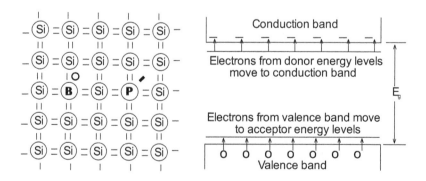

FIGURE 10.4 Acceptor and donor impurities in Si.

The Physics of Photovoltaic Cells

10.3.2 THE PN JUNCTION

10.3.2.1 Drift and Diffusion

When charged particles are placed in an electric field, they are exposed to an electrostatic force. This force accelerates the particles until they undergo a collision with another component of the material that slows them down. They then accelerate once again and collide again. The process continues, with the net result of the charge carrier's achieving an average velocity, either in the direction of the electric field for positive charges or opposite the electric field for negative charges. It should be noted that this average, or *drift* velocity, is superimposed on the thermal velocity of the charge carrier. Normally the thermal velocity is much larger than the drift velocity, but the thermal velocity is completely random so the net displacement of the charge carriers is zero. Another way to consider the thermal velocity is that at any instant, it is equally likely that a particle will be moving in any direction.

Drift current, then, is simply the component of current flow due to the presence of an electric field and is described by the familiar equation,

$$\vec{J} = \sigma \vec{E}, \quad (10.6)$$

This is simply the vector form of Ohm's law, where J is the current density in A/cm² when σ is expressed in $\Omega^{-1}\text{cm}^{-1}$ and E is measured in V/cm.

Diffusion is that familiar process by which the random thermal motion of particles causes them to ultimately distribute themselves uniformly within a space. Whenever particles are in thermal motion, they will tend to move from areas of greater concentration to areas of lesser concentration, simply because at any point, the probability of motion in all directions is equal. Suppose, for example, that regions A and B are adjoining, as shown in Figure 10.5. Suppose also that all particles in both regions are experiencing random thermal motion and that the concentration of certain particles, z, in region A is greater than the concentration of z-particles in region B. At any instant, half the z-particles in region A will be moving toward B, and half the z-particles in B will be moving toward A. Since there are more z-particles in A, the net motion of z-particles from A to B will continue until the concentrations in A and B are equal. If the particles are holes, this net movement from regions of greater concentration to regions of lesser concentration constitutes a flow of current that can be described in one dimension by the equation

$$J_p = -qD_p \frac{dp}{dx}, \quad (10.7a)$$

and if the particles are electrons,

$$J_n = qD_n \frac{dn}{dx}. \quad (10.7b)$$

In (10.7a) and (10.7b) the change in concentration with position is known as the concentration gradient and D_p and D_n are the hole diffusion constant and the electron diffusion constant, respectively. The minus sign in (10.7a) accounts for the fact that if the gradient is negative, the holes flow in the positive x-direction. The lack of the minus sign in (10.7b), accounts for the fact that if electrons flow in the positive x-direction, the associated current is in the negative x-direction. In each case, q = 1.6×10^{-19} coulomb.

At this point, all that is necessary is to make two additional observations. The first observation is that donor and acceptor atoms become donor and acceptor ions when they give up their electron or hole to the host material. These ions are fixed in position in the host by covalent bonds. The second observation is that in either n-type material or p-type material, any point in the bulk material will have charge neutrality. That is, the net charge present at any point is zero due to positive charges being neutralized by negative charges. However, when n-type and p-type materials are joined to form a pn junction, something special happens at the boundary.

10.3.2.2 Junction Formation and Built-In Potential

Although n-type and p-type materials are interesting and useful, the real fun starts when a junction is formed between n-type and p-type materials. The pn junction is treated in gory detail in most semiconductor device textbooks. Here, the need is to establish the foundation for the establishment of an electric field across a pn junction and to note the effect of this electric field on photo-generated EHPs.

Figure 10.6 shows a pn junction formed by placing p-type impurities on one side and n-type impurities on the other side. There are many ways to accomplish this structure. The most common is the diffused junction. To form a diffused pn junction, the host material is grown with impurities, so it will be either n-type or p-type. The material is either grown or sliced into an appropriate thickness. Then the material is

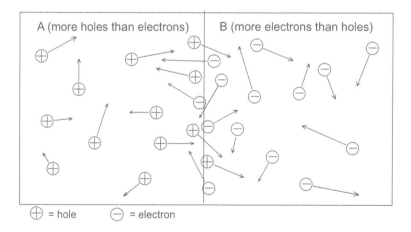

FIGURE 10.5 Random thermal motion and diffusion for electrons and holes.

The Physics of Photovoltaic Cells

heated in the presence of the opposite impurity, which is usually in the form of a gas. This impurity will diffuse into the host material at a level that exceeds the host impurity level, but will only penetrate a small distance into the host material, depending on how long the host material is left at the elevated temperature. The result is a layer of material of one dominant impurity on top of the remainder of the material, which is doped with the other dominant impurity.

When the two materials are brought together, the first thing to happen is that the conduction electrons on the n-side of the junction notice the scarcity of conduction electrons on the p-side, and the valence holes on the p-side notice the scarcity of valence holes on the n-side. Since both types of charge carriers undergo random thermal motion, they begin to diffuse to the opposite side of the junction in search of wide open spaces. The result is the diffusion of electrons and holes across the junction, as indicated in Figure 10.6.

When an electron leaves the n-side for the p-side, however, it leaves behind a positive donor ion on the n-side, right at the junction. Similarly, when a hole leaves the p-side for the n-side, it leaves a negative acceptor ion on the p-side. If large numbers of holes and electrons travel across the junction, large numbers of fixed positive and negative ions are left at the junction boundaries. These fixed ions, as a result of Gauss' law, create an electric field that originates on the positive ions and terminates on the negative ions. Hence, the number of positive ions on the n-side of the junction must be equal to the number of negative ions on the p-side of the junction.

The electric field across the junction, of course, gives rise to a drift current in the direction of the electric field. This means that holes will travel in the direction of the electric field and electrons will travel opposite the direction of the field, as shown in Figure 10.6. Notice that for both the electrons and the holes, the drift current component is opposite the diffusion current component. At this point, one can invoke Kirchhoff's current law to establish that the drift and diffusion components for each charge carrier must be equal and opposite, since there is no net current flow through the junction region. This phenomenon is known as the law of detailed balance. By

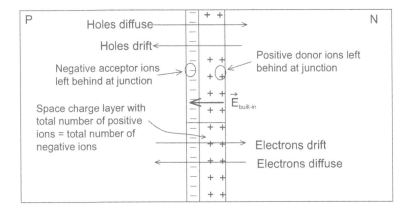

FIGURE 10.6 The pn junction showing electron and hole drift and diffusion.

setting the sum of the electron diffusion current and the electron drift current equal to zero and recalling from electromagnetic field theory that

$$E = -\frac{dV}{dx} \qquad (10.8)$$

it is possible to solve for the potential difference across the junction in terms of the impurity concentrations on either side of the junction. Proceeding with this operation yields

$$-q\mu_n n \frac{dV}{dx} + qD_n \frac{dn}{dx} = 0, \qquad (10.9a)$$

that can be rewritten as

$$dV = \frac{D_n}{\mu_n} \frac{dn}{n}. \qquad (10.9b)$$

Finally, recognizing the Einstein relationship, $D_n/\mu_n = kT/q$, which is discussed in solid-state physics textbooks, and integrating both sides from the n-side of the junction to the p-side of the junction, yields the magnitude of the built-in voltage across the junction to be

$$V_j = \frac{kT}{q} \ln \frac{n_{no}}{n_{po}} \qquad (10.10a)$$

It is now possible to express the built-in potential in terms of the impurity concentrations on either side of the junction by recognizing that $n_{no} \cong N_D$ and $n_{po} \cong (n_i)^2/N_A$. Substituting these values into (10.10a) yields, finally,

$$V_j = \frac{kT}{q} \ln \frac{N_A N_D}{n_i^2}. \qquad (10.10b)$$

At this point, a word about the region containing the donor ions and acceptor ions is in order. Note first that outside this region, electron and hole concentrations remain at their thermal equilibrium values. Within the region, however, the concentration of electrons must change from the high value on the n-side to the low value on the p-side. Similarly, the hole concentration must change from the high value on the p-side to the low value on the n-side. Considering that the high values are really high, that is, on the order of $10^{18}/cm^3$, while the low values are really low, that is, on the order of $10^2/cm^3$, this means that within a short distance of the beginning of the ionized region, the concentration must drop to significantly below the equilibrium value. Because the concentrations of charge carriers in the ionized region are so low, this region is often termed the *depletion region*, in recognition of the depletion of mobile charge carriers

The Physics of Photovoltaic Cells

in the region. Furthermore, because of the charge due to the ions in this region, the depletion region is also often referred to as the *space charge layer*. For the balance of this text, this region will simply be referred to as the junction. The next step in the development of the behavior of the pn junction in the presence of sunlight is to let the sunshine in and see what happens.

10.3.2.3 The Illuminated pn Junction

Equation (10.3) governs the absorption of photons at or near a pn junction. Noting that an absorbed photon releases an EHP, it is now possible to explore what happens after the generation of the EHP. Those EHPs generated within the pn junction will be considered first, followed by the EHPs generated outside, but near, the junction.

If an EHP is generated within the junction, as shown in Figure 10.7 (points B and C), both charge carriers will be acted upon by the built-in electric field. Since the field is directed from the n-side of the junction to the p-side of the junction, the field will cause the electrons to be swept quickly toward the n-side and the holes to be swept quickly toward the p-side. Once out of the junction region, the optically generated carriers become a part of the majority carriers of the respective regions, with the result that excess concentrations of majority carriers appear at the edges of the junction. These excess majority carriers then diffuse away from the junction toward the external contacts, since the concentration of majority carriers has been enhanced only near the junction.

The addition of excess majority charge carriers to each side of the junction results in either a voltage between the external terminals of the material or a flow of current in the external circuit or both. If an external wire is connected between the n-side of the material and the p-side of the material, a current, I_l, will flow in the wire from the p-side to the n-side. This current will be proportional to the number of EHPs generated in the junction region.

If an EHP is generated outside the junction region, but close to the junction (with "close" yet to be defined, but shown as point A in Figure 10.7), it is possible that, due

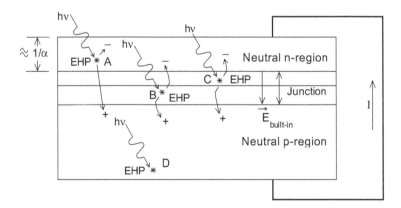

FIGURE 10.7 The illuminated pn junction showing desirable geometry and the creation of electron–hole pairs.

to random thermal motion, the electron, the hole or both will end up moving into the junction region. Suppose, for example, that an EHP is generated in the n-region close to the junction. Then suppose the hole, which is the minority carrier in the n-region, manages to reach the junction before it recombines. If it can do this, it will be swept across the junction to the p-side and the net effect will be the same as if the EHP had been generated within the junction, since the electron is already on the n-side as a majority carrier. Similarly, if an EHP is generated within the p-region, but close to the junction, and if the minority carrier electron reaches the junction before recombining, it will be swept across to the n-side where it is a majority carrier. So, what is meant by close?

Clearly, the minority carriers of the optically generated EHPs outside the junction region must not recombine before they reach the junction. If they do, then, effectively, both carriers are lost from the conduction process, as in point D in Figure 10.7. Since the majority carrier is already on the correct side of the junction, the minority carrier must therefore reach the junction in less than a minority carrier lifetime, τ_n or τ_p.

To convert these times into distances, it is necessary to note that the carriers travel by diffusion once they are created. Since only the thermal velocity has been associated with diffusion, but since the thermal velocity is random in direction, it is necessary to introduce the concept of minority carrier diffusion length, which represents the distance, on average, which a minority carrier will travel before it recombines. The diffusion length can be shown to be related to the minority carrier lifetime and diffusion constant by the formula

$$L_m = \sqrt{D_m \tau_m} \qquad (10.11)$$

where m has been introduced to represent n for electrons or p for holes. It can also be shown that on average, if an EHP is generated within a minority carrier diffusion length of the junction, the associated minority carrier will reach the junction. In reality, some minority carriers generated closer than a diffusion length will recombine before reaching the junction, while some minority carriers generated more than a diffusion length away from the junction will reach the junction before recombining.

Hence, to maximize photocurrent, it is desirable to *maximize the number of photons that will be absorbed either in the junction or within a minority carrier diffusion length of the junction*. The minority carriers of the EHPs generated outside this region have a higher probability of recombining before they have a chance to diffuse to the junction. If a minority carrier from an optically generated EHP recombines before it crosses the junction and becomes a majority carrier, it, along with the opposite carrier with which it recombines, is no longer available for conduction. Furthermore, the combined width of the junction and the two diffusion lengths should be several multiples of the reciprocal of the absorption constant, α, and the junction should be relatively close to a diffusion length from the surface of the material upon which the photon impinges, to maximize the collection of photons. Figure 10.7 shows this

The Physics of Photovoltaic Cells

desirable geometry. The engineering design challenge then lies in maximizing α, as well as maximizing the junction width and minority carrier diffusion lengths.

10.3.2.4 The Externally Biased PN Junction

To complete the analysis of the theoretical performance of the pn junction operating as a PV cell, it is useful to look at the junction with external bias. Figure 10.8 shows a pn junction connected to an external battery with the internally generated electric field direction included. If (10.10a) is recalled, taking into account that the externally applied voltage, with the exception of any voltage drop in the neutral regions of the material, will appear as opposing the junction voltage, the equation becomes

$$V_j - V = \frac{kT}{q} \ln \frac{n_n}{n_p}. \quad (10.12)$$

Note that the only difference between (10.12) and (10.10a) is that the electron concentrations on the n-side and on the p-side of the junction are no longer expressed as the thermal equilibrium values. This will be the case only when the externally applied voltage is zero. However, under conditions known as low injection levels, it will still be the case that the concentration of electrons on the n-side will remain close to the thermal equilibrium concentration. For this condition, (10.12) becomes

$$V_j - V = \frac{kT}{q} \ln \frac{N_d}{n_p}. \quad (10.13)$$

Since V_j can be calculated from (10.10b), the quantity of interest in (10.13) is n_p, the concentration of minority carriers at the edge of the junction on the p-side. Equation (10.13) can thus be solved for n_p with the result

$$n_p = N_D e^{-\frac{q(V_j - V)}{kT}} = N_D e^{\frac{-qV_j}{kT}} e^{\frac{qV}{kT}}. \quad (10.14)$$

Next, note that the thermal equilibrium value of the minority carrier concentration occurs when V = 0. If the excess minority carrier concentration is now defined as $n'_p = n_p - n_o$ and if n_o is subtracted from (10.13), the result is

$$n'_p(0) = n_p(0) - n_o(0) = N_D e^{\frac{-qV_j}{kT}} \left(e^{\frac{qV}{kT}} - 1 \right) = n_{po}\left(e^{\frac{qV}{kT}} - 1 \right). \quad (10.15)$$

What happens to these excess minority carriers? They diffuse toward a region of lesser concentration, which happens to be away from the junction toward the contact, as shown in Figure 10.9.

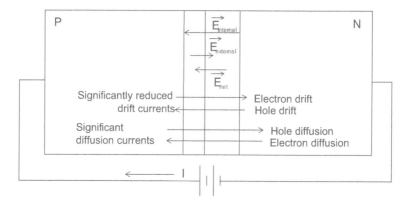

FIGURE 10.8 The pn junction with external bias.

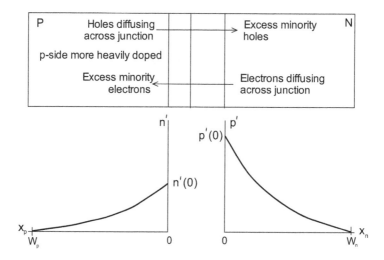

FIGURE 10.9 Excess minority carrier concentrations in neutral regions of pn junction device.

If the contact is a few diffusion lengths away from the junction, it can be shown, by solving the well-known continuity equation, that the distribution of excess minority electrons between the junction and the contact will be

$$n'_p(x_p) = n'_p(0)\cosh\frac{x_p}{L_n} - n'_p(0)\operatorname{ctnh}\frac{W_p}{L_n}\sinh\frac{x_p}{L_n}. \qquad (10.16)$$

A similar expression can be obtained for the concentration of excess minority holes in the neutral region of the n-side of the device. To obtain an expression for the total device current, (10.7) is used for each side of the device, noting that the gradient in the total carrier concentration is given by the gradient of the excess carrier concentration, since the spatial variation in the excess concentration will far exceed any spatial variation in the equilibrium concentration. Combining the results for electron and hole currents yields the total current in the device as it depends on the externally applied voltage along with the indicated device parameters. The result is, if A represents the cross-sectional area of the pn junction and adjoining regions,

$$I = I_n + I_p = qA\left(\frac{D_n n_{po}}{L_n}\operatorname{ctnh}\frac{W_p}{L_n} + \frac{D_p p_{no}}{L_p}\operatorname{ctnh}\frac{W_n}{L_p}\right)\left(e^{\frac{qV}{kT}} - 1\right). \quad (10.17)$$

Equation (10.17) is, of course, the familiar diode equation that relates diode current to diode voltage. Note that the current indicated in (10.17) flows in the direction opposite to the optically generated current described earlier. Letting qA (nasty expression) = I_o and incorporating the photocurrent component into (10.17) finally yields the complete equation for the current in the PV cell to be

$$I = I_\ell - I_o\left(e^{\frac{qV}{kT}} - 1\right). \quad (10.18)$$

Those readers who remember everything they read will recognize (10.18) to be the same as (3.1). Note that the current of (10.18) is directed out of the positive terminal of the device so that when the current and voltage are both positive, the device is delivering power to the external circuit.

10.4 MAXIMIZING PV CELL PERFORMANCE

10.4.1 INTRODUCTION

Equation (10.18) indicates, albeit in a somewhat subtle manner, that to maximize the power output of a PV cell, it is desirable to maximize the open-circuit voltage, short-circuit current and fill factor of a cell. Recalling the plot of (10.18) from Chapter 3, it should be evident that maximizing the open-circuit voltage and the short-circuit current will maximize the power output for an ideal cell characterized by (10.18). Real cells, of course, have some series resistance, so there will be power dissipated by this resistance, similar to the power loss in a conventional battery due to its internal resistance. In any case, recalling that the open-circuit voltage increases as the ratio of photocurrent to reverse saturation current increases, desirable design criteria is to maximize this ratio, provided that it does not proportionally reduce the short-circuit current of the device.

Fortunately, this is not the case, since maximizing the short-circuit current requires maximizing the photocurrent. It is thus instructive to look closely at the parameters that determine both the reverse saturation current and the photocurrent. Techniques for lowering series resistance will then be discussed.

10.4.2 Minimizing the Reverse Saturation Current

Beginning with the reverse saturation current as expressed in (10.17), the first observation is that the equilibrium minority carrier concentrations at the edges of the pn junction are related to the intrinsic carrier concentration through (10.4). Hence,

$$p_{no} = \frac{n_i^2}{N_D} \text{ and } n_{po} = \frac{n_i^2}{N_A}. \qquad (10.19)$$

So far, no analytic expression for the intrinsic carrier concentration has been developed. Such an expression can be obtained by considering Fermi levels, densities of states and other quantities that are discussed in solid-state devices textbooks. Since the goal here is to determine how to minimize the reverse saturation current and not to go into detail about quantum mechanical proofs, the result is noted here with the recommendation that the interested reader consult a good solid-state devices text for the development of the result. The result is

$$n_i^2 = 4\left(\frac{2\pi kT}{h^2}\right)^3 (m_n^* m_p^*)^{\frac{3}{2}} e^{-\frac{E_g}{kT}}, \qquad (10.20)$$

where m_n^* and m_p^* are the electron-effective mass and hole-effective mass in the host material and E_g is the bandgap energy of the host material. These effective masses can be greater than or less than the rest mass of the electron, depending on the degree of curvature of the valence and conduction bands when plotted as energy versus momentum as in Figure 10.2. In fact, the effective mass can also depend on the band in which the carrier resides in a material. For more information on effective mass, the reader is encouraged to consult the references listed at the end of the chapter.

Now, using (10.11) with (10.19) and (10.20) in (10.17), the following final result for the reverse saturation current is obtained.

$$I_o = \left(4qA\left(\frac{2\pi kT}{h^2}\right)(m_n^* m_p^*)^{\frac{3}{2}} e^{-\frac{E_g}{kT}}\right) \times \\ \left(\frac{1}{N_A}\sqrt{\frac{D_n}{\tau_n}}\operatorname{ctnh}\frac{\ell_p}{\sqrt{D_n \tau_n}} + \frac{1}{N_D}\sqrt{\frac{D_p}{\tau_p}}\operatorname{ctnh}\frac{\ell_n}{\sqrt{D_p \tau_p}}\right). \qquad (10.21)$$

Since the design goal is to minimize I_o while still maximizing the ratio $I:I_o$, the next step is to express the photocurrent in some detail so the values of appropriate parameters can be considered in the design choices.

10.4.3 Optimizing Photocurrent

In Section 10.3.2, the photocurrent optimization process was discussed qualitatively. In this section the specific parameters that govern the absorption of light and the lifetime of the absorbed charge carriers will be discussed, and a formula for the photocurrent will be presented for comparison with the formula for reverse saturation current. In particular, minimizing the reflection of the incident photons, maximizing the minority carrier diffusion lengths, maximizing the junction width and minimizing surface recombination velocity will be discussed. The PV cell designer will then know exactly what to do to make the perfect cell.

10.4.3.1 Minimizing Reflection of Incident Photons

The interface between air and the semiconductor surface constitutes an impedance mismatch, since the electrical conductivities and the dielectric constants of air and a PV cell are different. As a result, part of the incident wave must be reflected to meet the boundary conditions imposed by the solution of the wave equation on the electric field, E, and the electric displacement, D.

Those readers who are experts in electromagnetic field theory will recognize that this problem is readily solved by the use of a quarter-wave matching coating on the PV cell. If the coating on the cell has a dielectric constant equal to the geometric mean of the dielectric constants of the cell and of air and if the coating is one-quarter wavelength thick, it will act as an impedance-matching transformer and minimize reflections. Of course, the coating must be transparent to the incident light. This means that it needs to be an insulator with a bandgap that exceeds the energy of the shortest wavelength light to be absorbed by the PV cell. Alternatively, it needs some other property that minimizes the value of the absorption coefficient for the material, such as an indirect bandgap [4].

It is also important to realize that a quarter wavelength is on the order of 0.1 µm. This is extremely thin and may pose a problem for spreading a uniform coating of this thickness. And, of course, since it is desirable to absorb a range of wavelengths, the antireflective coating will be optimized at only a single wavelength. Despite these problems, coatings have been developed that meet the requirements quite well.

An alternative to antireflective coatings now commonly in use with Si PV cells is to manufacture the cells with a textured front surface, as shown in Figure 10.10.

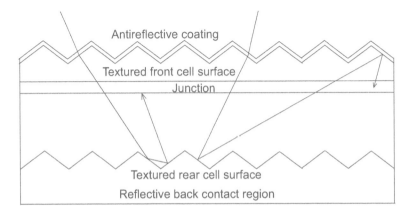

FIGURE 10.10 Maximizing photon capture with textured surfaces.

Textured front and back surfaces and their contribution to the capture of photons will be discussed later in this section. The bottom line is that a textured surface acts to enhance the capture of photons and also acts to prevent the escape of captured photons before they can produce EHPs. Furthermore, the textured surface is not wavelength dependent as is the antireflective coating.

10.4.3.2 Maximizing Minority Carrier Diffusion Lengths

Since the diffusion lengths are given by (10.11), it is necessary to explore the factors that determine the diffusion constants and minority carrier lifetimes in different materials. It needs to be recognized that changing a diffusion constant may affect the minority carrier lifetime, so the product needs to be maximized.

Diffusion constants depend on the scattering of carriers by host atoms as well as by impurity atoms. The scattering process is both material dependent and temperature dependent. In a material at a low temperature with a well-defined crystal structure, the scattering of charge carriers is relatively minimal, so they tend to have high mobilities. The Einstein relationship shows that the diffusion constant is proportional to the product of mobility and temperature [1]. Thus, once again, there is a trade-off. While increasing impurity concentrations in the host material *increases* the built-in junction potential, increasing impurity concentrations *decreases* the carrier diffusion constants.

While crystalline materials generally have relatively high carrier mobilities and diffusion constants, in polycrystalline or amorphous material, the lack of crystal lattice symmetry has a significant effect on the mobility and diffusion constant, causing a significant reduction in these quantities. However, if the absorption constant can be made large enough for these materials, the corresponding decrease in diffusion length may be compensated for by the increased absorption rate.

When an electron and a hole recombine, certain energy and momentum balances must be achieved. Locations in the host material that provide optimal recombination conditions are known as recombination centers. Hence, the minority carrier lifetime is determined by the density of recombination centers in the host material. One type of recombination center is a crystal defect, so as the number of crystal defects increases, the number of recombination centers increases. This means that crystal defects reduce the diffusion constant as well as the minority carrier lifetime in a material.

Impurities also generally make good recombination centers, especially those impurities with energies near the center of the bandgap. These impurities are thus different from the donor and acceptor impurities that are purposefully used in the host material, since donor impurities have energies relatively close to the conduction band and acceptor impurities have energies relatively close to the valence band.

Minority carrier lifetimes also depend on the concentration of charge carriers in the material. An approximation of the dependence of electron minority carrier lifetime on carrier concentration and location of the trapping energy within the energy gap is given by

$$\tau_n = \frac{n'\left[n+p+2n_i \cosh\frac{E_t-E_i}{kT}\right]}{CN_t(np-n_i^2)}, \quad (10.22a)$$

where C is the capture cross section of the impurity in cm³/sec, N_t is the density of trapping centers and E_t and E_i are the energies of the trapping center and the intrinsic Fermi level. In most materials, the intrinsic Fermi level is very close to the center of the bandgap. Under most illumination conditions, the hyperbolic term will be negligible compared to the majority carrier concentration and the excess electron concentration as minority carriers in p-type material will be much larger than the electron thermal equilibrium concentration. Under these conditions, for minority electrons in p-type material, (10.22a) reduces to

$$\tau_n \cong \frac{1}{CN_t}. \quad (10.22b)$$

Hence, to maximize the minority carrier lifetime, it is necessary to minimize the concentration of trapping centers and to be sure that any existing trapping centers have minimal capture cross sections.

10.4.3.3 Maximizing Junction Width

Since it has been determined that it is desirable to absorb photons within the confines of the pn junction, it is desirable to maximize the width of the junction. It is therefore necessary to explore the parameters that govern the junction width. Perhaps the reader recalls similar discussions in a previous electronics class.

An expression for the width of a pn junction can be obtained by solving Gauss' law at the junction, since the junction is a region that contains electric charge. The solution of Gauss' law, of course, is dependent on the ability to express the spatial distribution of the space charge in mathematical, or, at least, in graphical form. Depending on the process used to form the junction, the impurity profile across the junction can be approximated by different expressions. Junctions formed by epitaxial growth or by ion implantation can be controlled to have impurity profiles to meet the discretion of the operator. Junctions grown by diffusion can be reasonably approximated by a linearly graded model. The interested reader is encouraged to consult a reference on semiconductor devices for detailed information on the production of various junction impurity profiles.

The junction with uniform concentrations of impurities is convenient to use to obtain a feeling of how to maximize the width of a junction. The solution of Gauss' law for a junction with a uniform concentration of donors on one side and a uniform concentration of acceptors on the other side yields solutions for the width of the space charge layer on each side of the junction. The total junction width is then simply the sum of the widths of the two sides of the space charge layer. The results for each side are

$$W_n = \left[\frac{2\varepsilon N_A}{qN_D(N_A+N_D)}\right]^{\frac{1}{2}} (V_j - V)^{\frac{1}{2}} \qquad (10.23a)$$

and

$$W_p = \left[\frac{2\varepsilon N_D}{qN_A(N_D+N_A)}\right]^{\frac{1}{2}} (V_j - V)^{\frac{1}{2}}. \qquad (10.23b)$$

Before combining these two results, it is interesting to note that the width of the junction on either the p-side or the n-side depends on the ratio of the impurity concentrations on each side. Again, since Gauss' law requires equal numbers of charges on each side of the junction, the side with the smaller impurity concentration will need to have a wider space charge layer to produce enough impurity ions to balance out the impurity ions on the other side. The overall width of the junction can now be determined by summing (10.23a) and (10.23b) to get

$$W = \left[\frac{2\varepsilon(N_D + N_A)}{qN_A N_D}\right]^{\frac{1}{2}} (V_j - V)^{\frac{1}{2}} \qquad (10.24)$$

At this point, it should be recognized that the voltage across the junction due to the external voltage across the cell, V, will never exceed the built-in voltage, V_j. The reason is that, as the externally applied voltage becomes more positive, the cell current increases exponentially and causes voltage drops in the neutral regions of the cell, so only a fraction of the externally applied voltage actually appears across the junction. Hence, there is no need to worry about the junction width becoming zero or imaginary. In the case of PV operation, the external cell voltage will hopefully be at the maximum power point, which is generally between 0.5 and 0.6 V for silicon.

Next, observe that, as the external cell voltage increases, the width of the junction decreases. As a result, the absorption of photons decreases. This suggests that it would be desirable to design the cell to have the largest possible built-in potential to minimize the effect of increasing the externally applied voltage. This involves an interesting trade-off, since the built-in junction voltage is logarithmically dependent on the product of the donor and acceptor concentrations (see 10.10b), and the junction width is inversely proportional to the square root of the product of the two quantities. Combining (10.10b) and (10.24) results in

$$W = \left[\frac{2\varepsilon(N_D + N_A)}{qN_D N_A}\right]^{\frac{1}{2}} \left(\frac{kT}{q} \ln \frac{N_A N_D}{n_i^2} - V\right)^{\frac{1}{2}}. \qquad (10.25)$$

Note now that maximizing W is achieved by making either $N_A \gg N_D$ or by making $N_D \gg N_A$. For example, if $N_A \gg N_D$, then (10.25) simplifies to

The Physics of Photovoltaic Cells

$$W = \left[\frac{2\varepsilon}{qN_D}\right]^{\frac{1}{2}} \left(\frac{kT}{q} \ln \frac{N_A N_D}{n_i^2} - V\right)^{\frac{1}{2}}, \quad (10.26a)$$

or, if $N_D \gg N_A$, then

$$W = \left[\frac{2\varepsilon}{qN_A}\right]^{\frac{1}{2}} \left(\frac{kT}{q} \ln \frac{N_A N_D}{n_i^2} - V\right)^{\frac{1}{2}}. \quad (10.26b)$$

Another way to increase the width of the junction is to include a layer of intrinsic material between the p-side and the n-side as shown in Figure 10.11. In this *pin* junction, there are no impurities to ionize in the intrinsic material, but the ionization still takes place at the edges of the n-type and the p-type material. As a result, there is still a strong electric field across the junction and there is still a built-in potential across the junction. Since the intrinsic region could conceivably be of any width, it is necessary to determine the limits on the width of the intrinsic region.

The only feature of the intrinsic region that degrades performance is the fact that it has a width. If it has a width, then it takes time for a charge carrier to traverse this width. If it takes time, then there is a chance that the carrier will recombine. Thus, the width of the intrinsic layer simply needs to be kept short enough to minimize recombination. The particles travel through the intrinsic region with a relatively high drift velocity due to the built-in electric field at the junction. Since the thermal velocities of the carriers still exceed the drift velocities by several orders of magnitude, the width of the intrinsic layer needs to be kept on the order of about one diffusion length.

10.4.3.4 Minimizing Surface Recombination Velocity

If an EHP is generated near a surface, it becomes more probable that the minority carrier will diffuse to the surface. Since photocurrent depends on minority carriers'

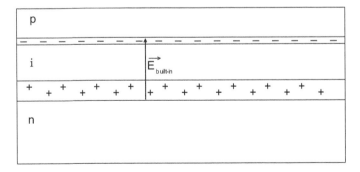

FIGURE 10.11 Introduction of an intrinsic layer to produce a PIN junction and increase overall junction width.

diffusing to the junction and ultimately across the junction, surface recombination of minority carriers before they can travel to the junction reduces the available photocurrent. When the surface is within a minority carrier diffusion length of the junction, which is often desirable to ensure that generation of EHPs is maximized near the junction, minority carrier surface recombination can significantly reduce the efficiency of the cell.

Surface recombination depends on the density of excess minority carriers, in this case, as generated by photon absorption and on the average recombination center density per unit area, N_{sr}, on the surface. The density of recombination centers is very high at contacts and is also high at surfaces in general, since the crystal structure is interrupted at the surface. Imperfections at the surface, whether due to impurities or crystal defects, all act as recombination centers.

The recombination rate, U, is expressed as number/cm²/sec and is given by

$$U = cN_{sr}m', \quad (10.27)$$

where m' is used to represent the excess minority carrier concentration, whether electrons or holes, and c is a constant that incorporates the lifetime of a minority carrier at a recombination center. Analysis of the dimensions of the parameters in (10.27) shows that the units of cN_{sr} are cm/sec. This product is called the *surface recombination velocity*, S. The total number of excess minority carriers recombining per unit time and subsequent loss of potential photocurrent, is thus dependent on the density of recombination centers at the surface and on the area of the surface. Minimizing surface recombination thus may involve reducing the density of recombination centers or reducing the density of minority carriers at the surface.

If the surface is completely covered by contact, then little can be done to reduce surface recombination if minority carriers reach the surface, since recombination rates at contacts are very high. However, if the surface is not completely covered by a contact, such as at the front surface, then a number of techniques have been discovered that will result in passivation of the surface. Silicon oxide and silicon nitrogen passivation are two methods that are used to passivate silicon surfaces.

Another method of reducing surface recombination is to passivate the surface and then only allow the back contact to contact the cell over a fraction of the total cell area. While this tends to increase series resistance to the contact, if the cell material near the contact is doped more heavily, the ohmic resistance of the material is decreased and the benefit of reduced surface recombination offsets the cost of somewhat higher series resistance. Furthermore, an E-field is created that attracts majority carriers to the contact and repels minority carriers. This concept will be explored in more detail in Section 10.5.2.

10.4.3.5 A Final Expression for the Photocurrent

An interesting exercise is to calculate the maximum obtainable efficiency of a given PV cell. Equation (10.3) indicates the general expression for photon absorption. Since the absorption coefficient is wavelength dependent, the general formula for overall absorption must take this dependence into account. The challenge in the design of

The Physics of Photovoltaic Cells

the PV cell and selection of appropriate host material is to avoid absorption before the photon is close enough to the junction but to ensure absorption when the photon is within a minority carrier diffusion length of either side of the junction.

Since direct bandgap materials tend to have larger absorption constants than indirect materials such as Si, it is relatively straightforward to capture photons in these materials. In Si, however, several clever design practices are used to increase absorption. Since some photons will travel completely across the cell to the back of the cell without being absorbed, if the back of the cell is a good reflector, the photons will be reflected back toward the junction. This is easy to do, since the back contact of the Si cell covers the entire back of the cell. However, rather than having a smooth back surface that will reflect photons perpendicular to the surface, the back surface is textured so the incident photons will be scattered at angles, thus increasing the path length.

The front of the cell, however, can be covered with an antireflective coating to maximize the transmission of photons into the material. Hence, a similar scheme is necessary to capture the photons in the material if, after bouncing off the back surface, they are still not absorbed. Once again, a textured surface will enhance the probability that a photon will undergo internal reflection, since the dielectric coefficient and, hence, the index of refraction of the host material, is greater than that of the antireflective coating. This is analogous to when ripples in water prevent a person below the surface from seeing anything above the surface. When the surface is smooth, it is possible to see objects above the water, provided that the angle of view is sufficiently close to the perpendicular. Figure 10.10, as introduced a few pages back, shows a cell with textured front and back surfaces and the effect on the travel of photons that enter the host material.

The foregoing discussion can be quantified in terms of cell parameters in the development of an expression for the photocurrent. Considering a monochromatic photon flux incident on the p-side of a p⁺n junction (the + indicates strongly doped), the following expression for the hole component of the photocurrent can be obtained. The expression is obtained from the solution of the diffusion equation in the neutral region on the n-side of the junction for the diffusion of the photon-created minority holes to the back contact of the cell.

$$\Delta I_{\ell p} = \frac{qAF_{ph}\alpha L_p}{\alpha^2 L_p^2 - 1}\left[\frac{S\cosh\frac{w_n}{L_p} + \frac{D_p}{L_p}\sinh\frac{w_n}{L_p} + (\alpha D_p - S)e^{-\alpha w_n}}{S\sinh\frac{w_n}{L_p} + \frac{D_p}{L_p}\cosh\frac{w_n}{L_p}} - \alpha L_p\right]. \quad (10.28)$$

It is assumed that the cell has a relatively thin p-side and that the n-side has a width, w_n. In (10.28), F_{ph} represents the number of photons per cm² per second per unit wavelength incident on the cell. The effect of the surface recombination velocity on the reduction of photocurrent is more or less clearly demonstrated by (10.28). The mathematical whiz will immediately be able to determine that small values of S maximize the photocurrent, and large values reduce the photocurrent, while one with average math skills may need to plug in some numbers.

Equation (10.28) is thus maximized when α and L_p are maximized and S is minimized. The upper limit of the expression then becomes

$$\Delta I_{\ell p} = -qAF_{ph}, \quad (10.29)$$

indicating that all photons have been absorbed and all have contributed to the photocurrent of the cell.

Since sunlight is not monochromatic, (10.28) must be integrated over the incident photon spectrum, noting all wavelength-dependent quantities, to obtain the total hole current. An expression must then be developed for the electron component of the current and integrated over the spectrum to yield the total photocurrent as the sum of the hole and electron currents. This mathematical challenge is clearly a member of the nontrivial set of math exercises and is not included as a homework problem. Yet, some have persisted in a solution to the problem and have determined the maximum efficiencies that can be expected for cells of various materials [5, 6]. Theoretical maximum efficiencies for cells of common materials are generally in the range between 24% and 29%, with 27% for crystalline Si, 27.5% for CdTe and 29% for Perovskite.

10.4.4 Minimizing Cell Resistance Losses

Any voltage drop in the regions between the junction and the contacts of a PV cell will result in ohmic power losses. In addition, surface effects at the cell edges may result in shunt resistance between the contacts. It is thus desirable to keep any such losses to a minimum by keeping the series resistance of the cell at a minimum and the shunt resistance at a maximum. With the exception of the cell front contacts, the procedure is relatively straightforward.

Most cells are designed with the front layer relatively thin and highly doped, so the conductivity of the layer is relatively high. The back layer, however, is generally more lightly doped to increase the junction width and to allow for longer minority carrier diffusion length to increase photon absorption. There must therefore be careful consideration of the thickness of this region to maximize the performance of these competing processes.

If the back contact material is allowed to diffuse into the cell, the impurity concentration can be increased at the back side of the cell, as illustrated in Figure 10.12. This is important for relatively thick cells, commonly fabricated by slicing single crystals into wafers. The contact material must produce either n-type or p-type material if it diffuses into the material, depending on whether the back of the cell is n-type or p-type.

In addition to reducing the ohmic resistance by increasing the impurity concentration, the region near the contact with increased impurity concentration produces an additional electric field that increases the carrier velocity, thus producing a further equivalent reduction in resistance. The electric field is produced in a manner similar to the electric field that is produced at the junction.

For example, if the back material is p-type, holes from the more heavily doped region near the contact diffuse toward the junction, leaving behind negative acceptor ions. Although there is no source of positive ions in the p-region, the holes that

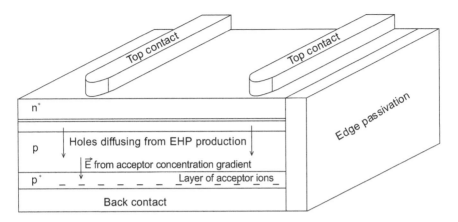

FIGURE 10.12 PV cell geometry for minimizing losses from cell series and shunt resistance

diffuse away from the contact create an accumulated positive charge that is distributed through the more weakly doped region. The electric field, of course, causes a hole drift current, which, in thermal equilibrium, balances the hole diffusion current. When the excess holes generated by the photoabsorption process reach the region of the electric field near the contact, however, they are swept more quickly toward the contact. This effect can be viewed as the equivalent of moving the contact closer to the junction, which, in turn, has the ultimate effect of increasing the gradient of excess carriers at the edge of the junction. This increase in gradient increases the diffusion current of holes away from the junction. Since this diffusion current strongly dominates the total current, the total current across the junction is thus increased by the heavily doped layer near the back contact.

At the front contact, another balancing act is needed. Ideally, the front contact should cover the entire front surface. The problem with this, however, is that if the front contact is not transparent to the incident photons, it will reflect them away. In most cases, the front contact is reflected. Since the front/top layer of the cell is generally very thin, even though it may be heavily doped, the resistance in the transverse direction will be relatively high because of the thin layer. This means that if the contact is placed at the edge of the cell to enable maximum photon absorption, the resistance along the surface to the contact will be relatively large.

The compromise, then, is to create a contact that covers the front surface with many tiny fingers, as shown in Figure 10.12. This network of tiny fingers, which, in turn, are connected to larger and larger fingers, is similar to the configuration of the capillaries that feed veins in a circulatory system. The idea is to maintain more or less constant current density in the contact fingers so that as more current is collected, the cross-sectional area of the contact must be increased. This subject is covered in more detail in the next chapter.

The alternative to trying to minimize the resistance of the front contact while avoiding too much shading of the cell by the front contacts can also be solved by eliminating the front contacts completely and, instead, attaching the "front" contacts to the back of the cell. Several methods, including back junction, emitter wrap through and metallization wrap through, have been developed [6] and these procedures are now commonly used in crystalline Si cell production. Note that this implementation involves some very sophisticated processing to isolate the conduction paths of the two sides of the cell to the contacts.

Another method of collecting more light on the cell is to allow light to enter the cell from the back of the cell. This involves using a transparent, as opposed to opaque, backplane for mounting the cells of the module. Bifacial modules have been around for a while and are now beginning to become more popular, especially in fixed-ground mount systems where the sun has direct access to the back of the modules in early summer morning and late summer afternoon hours. Bifacial module specification sheets indicate the possible gain in power production under bifacial operation.

Yet another method of avoiding the front contact shading the junction is found in some thin-film cells that use transparent conducting films across the top of the cell as the front conductor. This requires special materials and Si is not one of them.

Finally, shunt resistance is maximized by ensuring that no leakage occurs at the perimeter of the cell. This can be done by nitrogen passivation or simply by coating the edge of the cell with insulating material to prevent contaminants from providing a current path across the junction at the edges.

10.5 EXOTIC JUNCTIONS

10.5.1 Introduction

Thus far, only relatively simple junctions have been considered. PV cell performance can be enhanced significantly by incorporating a variety of more sophisticated junctions, including, but not necessarily limited to, graded junctions, heterojunctions, Schottky junctions, multijunctions and tunnel junctions. This section provides an introduction to these junctions so the reader will better understand the various junction types in the specific devices to be presented in the next chapter.

10.5.2 Graded Junctions

All the formulations to this point have related to the abrupt junction, in which the impurity concentration is constant to the junction and then abruptly changes to the opposite impurity concentration. While these junctions do exist, graded junctions are more common in materials for which the junction is fabricated by diffusion of impurities from the surface.

The significance of the graded junction is that the majority carrier transport beyond the junction is improved by an additional electric field component resulting from the decreasing impurity concentration from surface to junction, as shown in Figure 10.13. The origin of this electric field component is the same as the origin of the additional field near the back contact when the back metallization is allowed to diffuse into the back of the cell. Each impurity atom donates either an electron or a hole and becomes ionized. If there is a gradient in impurity concentration, then the

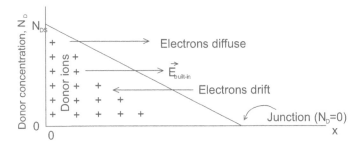

FIGURE 10.13 Diffusion and drift directions and electric field resulting from linearly decreasing impurity profile.

mobile carriers will diffuse in the direction of the gradient, leaving behind fixed impurity ions that serve either as the origin or termination of electric field lines.

An expression for the electric field in terms of the impurity concentration can be obtained by equating the diffusion current to the resulting drift current needed to balance the net current to zero when no external circuit is available for current flow. Assuming the majority carriers to be electrons in a heavily doped surface region, setting the sum of electron drift and diffusion current to zero results in

$$q\mu_n nE + qD_n \frac{dn}{dx} = 0, \tag{10.30}$$

which can be solved for E with the result

$$E = -\frac{D_n}{\mu_n} \frac{1}{n} \frac{dn}{dx} = -\frac{kT}{q} \frac{1}{N_D} \frac{dN_D}{dx}. \tag{10.31}$$

This additional E-field provides a drift component to the electron flow away from the junction toward the contact that effectively increases the velocity of the electrons toward the contact, resulting in lower resistive losses in the electron transport process toward the contact. Problem 10.12 provides an opportunity for calculation of the E-field present if the impurity concentration decreases linearly from surface to junction.

10.5.3 Heterojunctions

One of the problems identified with the optical absorption process is the absorption of higher energy photons close to the surface of the cell. Most of the EHPs generated by these higher energy photons are lost to recombination when they are created more than a diffusion length from the junction. This phenomenon can be mitigated to some extent by the use of a heterojunction. A heterojunction is simply a composite junction comprising two materials with closely matched crystal lattices, so the bandgap near the surface of the material is greater than the bandgap near the junction. The higher bandgap region will appear transparent to photons with lower energies, so these photons can penetrate the junction region where the bandgap is less than the incident photon energy. In the region of the junction, they can generate EHPs that will be collected before they recombine.

Heterojunctions are sometimes made between two n-regions or two p-regions as well as between n-region and p-region. The behavior of the heterojunction is dependent upon the crystal lattices, work functions, impurity doping profiles and energy band properties of each semiconductor material, to the extent that discussion of any particular junction would probably not be applicable to a different junction. Readers interested in specific junctions, such as Ge:GaAs or AlGaAs:GaAs, will be able to obtain more information from journal and conference publications.

In some cases, materials cannot be made either n-type or p-type. The use of a heterojunction with a material that can be made to complete a pn junction is another important use of heterojunctions. For example, a thin n-type CdS layer on top of a CIS structure will produce a pn junction effect, so CdS is often used as a part of a thin film structure to produce a thin, heavily doped n-region near the front surface of the cell.

10.5.4 Schottky Junctions

Sometimes, when a metal is in contact with a semiconductor material, an ohmic contact is formed and sometimes a rectifying contact is formed. It all depends on the relative positions of the work functions of the two materials. Figure 10.14 shows situations where the work function of the metal, $q\phi_m$, is greater than that of the n-type semiconductor, $q\phi_s$, and where the work function of the metal is less than that of the n-type semiconductor material. The work function is simply the energy difference between the

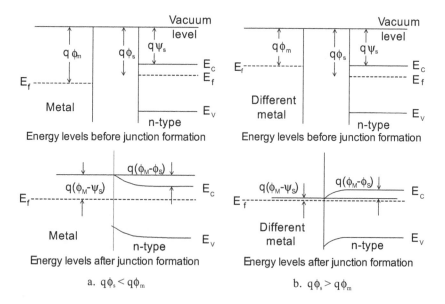

FIGURE 10.14 Energy levels in metal–semiconductor junctions.

The Physics of Photovoltaic Cells

Fermi level and the vacuum level and the Fermi level is that energy where the probability of occupancy by a mobile carrier is 0.5. For the semiconductor, the energy difference between the bottom of the conduction band and the vacuum level is designated as $q\phi_s$.

When the metal and semiconductor are joined, electrons flow either from metal to semiconductor or from semiconductor to metal, depending on which has the higher Fermi level. In the case where $q\phi_s < q\phi_m$, as indicated in Figure 10.14a, electrons will flow from semiconductor to metal in a manner similar to diffusion across a pn junction from n-type to p-type. As the electrons leave, they leave behind positive donor ions, just as in the pn junction. Since the materials are in contact, at the point of contact the probability of electron occupancy must be the same for each material, which requires the Fermi levels of each material to align. Since the semiconductor surface in contact with the metal is depleted of electrons, the Fermi level must move farther away from the conduction band, because, as the Fermi level moves closer to the conduction band, more electrons will appear in the conduction band and vice versa.

The donor ions on the n-side of the contact thus become the origin of an electric field directed from the semiconductor to the metal. This field causes a drift of electrons from metal to semiconductor to balance the diffusion from semiconductor to metal. Note that holes cannot diffuse from metal to semiconductor, since the metal supports only electrons. Thus, the built-in potential is only about half what it might be if holes could also diffuse across the junction.

In Figure 10.14b, where $q\phi_s > q\phi_m$, the electrons diffuse from the metal to the semiconductor, thus increasing the concentration of electrons on the semiconductor side of the junction. This diffusion, in effect, causes the equivalent of heavier doping of the n-type semiconductor and enables current to flow easily in either direction across the junction. This is therefore an ohmic junction, whereas the junction of Figure 10.14a is a rectifying junction. External contacts to a PV cell need to be ohmic to prevent unnecessary voltage drop at the contact, whereas the rectifying contacts are useful for other purposes.

For the rectifying junction, a positive external voltage from metal to semiconductor reduces the built-in field, enabling the electrons that have diffused to the metal to continue flowing in the external circuit, as shown in Figure 10.15a. Note that as the Fermi level on the semiconductor side is drawn closer to the conduction band, more electrons can diffuse to the metal, resulting in significant current flow from metal to semiconductor. The reduced built-in field reduces drift components in the opposite direction in a manner similar to the forward bias condition of a conventional pn junction.

When the external voltage is negative, the built-in field is as shown in Figure 10.15b. Although this might be expected to result in significant electron flow from metal to semiconductor, this does not occur, since the metal electrons must overcome the barrier between the metal Fermi level and the semiconductor conduction band. Only a few of the metal electrons are sufficiently energetic to do so, so the result is a relatively small reverse saturation current and thus a rectifying contact. The contact is rectifying even though the relatively strong E-field in the semiconductor region would be capable of enhancing drift current if replacement mobile charges were available.

For PV action, photons need to generate EHPs on the semiconductor side of the junction within a minority carrier diffusion length of the junction, so the minority

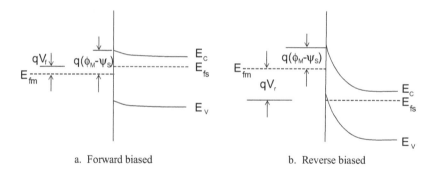

FIGURE 10.15 The rectifying metal-n-type semiconductor junction under forward-biased and reverse-biased conditions.

carrier can be swept out of the region to the other side of the junction by the built-in field.

It has been shown that, if the semiconductor material is n-type and $q\phi_m < q\phi_s$, the junction becomes ohmic, with bidirectional current conduction. When the semiconductor is p-type, the opposite is true. That is, for $q\phi_m > q\phi_s$, the contact is ohmic, and for $q\phi_m < q\phi_s$, the contact is rectifying.

These conditions must thus be met when ohmic contacts are to be made to semiconductor materials. This is why certain metals such as Mo, Al and Au are acceptable for ohmic contacts on some materials but not on others.

The Schottky barrier junction is relatively straightforward to fabricate but is not very efficient as a PV cell since it has a relatively smaller open-circuit voltage than a conventional pn junction due to diffusion currents flowing in only one direction across the junction. On the other hand, the lower voltage drop across the junction when it is conducting results in lower power dissipation, making the Schottky diode convenient for use as a bypass diode.

10.5.5 Multijunctions

Since photon energy is most efficiently absorbed when it is near the bandgap, a clever way to absorb more photons is to stack junctions of different bandgaps. By starting near the front surface with a relatively larger bandgap material, the higher energy photons can be absorbed relatively efficiently at this junction. Then a smaller bandgap junction, perhaps followed by an even smaller bandgap junction, will enable lower energy photons to be absorbed more efficiently. Since the junctions are in series, they must produce equal currents. This is an interesting challenge, since the current at any of the multiple junctions depends more strongly on the percentage of photons with energies close to the bandgap of a specific junction. In a terrestrial environment, as time of day changes, different wavelengths are present to a greater or lesser extent, resulting in an unbalance in currents generated among multiple junctions. On the other hand, in an extraterrestrial environment, the incident spectrum is constant, allowing for optimal performance of the multijunction structure.

The Physics of Photovoltaic Cells

FIGURE 10.16 Three junctions in tandem (series), showing opposing pn junctions.

Figure 10.16 shows three junctions in series, each of a somewhat different material to achieve three different bandgaps. The illustration may seem very logical, but there is an additional problem that is encountered when more than one junction is connected in series. That is simply the fact that, although the p-to-n direction of the first junction may be forward biased, this makes the n-to-p junction between the first and second pn junctions become reverse biased. The reverse bias across these junctions causes unnecessary voltage drops equivalent to the drop across blocking diodes that are sometimes inserted in strings of PV modules. Hence, a means must be devised to eliminate the effect of these reverse-biased junctions. Fortunately, the tunnel junction can eliminate this problem.

10.5.6 Tunnel Junctions

Tunnel junctions take advantage of the Heisenberg uncertainty principle, that is,

$$\Delta p \Delta x \geq \hbar, \tag{10.32}$$

where Δp and Δx are, respectively, the uncertainty in particle position and the uncertainty in particle momentum and $\hbar$ is Planck's constant divided by 2π. In a tunnel junction, the junction width is so extremely narrow, that with relatively small uncertainty in particle momentum, it cannot be specified with certainty that the particle is on one side or the other side of the junction. This phenomenon is known as quantum-mechanical tunneling. The process takes place without any change in energy, since the particle essentially tunnels through the potential barrier produced by the tunnel junction. Noting (10.24), it is evident that if impurity concentrations are very large on each side of the junction, the junction width will be quite small and the built-in potential will be relatively large, according to (10.10b). As the junction is forward biased, the junction width becomes even smaller, per (10.24). Tunneling occurs when the electrons in the conduction band on the n-side rise to energy levels adjacent to empty states on the p-side, provided that the junction is sufficiently narrow, as illustrated in the energy band diagram of Figure 10.17. Thus, the multiple pn junctions of the multijunction configuration can be accomplished by incorporating p^+n^+ tunnel junctions between each junction to eliminate the reverse-biased pn junctions. This is shown in Figure 10.18.

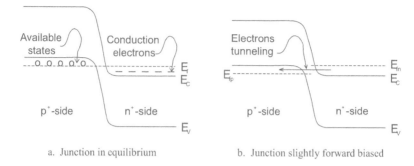

FIGURE 10.17 Tunnel junction showing conditions necessary for tunneling.

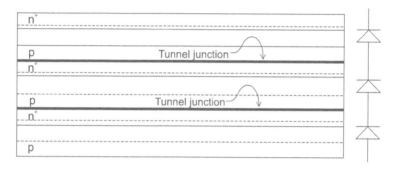

FIGURE 10.18 Separation of multiple pn junctions with tunnel junctions.

Homework Problems

10.1 What can be said about the bandgap of glass? Why is silicon opaque? For what range of wavelengths will silicon appear transparent?

10.2 Given the arguments about direct versus indirect bandgap materials and the ability of these materials to absorb photons offer an argument about the optimal bandgap structure of a material that will *emit* photons efficiently. Note that photon emission involves a transition of an electron from the conduction band to the valence band.

10.3 What would the bandgap need to be for an infrared LED? What about a red LED? What about a green LED? What about a blue LED? Would you expect the materials with these bandgaps to be direct bandgap materials or indirect bandgap materials?

10.4 Compare the average thermal velocity of an electron, v_{th}, at room temperature, if $½mv_{th}^2 = ½kT$, where m is the electron mass, k is Boltzmann's constant and T is the absolute temperature, with the drift velocity of an electron having a mobility of 10^3 cm²/volt-sec in an electric field of 1000 V/cm. The

The Physics of Photovoltaic Cells

drift velocity equals the product of the mobility and the electric field. At what electric field strength will the drift velocity equal the thermal velocity?

10.5 Use (10.9b) to solve for the concentration of electrons as a function of position between the neutral n-side edge of the junction and the edge of the neutral p-side of the junction. Plot this concentration function on a logarithmic scale along with the concentration of ionized impurities in the space charge layer.

10.6 Solve for the concentration of holes in the space charge layer. Show that in the space charge layer in thermal equilibrium, $np = n_i^2$ at every position, x.

10.7 Show that (10.10b) can also be obtained by setting the net hole current across the junction equal to zero.

10.8 If a material has μ_n = 8600 cm²/volt-sec, μ_p = 400 cm²/volt-sec and τ_n = τ_p = 10 ns, would you design a PV cell with the light incident on the n-side or the p-side of the material? Approximately what should be the distance between the edge of the space charge layer and the back contact? Explain why.

10.9 Show that (10.24) results from adding (10.23a) to (10.23b).

10.10 Show that combining (10.10b) and (10.24) results in (10.25).

10.11 Assume $N_A = 10^{20}$ cm⁻³ and $N_D = 10^{17}$ cm⁻³. Assume room temperature.
   a. Calculate the built-in junction potential for Si and GaAs.
   b. Calculate the junction width for Si and for GaAs under short-circuit conditions.

10.12 Use (10.30) to calculate the electric field in the n-region adjacent to the junction if the donor concentration is given by $N_D(x) = N_S(1-bx)$, where x is the distance measured from the surface of the cell and b is a constant.

10.13 Calculate the widths of the following abrupt junctions in Si at room temperature with no external bias applied.
   a. $N_A = 10^{16}$ and $N_D = 10^{20}$.
   b. $N_A = 10^{21}$ and $N_D = 10^{21}$.

   Next, calculate the widths of each junction if a forward bias of 0.3 V is applied, with all the external voltage appearing across the junction. Sketch the energy band picture for part b.

REFERENCES

[1] Yang, E. S., *Microelectronic Devices*, McGraw-Hill, New York, 1988.
[2] Granata, J., Sites, J. R. and Tuttle, J. R. "14th NREL PV Program Review," *AIP Conference Proceedings 394*, Lakewood, CO, 1996, 621–630.
[3] Streetman, B. G., *Solid State Electronic Devices*, 4th ed., Prentice Hall, Englewood Cliffs, NJ, 1995.

[4] Hu, C. and White, R. M., *Solar Cells: From Basic to Advanced Systems*, McGraw-Hill, New York, 1983.
[5] Zweibel, K., *Harnessing Solar Power*, Plenum Press, New York, 1990.
[6] Gulomov, Jasurbek, Accouche, Oussama, Aliev, Rayimjon, Ghandour, Raymond and Gulomova, Irodakhon, "Investigation of n-ZnO/p-Si and n-TiO 2/p-Si Heterojunction Solar Cells: TCAD + DFT," *IEEE Access*, vol. 11, pp. 38970–38981, 2023: https://doi.org/10.1109/ACCESS.2023.3268033

11 Evolution of Photovoltaic Cells and Systems

11.1 INTRODUCTION

In the last chapter, the basic theory of photovoltaic (PV) cells was presented without regard to any specific cell technology. This chapter will cover some of the fabrication processes associated with current cells along with discussions of the operation of a variety of cells, some of which are commonly in use, some of which have gone by the wayside and others of which are still in the experimental phases. But the world of PV is experiencing more than changes in cell technology. It is also in the midst of a systems paradigm shift as large-scale system deployment is now well underway.

Progress in PV research and development is moving so rapidly that by the time this chapter is in print, some of it will likely be outdated. As an example of the extent of PV research currently underway, the 2023 IEEE PV Specialists Conference had more than 1000 participants and dozens of topics, and the PVSC is only one of many worldwide PV conferences. Hence, in addition to introducing the reader to a few of the technologies of 2023, it is also the goal of this chapter to provide the reader with the intellectual tools needed to read and understand current and future literature. Many of the sources of information on PV technology are listed in the reference section of this chapter.

In general, cell fabrication begins with the refining and purification of the cell base materials. After extremely pure materials are available, then the pn junction or its equivalent must be formed. In some multiple-layer cells, more than one junction is formed along with various fascinating isolation steps.

The crystalline silicon cell has made its mark on history. Whether it will continue to make its mark in the future will depend on continuing to reduce the amount of energy consumed in the production of the cell to maintain its cost competitiveness over other technologies. Its fabrication and characteristics will be discussed first. During the discussion of the single-crystal silicon cell, important processes, such as crystal growth and diffusion will be discussed. These basic processes in many cases are applicable to the fabrication of other types of PV cells. Crystalline cells are generally considered to be the first generation of PV cells and thin films are considered to be the second generation. Perhaps the most promising near-term PV cells will consist of thin films, although the authors have been observing these forecasts since the 1990s. The third generation, yet to be commercialized, will have production costs in the range of $0.20–$0.50/watt and will have efficiencies between 31% and 74%, according to Martin Green in 2009 [1]. Certain thin film materials have direct bandgaps with energies near the peak of the solar spectrum, along with relatively high absorption constants and the capability of being fabricated with pn junctions. These films are not single-crystal devices, so there are limitations to carrier mobilities and subsequent device performance. However, in spite of the non-single-crystal structures, laboratory cell conversion efficiencies of 23.6% have been achieved [2].

Thin films are advantageous because only a minimal amount of material is required to deposit a film with a thickness of 1 or 2 μm. Figure 11.1 [3] shows the photon current versus optical path length for common thin film materials. Note that within 1 to 2 μm all the materials approach photon current saturation, whereas crystalline Si requires significantly greater thickness for full photon absorption. Examples of thin film PV cells that are currently in commercial production are amorphous silicon (a-Si), copper indium gallium diselenide (CIGS) and cadmium telluride (CdTe). Gallium arsenide has been used for high-performance crystalline cells and has also been the subject of thin film experimentation. On the other hand, the last National Renewable Energy Laboratories' (NREL) efficiency report on amorphous Si was in 2017 at 14% [2]. And, with the dawn of the age of nanotechnology, PV applications have been given more than a cursory bit of attention.

The availability of cell component materials is also of interest if many gigawatts of PV power are to be obtained from any technology. To add some perspective to the overall PV technology picture, IEA has reported that 2025 worldwide electrical generation capacity has been estimated to be 8601 GW [4], with 3000 GW from all renewable sources, including hydro. Other thin film materials, although scarce, are sufficiently abundant for the production of many gigawatts of PV cells at a reasonable cost, provided that certain production shortcomings that limit cell performance can be overcome.

The current and historic benchmarks for essentially all PV cells have been tracked by the NREL since 1976. In 2023, the NREL Best Research Cell Efficiencies chart includes six varieties of Multijunction cells, three versions of single-junction GaAs

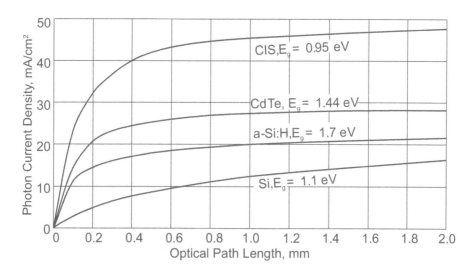

FIGURE 11.1 Photon current versus optical path length for thin film materials, compared with crystalline silicon, standard test conditions [3].

cells, five varieties of crystalline Si cells, four thin film technologies and eight different emerging technologies [2]. Since crystalline Si still has by far the largest market share, these cells will be discussed in some detail. Since all of the remaining technologies at the time of this writing make up about 4% of the total PV cell market, they will be discussed in a bit less detail, but enough to give the reader some understanding of some of them.

11.2 SILICON PV CELLS

The first silicon PV cells were of the single-crystal variety. Single crystal cells are the most efficient and most robust of the silicon PV cell family but are also the most energy intensive in their production. For this reason, other varieties of silicon cells have been developed. Polycrystalline cells are somewhat less efficient but are less energy intensive. Amorphous silicon cells are even less energy intensive but have significantly lower conversion efficiencies. However, since amorphous silicon is a direct bandgap material with an energy gap larger than pure silicon, it has a much larger absorption constant than crystalline silicon, with peak absorption at a wavelength closer to the peak of the solar spectrum, and is thus a suitable material for thin film cells. The instability mechanism present in early versions of amorphous silicon is reasonably well understood and means for overcoming the instability are now in common use. Additional efficiency increases have been achieved for amorphous silicon by the use of multijunction cell structures. Thin Si cells, such as crystalline silicon on glass, represent a compromise between crystalline cells and amorphous cells. These relatively recent introductions to the Si cell family use novel techniques for photon capture and minimization of top surface area blocked by contacts. Good efficiencies have been achieved in lab scale devices.

11.2.1 PRODUCTION OF PURE SILICON

The first thing that must be done in the production of a silicon cell is to find some silicon. Although nature has not seen fit to leave behind large quantities of relatively pure silicon in a form similar to the large deposits of carbon that have been mined for many years and burned in furnaces and worn on fingers, silicon has been provided in abundant quantities mixed with oxygen. Indeed, more than half the crust of the earth is composed of this silica compound, so all that is needed is to mine it and remove the oxygen and any other impurities. Generally, quartz is the preferred starting point in the production of pure silicon due to fewer impurities in this material.

Of course, this is not done with zero energy. Separation of oxygen from silicon is quite energy intensive, requiring the reduction of SiO_2 with carbon in a carbon arc furnace at a temperature of approximately 2000°C. Regrettably, the CO_2 byproduct of this process is a greenhouse gas, so the production of pure Si for PV cells begins with the production of a greenhouse gas along with the silicon. Fortunately, over the lifetime of the cell, the amount of CO_2 from the reduction process is significantly less than the amount of CO_2 that would be produced if fossil fuels were used instead of PV cells to generate the same amount of electrical energy. Furthermore, as carbon capture technology advances, it is possible that the CO_2 from the production of Si will ultimately be sequestered and kept out of the atmosphere. The energy cost of this step, once at approximately 50 kWh/kg of metallurgical grade (99.0% pure) silicon, had dropped to 11 kWh/kg by 2022 [6] and is expected to continue to decrease as more efficient processing steps continue to be developed.

A purity of 99% means impurity levels of 1 part in 100. This means that additional refining needs to take place, since the purity of solar-grade silicon needs to be in the neighborhood of 1 part in 10^6, depending on how efficient the cell is to be. Unwanted impurities lead to crystal defects and trapping levels that affect the cell output. Hence, the next step in the refining process is to remove these impurities. The traditional chemical method has been to react the metallurgical grade silicon with either hydrogen or chlorine to form either SiH_4 (silane) or $SiHCl_3$ (trichlorosilane). The more common reaction has been to combine the Si with HCl in a fluidized bed reaction. This reaction, however, produces a combination of Si-H-Cl compounds, such as SiH_2Cl_2, $SiHCl_3$ and $SiCl_4$ along with chlorides of impurities. Each of these liquids has its own boiling point, so it is relatively straightforward to use fractional distillation to separate the $SiHCl_3$ from the other components. Although it looks simple on paper, the energy required in this step is relatively high.

The next step is to remove the silicon from the trichlorosilane. This is done by reacting the trichlorosilane with hydrogen at a high temperature, producing polycrystalline silicon and HCl. Once again, improvements in process efficiency have led to reductions in the energy cost, which in 1999 was close to 200 kWh/kg of silicon. At this point, the silicon achieves the electronic grade level of purity (99.999999%).

An alternative metallurgical process is now becoming popular that yields solar grade silicon with somewhat more impurities than electronic grade material, but still more than adequate for PV cells. This process is currently capable of producing acceptable material at an energy cost that has been decreasing steadily to a 2023 value of approximately 20 kWh/kg [7].

With the steady decrease over time and anticipated continued future decrease, crystalline silicon has been able to compete with all other technologies over the past 30 years or so and still maintain nearly 96% of the overall PV technology market share. By 2023, the energy payback time (EPT) for crystalline silicon modules has dropped to 3.3 years and is expected to drop to one year within a few years [8].

11.2.2 Single Crystal Silicon Cells

11.2.2.1 Fabrication of the Wafer

To produce single crystal silicon from the polycrystalline material, the silicon must be melted and recrystallized. This is done by dipping a silicon seed crystal into the melt and slowly withdrawing it from the melt with a slight twisting motion. As the silicon is pulled from the melt, it reaches a somewhat higher purity level since the remaining impurities tend to remain behind in the melt. Under properly controlled conditions for solidification and replenishment of the melt, crystals as large as 11.8 in (30 cm) in diameter and 6 ft (1.83 m) long can be grown [9]. To produce n-type or p-type crystals, boron or arsenic can be introduced to the melt in whatever quantities are needed to produce the desired doping levels. Other type III or type V materials may also be introduced, but generally the diffusion constants or activation energies of these materials are less desirable from a PV performance perspective.

Special diamond wire saws are then used to cut the ingots into wafers. Since circular wafers mounted in a module leave a large amount of empty space between the wafers, the edges of the wafers are trimmed to make the wafers closer to square. The wafers are approximately 0.01 in (0.254 mm) thick and are quite brittle. The sawing causes significant surface damage, so the wafers must next be chemically etched to restore the surface. To achieve the textured finish as described in the last chapter, it is possible to use a preferential etching process, so that after etching, the wafer surface is textured and relatively defect free. Again, the wafering process is quite energy intensive and leaves behind kerf loss that then needs to be recycled.

11.2.2.2 Fabrication of the Junction [10]

The next step in the production of the single crystal silicon cell is to create the pn junction. In electronic semiconductors, pn junctions can be created by diffusion, epitaxial growth and ion implantation. These methods work fine for junctions having areas in the 10^{-12} m^2 range, but for PV cells with junction areas in the 10^{-2} m^2 range, the time it takes to produce a given area of junction becomes very important. Hence, due to cost constraints, pn junctions in crystalline silicon PV cells are made primarily by diffusion.

Impurity atoms diffuse into silicon in a manner similar to the diffusion of holes and electrons across regions of non-uniform impurity density. The concentration as a function of distance from the surface, x, and time is determined by the solution of the familiar diffusion equation

$$\frac{\partial N}{\partial t} = D \frac{\partial^2 N}{\partial x^2}, \tag{11.1}$$

where N represents the density of the impurity and D is the diffusion constant associated with the specific impurity. It should be noted that D is highly temperature dependent, especially close to the melting temperature of the host material. At room temperature, diffusion is negligible, but in the neighborhood of 1000°C, diffusion becomes appreciable.

The solution to (11.1) depends upon the boundary conditions imposed on the impurity. Most diffusion processes begin by holding the surface concentration of impurities constant for a fixed time. Then the impurity source is removed so the total number of impurities now remains constant, equal to the number of impurity atoms that diffused into the host during the constant surface concentration step of the process. The first step, known as the *predeposition*, leads to the well-known complimentary error function solution, that is,

$$N(x,t) = N_0 \operatorname{erfc} \frac{x}{2\sqrt{Dt}}, \tag{11.2}$$

where N_0 is the concentration of impurities at the surface of the wafer.

Suppose the substrate is uniformly doped p-type material and the impurities being diffused are n-type. The net impurity concentration at any depth, x, then, is the difference between the n-type and the p-type concentrations, or,

$$N(x,t) = N_o \, \text{erfc} \frac{x}{2\sqrt{Dt}} - N_A. \qquad (11.3)$$

When the net impurity concentration is positive, the material is n-type, and when the net impurity concentration is negative, the material is p-type. The junction is located where the net impurity concentration is zero, which represents the transition between n-type and p-type. The depth of the junction can thus be determined by setting $N(x,t) = 0$ and solving for x, yielding the result

$$x = 2\sqrt{Dt} \, \text{erfc}^{-1} \frac{N_A}{N_o}. \qquad (11.4)$$

This will not normally be the final junction depth, since after the predeposition step, the impurity concentration at the wafer surface is close to the solid solubility limit for the host material, which shortens the minority carrier diffusion length in this region. This preliminary junction depth estimate is useful, however, because it leads to a determination of the amount of time needed in the predeposition step to produce a junction at a depth less than the final desired junction depth, which will ultimately be five to ten times the initial junction depth.

To reduce the surface concentration, a *drive-in* diffusion is carried out next. The drive-in diffusion is simply the continued application of heat to the material after the supply of the surface impurity is removed. The result is that the impurity continues to diffuse from the region of greater concentration to the region of lesser concentration. The impurity concentration is given, to a good approximation, by the solution of the diffusion equation under conditions of constant impurities per unit area. The assumption is that all of the impurities diffused into the material during the predeposition step remain at the surface of the material. The number of impurities per unit area can be determined by integrating (11.3) from the surface into the material. While the actual distance of travel of the impurities is less than 1 μm, the integration is performed from 0 to infinity to yield

$$Q(t) = \int_0^\infty N(x,t) \, dx = 2N_o \sqrt{\frac{Dt}{\pi}}. \qquad (11.5)$$

After the drive-in step, the net impurity concentration is represented fairly accurately by the Gaussian distribution function,

$$N(x,t) = \frac{Q}{\sqrt{\pi Dt}} e^{-\frac{x^2}{4Dt}} - N_A. \qquad (11.6)$$

This expression can now be solved for the junction depth by again setting $N(x,t) = 0$. The solution is left as an exercise for the reader (Problem 11.2). It is thus a straightforward matter to control the diffusion of impurities into the wafer to produce a junction

Evolution of Photovoltaic Cells and Systems 359

at any desired depth from the surface. Since it is desirable to have the junction within a minority carrier diffusion length of the surface, the depth can be set to achieve this goal. Note that (11.3) through (11.6) also enable the cell designer to control the impurity concentration profile on the n-side of the junction. Figure 11.2 shows the impurity concentrations after the two diffusion steps have been completed.

11.2.2.3 Contacts

After creating the pn junction, the next step is to affix contacts to the cell. If the intent is to fabricate contacts that will last as long as the rest of the cell, then alligator clips are somewhat inadequate. The contacts must be low resistance and ohmic and must maintain their physical and electrical integrity through extreme temperature cycling. The contact material must be compatible with bonding of connecting wires/ribbons so the cell can be integrated into a module.

For older cells, the back contact covered the entire cell and was commonly made of evaporated aluminum that was annealed by heating the material after the evaporation step was complete. The annealing process caused the aluminum to diffuse slightly into the silicon, creating a strong bond that would not break under thermal cycling. Since aluminum is a group III element, it also acted as a p-type donor and produced a heavily doped p-region adjoining the contact. This heavily doped p-region created an impurity gradient that produced a resulting electric field that accelerated holes toward the back contact.

The front contact is considerably more challenging for several reasons. First of all, if the front contact is opaque to light, then it will block photons from entering the cell and being absorbed. This means the area of the front contact must be minimized. On the other hand, if the area is too small, then the resistance of the contact increases. Since the junction is only a small distance below the front surface, if the contact material is annealed, it is possible that it will diffuse into the silicon and short out the junction. For that matter, even if the contact does not diffuse across the junction initially, this can occur over time under operation at higher temperatures

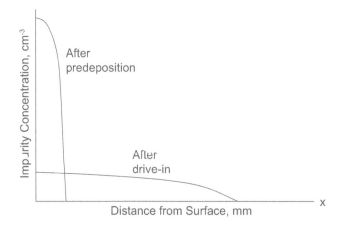

FIGURE 11.2 Concentrations of impurities after two diffusion steps.

and shorten the lifetime of the cell. Hence, the design constraints on the front contact are that it must be small, ohmic and present low series resistance while not threatening the junction.

After affixing the contacts to the material, an antireflective coating may be applied to the cell, normally by evaporation, since the coating must be so thin. A quarter wavelength coating has a thickness of approximately 0.15 μm. These coatings are commonly used for photographic lenses to increase their speed by reducing reflection and simultaneously increasing transmission. As long as the photon wavelength is relatively close to the quarter-wavelength constraint, transmission in excess of 90% can be achieved for the cell surface. However, because antireflective coatings optimize transmission at only one wavelength, textured cell surfaces have become more common for enhancing light trapping over the full spectrum.

Another recent improvement in cell design that has resulted in efficiency increases is the passivation emitter rear contact (PERC) process [11]. In this process, the front surface of the cell incorporates all the traditional features that have been used to increase cell efficiency, including textured surfaces. However, at the back contact, it was learned that by exposing the entire back contact to the bulk of the cell, some carriers generated were lost to recombination at the back contact due to defects at the back surface. By depositing a thin, insulating, passivating layer between the back contact and the bulk of the cell and opening smaller conduction channels to the rear contact, a significant amount of this minority carrier recombination can be eliminated. By optimizing the ratio of passivation area to the area of the channels that provide a majority carrier path to the rear contact, improved cell performance was achieved. Still, overall cell performance is limited by the presence of the front contacts. In fact, these very design constraints prompted researchers to develop a method of moving the front contacts to the back of the cell. This may sound like an oxymoron, but since the purpose of the front contact is to collect charge carriers injected across the junction to the top of the cell, moving the front contacts to the back involved developing a method to force these carriers to the back of the cell such that both positive and negative contacts could live harmoniously together at the back of the cell without interfering with each other via short circuits or any other process that might diminish cell efficiency.

11.2.2.4 A High-Efficiency Si Cell With All Contacts on the Back [12]

In 2009, cell efficiencies of 22% were achieved and 20.3% efficiency was achieved for modules that used the SunPower A-300 monocrystalline Si cell [1]. Basically, the cell consists of a substrate of n-type Si that has alternating positive and negative contacts to alternating n^+- and p-type contact areas that are diffused into the substrate. As EHPs are generated in the substrate, with a few being generated in the pn junction regions, the electrons are attracted to the n-type contact and the holes are attracted to the p-type contact as a result of the E-fields across the pn junction and between the n^+ contact and the substrate.

Note that the region between the n^+ contact and the weaker n-type substrate essentially constitutes a graded junction, where electrons flow from the more heavily doped contact region to the less heavily doped substrate. As a result, net positive charges are left behind in the n^+ region such that an electric field is created that is directed toward the substrate from the contact region. This field enhances the drift of electrons from the photon-generated EHPs from the substrate to the contact region.

The p-regions create pn junctions with the substrate with resulting E-fields directed from the substrate toward the p-region, as is the case for conventional pn junctions. The substrate doping is weaker than the p-region doping, so that the junction region is predominantly in the substrate. This field enhances the drift of minority holes from photon-generated EHPs in the substrate, across the junction into the p-regions. The cell surface is textured to optimize photon capture. The lack of front contacts on the cell allows a greater level of photon penetration into the substrate layer, since the contacts on the surface are not reflecting.

Fast forward to 2023, where several variations of rear contact (interdigitated back contact, IBC) modules have become particularly popular, with verified efficiencies between 22.3% and 23.6%. Ten manufacturers are now producing their own versions of these modules, with residential module power ratings between 400 and 460 W [13]. At present, the most popular back contact monocrystalline silicon technology is the N-type TOPcon technology, which is more efficient than traditional front-contact technologies.

11.2.2.5 Modules

The cells must then be mounted in modules, with interconnecting wires or metal foil "ribbons." The interconnects are generally either ultrasonically bonded or soldered to the cell contacts. The cells are then mounted to the module base and encapsulated with a glass or composite. The encapsulant must be chosen for long life in the presence of ultraviolet radiation as well as possible degradation from other environmental factors. Depending on the specific module environment, such as blowing sand, salt spray, acid rain or other not-so-friendly environmental component, the encapsulant must be chosen to minimize scratching, discoloration, cracking or any other damage that might be anticipated. Earlier encapsulants, such as ethylene vinyl acetate (EVA), tended to discolor as a result of exposure to high levels of ultraviolet radiation and higher temperatures. Other encapsulants had a tendency to delaminate under thermal stresses. By 1998, it was observed that modern materials now seem to have overcome these problems [14].

Bifacial modules have transparent backplanes that allow photons to enter the module from both directions. Presently, at least a dozen major manufacturers are ramping up production of bifacial modules with an equal number of variations in technology for maximizing photon capture from the rear of the module [15]. Depending upon how and where these modules are installed, it has been estimated that generation capacity can be enhanced up to 25% by capturing these mostly diffuse photons at the rear of PV Modules. This is an important factor in marketing of bifacial modules, since they generally have a higher first cost than monofacial units. However, reliable simulations now can determine whether systems using bifacial modules will have a lower LCOE than monofacial units. In fact, it has also been estimated that bifacials will capture 70% of the market share by 2030.

After the cells are encapsulated in modules, they are ready to produce electricity very reliably for a long time, normally in excess of 25 years. Because of the intense effort to bring down the energy overhead and overall cost, while increasing lifetimes and module efficiencies to more than 21%, crystalline Si modules have captured 96% of the global market share as of 2023.

11.3 GALLIUM ARSENIDE CELLS

11.3.1 Gallium Arsenide

The 1.43 eV direct bandgap, along with a relatively high absorption constant, makes GaAs an attractive PV material. Historically high production costs once limited the use of GaAs PV cells to extraterrestrial and other special purpose uses, such as in concentrating collectors. Recent advances in concentrating technology, however, enable the use of significantly less active material in a module, such that cost-effective, terrestrial devices are now commercially available. In this section, the basic GaAs cell and its components will be discussed. Concentrating PV will be discussed in more detail in Section 11.6.5.

11.3.2 Fabrication of the Gallium Arsenide Cell

Crystalline gallium arsenide is somewhat more difficult to form than silicon, since gallium and arsenic react exothermally when combined. The most common means of growing GaAs crystals is the liquid encapsulated Czochralski method. In this method, the GaAs crystal is pulled from the melt. The melted GaAs must be confined by a layer of liquid boric oxide. The biggest challenge is to create the GaAs melt in the first place. As of 1990, several means had been developed, such as first melting the Ga, then adding the boric oxide and then injecting the As through a quartz tube [16].

Most modern GaAs cells, however, are prepared by epitaxial growth of a GaAs film on a suitable substrate. In one basic GaAs cell structure, the cell begins with the growth of an n-type GaAs layer on a substrate, typically Ge, but more recently, single crystal GaAs, using the MOCVD process followed by an epitaxial lift-off (ELO) step [17]. Then a p-GaAs layer is grown to form the junction and collection region. The top layer of p-type GaAlAs has a bandgap of approximately 1.8 eV. This structure reduces minority carrier surface recombination and transmits photons below the 1.8 eV level to the junction for more efficient absorption.

A number of other GaAs structures have been reported, including cells of other III–V compounds. One of these cells consists of a cascaded AlInP/GaInP/GaAs structure grown by molecular beam epitaxy [18] and another multi-layer cell consists of an InP cell fabricated with the organo-metallic vapor phase epitaxy process [19]. The GaAs cell has 15 layers and the InP cell has nine layers.

The epitaxial growth process involves passing appropriate gases containing the desired cell constituents over the surface of the heated substrate. As the gases contact the substrate, the H or CH_3 attached to the In or P or Ga are liberated and the In, P or Ga attaches to the substrate. Hence, to grow a layer of p-type InP, a combination of trimethyl indium, phosphine (for P) and diethyl zinc (for acceptor impurities) are mixed in the desired proportions and passed over the heated substrate for a predetermined time until the desired layer thickness is obtained. The process is repeated

Evolution of Photovoltaic Cells and Systems 363

with different mixes of gases to form the other layers at the desired thickness. GaAs cells remain expensive to fabricate and are thus used primarily for extraterrestrial applications and in concentrating systems.

11.3.3 CELL PERFORMANCE

Cells fabricated with III–V elements are generally extraterrestrial quality. In other words, they are expensive, but they are high-performance units. Efficiencies in excess of 25% are common for single junction cells and efficiencies of cells fabricated on more expensive GaAs substrates had exceeded 34% by 2009 [2]. An important feature of extraterrestrial quality cells is the need for them to be radiation resistant. Cells are generally tested for their degradation resulting from exposure to healthy doses of 1 MeV or higher energy protons and electrons. Degradation is generally less than 20% for high exposure rates.

Extraterrestrial cells are sometimes exposed to temperature extremes, so the cells are also cycled between −170°C and +96°C for as many as 1600 cycles. The cells also need to pass a bending test, a contact integrity test, a humidity test and a high temperature vacuum test, in which the cells are tested at a temperature above 140°C in vacuum for 168 hours [20]. Fill factors in excess of 80% have been achieved for GaAs cells. Single cell open-circuit voltages are generally between 0.8 and 0.9 V.

The reason for stacking cells is simply that, for any particular material, wavelengths absorbed with energies greater than the bandgap lose the difference in energy to thermal losses in the conduction band of the material. Thus, if a stack of cells can be created that has a relatively high bandgap material at the top, this material will convert the higher energy photons into EHPs with lower thermal losses. The next layer below the top will have a bandgap slightly smaller than the top layer, etc., until the bottom layer is reached. Of course, it a non-trivial exercise to maintain a crystalline structure all the way down to the last pn junction of the cell. Another challenge is establishing equal cell current production in each layer, which involves adjustment of layer thickness. This problem is similar to series connections of cells in modules. The worst cell dominates the entire module performance as discussed in earlier chapters. In addition, for other reasons, these cells are typically not designed for terrestrial use.

The design of stacked cells depends upon the air mass under which the device is intended to operate. To ensure equal photon-generated current in each absorption layer, layer thickness needs to be adjusted for the air mass under which operation is anticipated, because air masses do not attenuate the entire spectrum proportionately, as noted in Chapter 2.

In particular, if a cell absorbs efficiently at $\lambda = 0.9$ μm, significantly more photons arc available at this wavelength at AM0 than at AM1. Thus, for operation at AM1, the absorber width would need to be increased to generate a photocurrent comparable to that which a narrower layer would generate at AM0. This consideration is particularly important when optimizing cell performance at AM0, but can also be relevant for cells designed for use in polar regions where exposures to AM 1.5 or AM 2.0 may occur for long periods of cell operation. The highest lab cell efficiency for GaAs cells (47.6%) has been obtained with a four-junction concentrating cell [2].

Recently, interest has been shown in growing epitaxial III–V compounds on a crystalline Si cell as substrate material, resulting in a multijunction device. Several experiments were reported on at the 2016 43rd Photovoltaic Specialists Conference. Meanwhile, because of its specialty applications, GaAs remains less than 1% of the global PV market share.

11.4 CIGS CELLS

11.4.1 Introduction

The first CIS PV cell, without the Ga, was reported in 1974 by a group at Bell Laboratories [21]. Copper indium diselenide was chosen as a potential PV material because of its attractive direct bandgap (1.0 eV), its very high optical absorption coefficient and its potentially inexpensive preparation. Furthermore, the cell components are available in adequate quantities and the manufacturing, deployment and decommissioning of the technology fall within acceptable environmental constraints.

11.4.2 Fabrication of the CIS Cell

While it is possible to produce both n-type and p-type CIS, homojunctions in the material are neither stable nor efficient. A good junction can be made, however, by creating a heterojunction with n-type CdS and p-type CIS.

The ideal structure uses near-intrinsic material near the junction to create the widest possible depletion region for collection of generated EHPs. The carrier diffusion length can be as much as 2 µm, which is comparable with the overall film thickness. A cell structure that was in popular use in 2004 consisted of a soda glass substrate with very thin layers of ZnO and CdS and then CIGS and a Mo back contact [22]. Since then, improvements in the TCO and processing steps have led to more efficient cells, but the basic cell structure is still the subject of incremental improvements. Again, CIGS technology is advancing rapidly as a result of the Thin-Film Photovoltaics Partnership Program and the U.S. Department of Energy SunShot Program [23], so by the time this paragraph is read, the ZnO/CdS/CIGS/Mo structure may be only suitable for history books and general discussion of the challenges encountered in thin-film cell development.

Nearly a dozen processes have been used to achieve the basic CIGS cell structure. The processes include rf sputtering, reactive sputtering, chemical vapor deposition, vacuum evaporation, spray deposition and electrodeposition. Sometimes these processes are implemented sequentially and sometimes they are implemented concurrently.

In the physical vapor deposition (PVD) process, which was used to achieve a record laboratory cell efficiency, the constituent elements are deposited under a relatively high vacuum of 10^{-6} Torr. In the PVD process, the four elements can be simultaneously evaporated, they can be sequentially evaporated followed by exposure to Se or they can be sequentially evaporated in the presence of Se. The soda lime glass substrate is maintained between 300 and 600°C during the evaporation process.

The front ohmic contact is straightforward, and ZnO generally works well. The trick is to achieve sufficiently high conductivity without absorbing any of the

Evolution of Photovoltaic Cells and Systems

incident photons. Often the ZnO is applied in two layers. The layer on the glass is high-concentration n-type, with a very thin intrinsic layer in contact with the CdS. The more heavily doped layer has high conductivity and the narrow intrinsic layer acts as a passivation layer between the thin CdS layer and the TCO, but is narrow enough to allow efficient transport of electrons to the TCO.

11.4.3 Cell Performance

When new technology PV cells are developed, normally small area cells are fabricated first to determine whether it appears practical to extend the technology to larger area cells and modules. In 2008, the highest performance achieved for a CIGS laboratory cell was 19.4%. Scaling up to the production of modules with areas up to 3459 cm^2 resulted in decrease in efficiency to 13.5% [1]. In 2015 a 22.3% efficiency was achieved for a lab cell and the 2014 record of 23.3% was achieved for a concentrator lab cell [2]. The increased efficiency resulted from improvements in the CIS absorber layer and in the junction formation process [24].

Clearly, the challenge in module development is to overcome the factors that result in degradation of cell performance and clearly these challenges are being undertaken ambitiously. To do so requires understanding of the factors that cause degradation. Some of these considerations include general device design, contact grid design, antireflection coatings and the sheet resistance of window layers.

An example of the trade-offs involved in scaling up a technology is the TCO. For a laboratory scale device with an area of 1 cm^2 or so, a TCO layer can be relatively thin, with a sheet resistivity of 15 Ω/square, since the relatively small current from the cell will experience minimal voltage drop through this contact. The thin window absorbs a minimal amount of incident radiation so the CIS absorber layer can achieve maximum conversion efficiency. However, as the cell is made larger, the sheet resistance of the TCO must be reduced to prevent voltage drop at the contact and corresponding degradation of fill factor and cell efficiency. The price paid for lower sheet resistance is a greater amount of absorption of incident photons by the transparent contact.

Series connection of cells can also cause cell performance degradation. Figure 11.3 [25] shows how individual cells have been replaced by minimodules

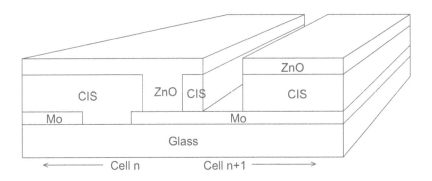

FIGURE 11.3 Siemens monolithic method of series cell connection [36].

that are monolithically connected. This monolithic connection process eliminates the need for separate fabrication processes for interconnecting cells.

After deposition of the Mo, it is scribed to separate adjacent cells, creating cells with a length of about 4 ft and a width of a fraction of an inch. Next, after deposition of the CIS cell components, the CIS is scribed. Finally, after deposition of the TCO, the TCO and CIS are scribed. While this process may appear to be straightforward, it can also be appreciated that the first cut needs to remove all of the Mo, but none of the glass. The second and third cuts must leave the Mo intact. Considering thickness of a few μm or less for these layers, the process falls somewhat short of being classified as trivial.

By incorporating Ga into the CIS mixture, the bandgap of the material can be increased beyond 1.1 eV. This movement of the bandgap energy closer to the peak of the solar spectrum increases conversion efficiency in this wavelength region and leaves lower energy photons for capture by the free carrier absorption process in the transparent conducting oxide (TCO) layer while the higher energy photons are converted to EHPs. The result is increasing the cell open-circuit voltage from approximately 0.4 V to as high as 0.68 V, with fill factors approaching 80% for laboratory cells. Experiments with adding sulfur to the selenium have also resulted in cell performance improvements. Some of the various CIS-related materials that have been proposed for tandem cells along with their bandgaps are shown in Table 11.1 [26].

Exposure to more intense light has actually caused efficiencies to increase slightly (concentrator cell). Exposure to elevated temperatures has resulted in loss of efficiency, but light soaking restored the modules to original efficiency levels. In 2023, the highest CIGS lab cell efficiency was 23.6% [2]. But the CIGS worldwide market share was less than 1%.

TABLE 11.1
Bandgaps of CIS-related materials

Material	Bandgap
$CuInSe_2$	1.05 eV
$AgInSe_2$	1.24 eV
$CuInS_2$	1.56 eV
$CuGaSe_2$	1.67 eV
$AgGaSe_2$	1.69 eV
$AgInS_2$	1.95 eV
$CuGaS_2$	2.33 eV
$AgGaS_2$	2.56 eV

Source: Tarrant, D. E. and Gay, R. R., *Research on Large Area CuInSe2 Modules, Final Technical Report*, NREL, Golden, CO, April 1995.

11.5 CADMIUM TELLURIDE CELLS

11.5.1 Introduction

In theory, CdTe cells have a maximum efficiency limit close to 25%. The material has a favorable direct bandgap and a large absorption constant, allowing for cells of a few μm thickness. By 2001, efficiencies approaching 17% were being achieved for laboratory cells, and module efficiencies had reached 11% for the best large area (8390 cm^2) module [27]. Efforts were then focused on scaling up the fabrication process to mass produce the modules, with the result of achieving a production cost of less than $1.00/watt in 2008. This cost included an escrow account to be used for recycling the materials at the end of module life [28].

Although tellurium is not as abundant as other cell components, cadmium and tellurium are both available in sufficient quantities for the production of many gigawatts of arrays. The amount of cadmium poses a possible fire hazard and some concern at the time of decommissioning of the modules, as indicated in Chapter 9. Analysis of the concerns, however, has shown that the Cd of the cells would be recycled at decommissioning time and that the danger of burns from any fire far exceeds the danger of contact with any Cd released from heating of the modules. Means for recycling CdTe modules exist that result in recovery of glass, $CdCO_3$, electrolytically refined Te and clean EVA at a cost of less than $0.04 per watt [29].

11.5.2 Production of the CdTe Cell

A typical CdTe cell begins with a glass superstrate with a transparent, conducting, oxide layer about 1 μm thick, a thin CdS buffer layer about 0.1 μm thick, a CdTe layer a few μm thick and a rear contact of Au, Cu/Au, Ni, Ni/Al, ZnTe:Cu or (Cu, HgTe) [30]. The TCO layer has been fabricated with SnO, InSnO and Cd_2SnO_4. The Cd_2SnO_4 has been shown to exhibit better conductivity and better transparency [31] than the other TCOs and may end up as the preferred TCO. The Cd_2SnO_4 layer is deposited by combining CdO and SnO_2 in a 2:1 proportion in a target material, which is then deposited by means of radio frequency magnetron sputtering onto the glass superstrate.

The thin n-type CdS layer has been deposited by the metal organic chemical vapor deposit (MOCVD) process as well as by other thin film deposition techniques. The layer needs to be annealed prior to deposition of the CdTe to reduce CdS surface roughness, thus reducing defects on the CdS/CdTe boundary. This is normally accomplished in air at approximately 400°C for about 20 minutes.

An interesting challenge existed with regard to the CdS layer, since the layer is so thin. During deposition of the CdTe or subsequent heat treatment of the cell, intermixing of the CdS and CdTe can occur at the boundary. This can result in junctions between the CdTe and the TCO layer, which causes significant reductions in cell open-circuit voltage. Several methods of minimizing this mixing were proposed [32] and this problem has now been overcome. Recently, it has been discovered that replacing the CdS layer with CdSO as the n-window layer, improves the bandgap and results in a higher cell efficiency [33].

Numerous methods have been used to deposit the CdTe layer, including atmospheric pressure chemical vapor deposition (APCVD), atomic layer epitaxy (ALE), close-spaced sublimation (CSS), electrodeposition (ED), laser ablation, PVD, screen printing (SP), spray, sputtering and MOCVD [27]. The CdTe layer is subjected to a heat treatment in the presence of $CdCl_2$ for about 20 minutes at about 420°C (788°F). This treatment enhances grain growth in the CdTe layer to reduce grain boundary trapping effects on minority carriers. Nonheat-treated CdTe cells tend to have open-circuit voltages less than 0.5 V, while after heat treating, V_{OC} can exceed 0.8 V.

Another important step in optimizing cell performance is to ensure stability of the back contact. Before the back contact is applied, the CdTe is etched with nitric-phosphoric, resulting in a layer of elemental Te at the back CdTe surface. The elemental Te produces a more stable contact between the p-type CdTe and the back metal [34].

Final processing of modules involves encapsulation of the back of the cell with a layer of EVA between the metallization and another layer of glass.

11.5.3 Cell Performance

No fewer than nine companies have shown an interest in commercial applications of CdTe. As of 2001, depending on the fabrication methodology, efficiencies of close to 17% had been achieved for small area cells (≈ 1 cm^2) and 11% on a module with an area of 8390 cm^2 [26]. Furthermore, sufficient experience had been logged with large-scale production to identify areas in which improvement was needed [34]. These areas included improving the design, operation and control of a CdTe reactor; increasing the understanding of the fundamental properties of CdTe films; maintaining uniformity of materials and device properties over large areas, including the interdiffusion of CdTe and CdS and the back contact; obtaining stability of the back contact and addressing any environmental concerns over Cd. Since Te availability may limit cell production, increasing performance with thinner layers of CdTe is also desirable. Reducing the thickness of the CdTe layer to 0.5 μm will allow for four to five times the cell area, provided that efficiency can be maintained or increased. Just as in the case of other thin-film arrays, area-related costs limit the minimal cost per watt of CeTe arrays, so an increase in efficiency remains important to minimize the overall cost per watt.

Fill factors for lab cells have been obtained in the 65–75% range [35], with the slope of the cell J-V curve at V_{OC} in the range of 5 Ω-cm^2. Short-circuit current densities upward from 25 mA/cm^2 were obtained by using a thin buffer layer of CdS along with a thin insulating TCO layer between the heavily doped TCO and the CdS layer [26].

Experimental CdTe arrays in sizes up to 25 kW were deployed in California, Ohio, Tunisia, Colorado and Florida for testing purposes [26]. Soon after it was shown that no degradation was observable after two years, production-scale manufacturing began. CdTe modules are now being manufactured and marketed at the rate of more than 9 GW annually [36] and the magic $1.00/watt production cost barrier has now been broken for these modules, with 2015 production costs for 16% efficient modules at approximately $0.51/W [23]. In 2023, CdTe modules claimed

3% of the worldwide market, with maximum lab cell efficiencies of 22.3% and average module efficiencies of 19%, which is close to crystalline Si [37].

11.6 EMERGING TECHNOLOGIES

11.6.1 Intermediate Band Solar Cells

In all cells described to this point, absorption of a photon has resulted in the generation of a single EHP. If an intermediate band material is sandwiched between two ordinary semiconductors, it appears that it may be possible for the material to absorb two photons of relatively low energy to produce a single EHP at the combined energies of the two lower energy photons. The first photon raises an electron from the valence band to the intermediate level, creating a hole in the valence band, and the second photon raises the electron from the intermediate level to the conduction band. The trick is to find such an intermediate band material that will "hold" the electron until another photon of the appropriate energy impinges upon the material. Such a material should have half its states filled with electrons and half empty to optimally accommodate this electron transfer process. It appears that III–V compounds may be the best candidates for implementation of this technology. The theoretical maximum efficiency of such a cell is 63.2% [38].

11.6.2 Hot Carrier Cells

The primary loss mechanism in PV cells is the energy lost in the form of heat when an electron is excited to a state above the bottom of the conduction band of a PV cell by a photon with energy greater than the bandgap. The electron will normally drop to the lowest energy available state in the conduction band, with the energy lost in the process being converted to heat. Hence, if this loss mechanism can be overcome, the efficiency of a cell with a single junction should be capable of approaching that of a supertandem cell. One method of preventing the release of this heat energy by the electron is to heat the cell, so the electron will remain at the higher energy state. The process is called *thermoelectronics* and was being investigated in 2002 [39].

11.6.3 Optical Up- and Down-Conversion

An alternative to varying the electrical bandgap of a material is to reshape the energies of the incident photon flux. Certain materials have been shown to be capable of absorbing two photons of two different energies and subsequently emitting a photon of the combined energy. Other materials have been shown to be capable of absorbing a single high-energy photon and emitting two lower-energy photons. These phenomena are similar to up-conversion and down-conversion in communications circuits at radio frequencies.

By the use of both types of materials, the spectrum incident on a PV cell can be effectively narrowed to a range that will result in more efficient absorption in the PV cell. An advantage of this process is that the optical up-and-down-converters need not be a part of the PV cell. They simply need to be placed between the photon source and the PV cell. In tandem cells, the down-converter would be placed ahead of the top cell and the up-converter would be integrated into the cell structure just ahead of the bottom cell.

11.6.4 Organic PV Cells

Even more exotic than any of the previously mentioned cells is the organic cell. In the organic cell, electrons and holes are not immediately formed as the photon is absorbed. Instead, the incident photon creates an *exciton*, which is a bound EHP. To free the charges, the exciton binding energy must be overcome. This dissociation occurs at the interface between materials of high electron affinity and low ionization potential [40]. One of the original cell structures consisted simply of an organic material, usually a polymer/plastic, sandwiched between a high-work function metal and a lower-work function metal, to create an electric field across the organic absorber [41].

One of the challenges in the development of organic PV (OPV) cells is getting the charge carriers out of the absorber and to the contacts, which is more or less equivalent to getting the charge carriers out of a pn junction and into the bulk regions of a conventional PV cell. The absorbers that have been tried have generally absorbed most of the incident photons in a submicron thickness but have been hampered by very short carrier lifetimes and correspondingly short diffusion lengths. As a result, a large majority of the charges recombine within the absorber before they can travel to contact [41]. The result is a solar cell with an efficiency of about 1%.

Researchers have found that, by using two different organic materials to form the active layer, transport properties and thus cell efficiency can be improved. The first layer consists of a heterojunction where the second junction material, usually fullerene, will act as an n-type semiconductor and the polymer layer will act as a p-type semiconductor. Thus, when a photon creates an EHP or an exciton, the electron is accepted by the fullerene and the hole is accepted by the polymer, thus separating the exciton and moving the electron and hole toward the contacts. This structure is still limited by short diffusion lengths, so to overcome this problem, the heterojunction is distributed throughout the absorber to enable separation of the charge carriers throughout the absorber. As of 2023, the best reported OPV lab cell efficiency was 19.2% [2].

OPV systems are currently in the advanced demonstration phase, with some products for sale. Most applications have been directed toward building integrated PV, in which, for example, the material has been integrated into cement and metal structural components, as well as glass and irregularly shaped fabrics used in backpacks and tents [42].

11.6.5 Concentrating PV

Concentrating PV (CPV) cells and systems are designed to focus large amounts of sunlight onto a small area cell. Work has been done up to 1000 suns, which is 1000 times the intensity of the sun, or, 1,000,000 W/m². This range of concentration is generally broken down into a 300–1000 suns range (high concentration, or HCPV) and < 100 suns (low concentration, or LCPV). HCPV cells are typically III–V compounds with multi-junctions and LCPV cells are typically Si-based cells, which may be single junction or sometimes heterojunction devices.

As mentioned earlier, since III–V cells are more expensive, it is advantageous to use smaller cells with less expensive concentrating structures to focus the incident sunlight on the cells. For high concentration, however, this requires using the direct beam of sunlight via double-axis trackers, which means the technology is most cost-effective in areas such as deserts that experience a large fraction of direct sunlight. In fact, standard test conditions for concentrating cells have been adopted using 800 W/m^2 as standard incident radiation, since the diffuse component of sunlight is eliminated. For LCPV either single-axis or double-axis trackers are used.

CPV has advantages and disadvantages. Perhaps one of the most attractive features of HCPV is the fact that the dual-axis trackers must be spaced apart far enough to avoid shading among the trackers. This allows for the land between the tracking stands to be used for agriculture, since the racking is generally implemented with a single mounting pole for each tracker. Although the tracking must be quite precise, recent significant progress in tracker design has made this possible.

Another important feature of tracking is that it results in maximum possible energy production throughout the day, especially during late afternoon utility peaking hours. Furthermore, III–V cells tend to be less degraded by ambient temperatures, since their output power has a smaller temperature dependence than Si devices. As the efficiency of concentrator cells continues to increase, as discussed in earlier sections, the cost-effectiveness of HCPV is also enhanced. As of 2015, an HCPV module efficiency of 38.9% was achieved. As of 2023, the efficiency had increased to 47.6% for a four-junction cell [2].

One disadvantage of CPV is that they only perform well in regions having direct sunlight of at least 2000 kWh/m^2/yr, with 2500 kWh/m^2/yr as a preferred level, which pretty much limits their use to desert environments.

Presently, most CPV systems use Fresnel lenses to focus the direct solar beam onto the multijunction III–V cell. The most commonly used cell in recent large commercial (MW) systems is the triple-junction, lattice-matched, $Ga_{0.50}In_{0.50}P/Ga_{0.99}In_{0.01}As/Ge$ cell. The limiting factor on this cell is that the Ge layer absorbs nearly twice as much of the incident radiation as the other layers, creating a mismatch of current sources in series.

Another disadvantage of the cell assembly is that dust and sand interfere with the passive cooling of the assembly, leading to temperature rise and efficiency losses. It has also been observed that when CPV modules are mounted on large trackers, the trackers bend, resulting in angular distortion between cells and thus overall less than optimal photon capture.

For CPV, it is important to distinguish between CPV and concentrating solar. Concentrating solar is perhaps familiar as a solar power tower, in which a field of mirrors is carefully controlled to focus sunlight on a boiler at the top of a structure in the center of the mirror field. The boiler can store hot material, such as salt, for use in generating electricity by conventional means during nighttime hours.

11.6.6 Perovskites

Perovskites, hardly on the radar screen in 2008, are an exciting new technology. The efficiency of laboratory cells has increased from about 14% in 2013 to 22.1% in 2016 and 26.1% in 2023 [2]. Even more impressive is a perovskite-Si tandem structure for which an efficiency of 33.7% [NREL] has been demonstrated.

The basic perovskite structure is shown in Figure 11.4 [44]. A close look shows that this structure might be classified as a body-centered, face-centered, cubic structure, since the A-component appears in the body center, the B-components appear at the cell corners and the X-components appear in the centers of the cube faces. Typically, the A-component is a large positively charged ion, which might be organic. The B-component is a somewhat smaller, positively-charged ion and the C-component is a small, negatively-charged ion. The basic structure is recognized as ABX_3, where A has a single positive charge, B has a double positive charge and X has a single negative charge [44].

From the figure, one can deduce that all of the A-component belongs to the cell, thus adding a charge of +2. The B-components, however, are each shared with seven other cells that join at the corners, so that each contributes only 1/8 of its charge to the cell and 7/8 to the other seven cells. Thus, the eight corner (B) components contribute a total of +1 charge to the cell. The six X-components are each shared with one adjacent cell, such that half their total charge is shared with adjoining cells and the remaining half can be claimed by the original cell, thus resulting in a contribution of −3 charge to the cell, resulting in a zero net charge for the cell.

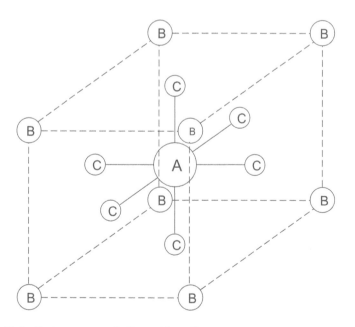

FIGURE 11.4 Basic structure of a Perovskite cell.
(Adapted from [45].)

For this structure to act as a photon absorber, it has been found that good absorption efficiency can be achieved by choosing an organic cation, such as methylammonium $(CH_3NH_3)^+$ to fill the A-position, lead Pb^{2+} for the B-positions and a halide or halide combination, such as I^-, Cl^-, Br^- or a combination of them, such as $xI^-(1-x)Cl^-$, for the X-positions [44]. The exact absorption process is still not well understood [44, 45]. For example, it has not yet been confirmed that the process involves the creation of excitons, such as in organic devices, or electron–hole pairs, as in traditional devices. What is understood is that the sort of charge-selective interface layers as used in organic devices also act as efficient collectors of electrons and holes from the perovskite layer.

Figure 11.5 [44] shows one implementation of a perovskite cell. Depending on specific cell components, the absorber layer can be a fraction of a μm thick and still harvest a large fraction of the incident photons. One group of researchers has found that electron–hole diffusion lengths can exceed 1 μm in an organometal trihalide perovskite absorber if the absorber is of the mixed halide triiodide variety [45]. With a perovskite layer thickness of ≈400 nm, this ensures that a very large fraction of carriers generated in the absorber layer will find their way to the charge selective interface layers, as opposed to traditional OPV cells that have very short diffusion lengths for charge carriers.

Since day one of perovskite development, stability has been the main issue that has kept them out of the marketplace. Doubling the efficiency of the basic cell in ten years had previously been very rare for any other technology. Then, in 2018, the perovskite/Si monolithic tandem cell was introduced at an efficiency close to 23.5%. The basic structure of this cell begins with glass, followed by an encapsulant, the rear electrode, a contact stack, a layer of crystalline silicon, another contact stack, an interconnection contact stack, a perovskite layer, metallic electrodes similar to those of a silicon cell, a top encapsulant and another layer of glass [46]. This collection of layers is an impressive collection of pn junctions, heterojunctions, tunnel junctions and other associated layers.

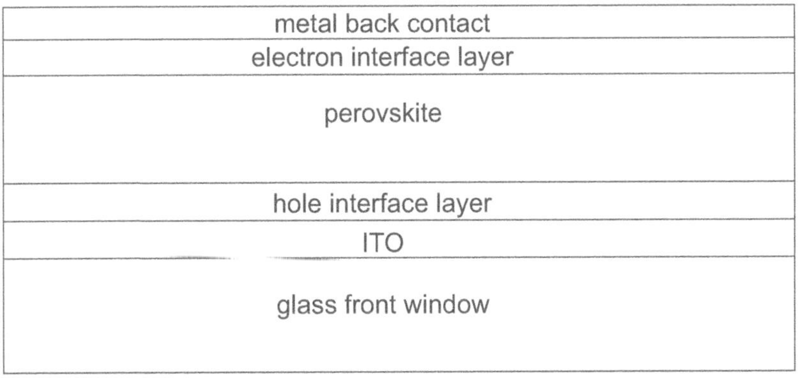

metal back contact
electron interface layer
perovskite
hole interface layer
ITO
glass front window

FIGURE 11.5 Perovskite cell layers.

(Adapted from [45].)

The active layers of the cell are the crystalline silicon and the perovskite. The perovskite layer is near the top of the cell, to convert the higher energy incident photons to charge carriers, since it has a bandgap of $\cong$ 1.7 eV. The crystalline silicon layer with its 1.1 eV bandgap then converts energies between the two bandgap energies. As with all tandem cells, something needs to be done between the two cells to facilitate charge transfer between the top and bottom contacts without interference by reverse-biased heterojunctions. This will normally be achieved with tunnel junctions. In addition, since the cells are in series, the ideal structure will result in equal current generation by each material. Since the Perovskite layer has a direct bandgap, it is expected to be a more efficient converter and is thus thinner than the crystalline silicon layer. The rear electrode needs to consist of a suitable work function material to achieve an ohmic connection and the top layer is a transparent conducting oxide layer with top metallic electrodes.

By 2023 the efficiency of lab cells had increased to 33.7% [2] and several companies began planning for mass production, since a significant improvement in stability had also been achieved. It appears now to be very likely that by the time this book is published, these cells may be available commercially in module form. Stay tuned.

11.7 MICRO GRIDS

Currently, utility-interactive inverters must incorporate anti-islanding algorithms so they will shut down if the grid shuts down, regardless of whether other PV inverters are connected to the grid that might tend to fool the inverter into thinking they are the utility grid. But what if a situation might exist where it is important to maintain a PV island in the midst of a utility outage? One might imagine a remote town or village with a single feeder connection to the utility grid. If the grid goes down, should the entire town go down, or, alternatively, be transferred to individual or collective non-utility generation? What about setting up PV systems such that islanding is allowed for the systems in the town, but still disconnects the town or neighborhood from the utility feed until utility power is restored?

In fact, exactly how utility power should be restored is now in question. As renewable energy sources with inverters compliant with IEEE-1547–2018 are now exceeding 10% of total generation capacity, they are capable of being controlled such that they can be incorporated back into the grid almost instantaneously. All it takes is one inverter to be the lead source for synchronizing the grid voltage and others can then be added very quickly without waiting for a rotating machine to get exactly up to speed and up to phase. Smart load management can control loads to coincide with available grid power. Add to this significant amounts of energy storage, and the process becomes even smoother.

11.8 SUMMARY

Regardless of the technology or technologies that may result in low-cost, high-performance PV cells, it must be recognized that the life cycle cost of a cell depends on the cells having the longest possible, maintenance-free lifetime. Thus, along with the development of new technologies for absorbers, the development of reliable encapsulants and packaging for the modules will also merit continued research and development activity.

Every year engineers make improvements on products that have been in existence for many years. Automobiles, airplanes, electronic equipment, building materials and many more common items see improvements every year. Even the yo-yo, a popular children's toy during the 1940s and 1950s, came back with better-performing models. Hence, it should come as no surprise to the engineer to see significant improvements and scientific breakthroughs in the PV industry well into this millennium. The years ahead promise exciting times for the engineers and scientists working on the development of new PV cells and system technologies, provided that the massive planning and execution phases can be successfully undertaken.

Homework Problems

11.1. Assume a PV module is to have dimensions of 39.4 in × 79.2 in. Also assume that 8-in round cells are available.
   a. Calculate the percentage of the module area that can be covered with circular cells.
   b. If the cells operate with 21% efficiency, and if there is no mismatch loss among cells, calculate the overall module efficiency if the round cells are used.
   c. Assume the cells are "squared up" by sawing off the edges to obtain cells that measure 6 in between the straight sides. Calculate the percentage fill of these "squared up" cells and repeat the module efficiency calculation.
   d. What is the percentage loss of material in the "squaring up" process?
11.2. Determine the expression for the depth of the junction after the drive-in diffusion step.
11.3. Sketch the impurity distribution profile between the junction and the back contact of a single crystal silicon cell to show how the annealed aluminum of the back contact creates an accelerating E-field.
11.4. What volume and weight of tellurium is needed to produce a square meter of CdTe thin film with a thickness of 1.0 μm? Assume the Cd and Te occupy equal volumes within the film.
11.5. What volume and weight of indium is required to fabricate a 1 MW PV CIS array if the CIS layer thickness is 1.0 μm, assuming that 18% of the layer volume is due to the In. Assume an array efficiency of 19% and standard test conditions.
11.6. The energy payback time (EPBT) for a PV technology is defined as the time needed for the completed cell to generate an amount of electrical energy equal to the energy required for producing the cell. Refer to the current literature and tabulate the EPBT for the most efficient crystalline Si cells.
11.7. Calculate or look up the EPBT for CIGS cells, CdTe cells, GaAs cells and Si-Perovskite tandem cells If you do the calculation, list your assumptions.

11.8. Look for modules with the highest efficiency you can find with not more than two cells in tandem and not concentrating. Determine the number of modules it will take and the area it will take to build a 100 kW STC PV array with these modules. Be sure to list the module data as a part of your solution.

REFERENCES

[1] Green, M., "Third Generation Photovoltaics: Assessment of Progress Over the Last Decade," *Proceedings of the 34th IEEE PV Specialists Conference*, 2009.
[2] National Renewable Energy Laboratories, Best Research-Cell Efficiencies, Lab Cell Efficiency Record History for All Technologies Up to 2023: www.nrel.gov/pv/assets/pdfs/best-research-cell-efficiencies.pdf
[3] Tarrant, D. E. and Gay, R. R., *Research on High Efficiency, Large Area CuInSe2-Based Thin Film Modules, Final Technical Report*, NREL, Golden, CO, April 1995.
[4] *Estimates of Future Electrical Generation Capacity: Statista*, IEA, 5–9. Statista Research Department, Jun 28, 2024. https://www.statista.com/statistics/859178/projected-world-electricity-generation-capacity-by-energy-source/
[5] Markvart, T., ed., *Solar Electricity*, John Wiley & Sons, Chichester, UK, 1994.
[6] Ballif, Christophe, Boccard, Mathieu, Haug, Franz-Josef, Verlinden, Pierre Jacques. March 2022. "2022 Status Report on Production of Pure Silicon: Status and Perspectives of Crystalline Silicon Photovoltaics in Research and Industry," *Nature Reviews Materials*. https://www.researchgate.net/publication/359077539_Status_and_perspectives_of_crystalline_silicon_photovoltaics_in_research_and_industry
[7] Hallam, Brett, Kim, Moonyong, et al., "A Polysilicon Learning Curve and the Material Requirements for Broad Electrification with Photovoltaics by 2050," July 21, 2022: https://doi.org/10.1002/solr.202200458
[8] Information on Energy Payback Time for Crystalline Si: www.nrel.gov/docs/fy04osti/35489.pdf
[9] Specifics on Silicon Wafer Fabrication and Size: https://en.wikipedia.org/wiki/Wafer_(electronics)
[10] Streetman, B. G., *Solid State Electronic Devices*, 4th ed., Prentice Hall, Englewood Cliffs, NJ, 1995.
[11] Satpathy, Rabindra and Pamuru, Venkateswarlu, *Manufacturing of Crystalline Silicon Solar PV Modules*, Solar PV Power, 2021.
[12] *Sunpower Discovers the "Surface Polarization Effect" in High Efficiency Solar Cells*, Sunpower Corp Internal Document, Personal Communication with SunPower Representative, May, 2009.
[13] More Information on Back Contact Cell Production in 2023: www.cleanenergyreviews.info/blog/most-efficient-solar-panels
[14] Rodriguez, Laura. Jan 23, 2022. Promotional Article on Bifacial Modules: Bifacial Modules: A Comprehensive Guide on Financial and Technical Performance of the Next Hot Thing in Solar—RatedPower. https://ratedpower.com/blog/bifacial-modules/
[15] Bohland, J., "Accelerated Aging of PV Encapsulants by High Intensity UV Exposure," *Proceedings of the 1998 Photovoltaic Performance and Reliability Workshop*, Cocoa Beach, FL, November 3–5, 1998.
[16] Williams, R., *Modern GaAs Processing Methods*, Artech House, Norwood, MA, 1990.
[17] Moon, S., et al., "Highly Efficient Single-Junction GaAs Thin-Film Solar Cell on Flexible Substrate," *Scientific Reports*, vol. 6, p. 30107, 2016. https://doi.org/10.1038/srep30107.

[18] Lammasniemi, J., et al., *Proceedings of the 26th IEEE Photovoltaic Specialists Conference*, 1997, 823–826. IEEE. https://ieeexplore.ieee.org/document/653910?denied=
[19] Hoffman, R., et al., *Proceedings of the 26th IEEE Photovoltaic Specialists Conference*, 1997, 815–818. https://ieeexplore.ieee.org/document/653910?denied=
[20] Brown, M. R., et al., *Proceedings of the 26th IEEE Photovoltaic Specialists Conference*, 1997, 805–810. https://ieeexplore.ieee.org/document/653910?denied=
[21] Shay, J. L., Wagner, S. and Kasper, H. M., *Applied Physics Letters*, 27, No. 2, 1975, 89. https://pubs.aip.org/aip/apl/article-abstract/27/2/89/44586/Efficient-CuInSe2-CdS-solar-cells?redirectedFrom=fulltext
[22] Konagai, M., *Proceedings of the 29th IEEE Photovoltaic Specialists Conference*, 2002, 38–43. https://publica.fraunhofer.de/entities/mainwork/adfd7ccf-5de7-462a-8c3a-e9c1848f0d54/details
[23] Woodhouse, Michael, Jones-Albertus, Rebecca, Feldman, David, et al., *On the Path to Sunshot, The Role of Advancements in Solar Photovoltaic Efficiency, Reliability, and Costs*, National Renewable Energy Laboratory, Golden, CO. NREL/TP-6A20-65872.
[24] Kamada, R., Yagioka, T., Adachi, S., et al., *Proceedings of the 2016 IEEE 43rd Photovoltaic Specialists Conference*, 2016. https://ieeexplore.ieee.org/xpl/conhome/7701171/proceeding
[25] Tarrant, D. E. and Gay, R. R., *Thin-Film Photovoltaic Partnership Program—CIS-based Thin Film PV Technology. Phase 2 Technical Report, October 1996–October 1997*, NREL report NREL/SR-520-24751, Golden, CO, May 1998.
[26] Tarrant, D. E. and Gay, R. R., *Research on High Efficiency, Large Area CuInSe2 Modules, Final Technical Report*, NREL, Golden, CO, April 1995.
[27] Kazmerski, L. L., *Proceedings of the 29th IEEE Photovoltaic Specialists Conference*, 2002, 21–27. https://publica.fraunhofer.de/entities/mainwork/adfd7ccf-5de7-462a-8c3a-e9c1848f0d54/details
[28] Fthenakis Session on Recycling First Solar Person at *Proceedings of the 29th IEEE Photovoltaic Specialists Conference*, 2002, Fthenakis Presentation. https://publica.fraunhofer.de/entities/mainwork/adfd7ccf-5de7-462a-8c3a-e9c1848f0d54/details
[29] Bohland, J., et al., *Proceedings of the 29th IEEE Photovoltaic Specialists Conference*, 2002, 355–358. https://publica.fraunhofer.de/entities/mainwork/adfd7ccf-5de7-462a-8c3a-e9c1848f0d54/details
[30] Ullal, H. S., Zweibel, K. and von Roedern, B., *Proceedings of the 29th IEEE Photovoltaic Specialists Conference*, 2002, 301–305. https://publica.fraunhofer.de/entities/mainwork/adfd7ccf-5de7-462a-8c3a-e9c1848f0d54/details
[31] Wu, X., Sheldon, P., et al., *Proceedings of the 29th IEEE Photovoltaic Specialists Conference*, 2002, 347–350. https://publica.fraunhofer.de/entities/mainwork/adfd7ccf-5de7-462a-8c3a-e9c1848f0d54/details
[32] McCandless, B. E. and Birkmire, R. W., *Proceedings of the 29th IEEE Photovoltaic Specialists Conference*, 2002, 307–312. https://publica.fraunhofer.de/entities/mainwork/adfd7ccf-5de7-462a-8c3a-e9c1848f0d54/details
[33] Chen, Y., Peng, S., Tan, X., et al., *Proceedings of the 43rd IEEE Photovoltaic Specialists Conference*, 2016. https://ieeexplore.ieee.org/xpl/conhome/7701171/proceeding
[34] Levi, D. H., et al., *Proceedings of the 26th IEEE Photovoltaic Specialists Conference*, 1997, 351–354. https://ieeexplore.ieee.org/document/653910?denied=
[35] Birkmire, R. W., *Proceedings of the 26th IEEE Photovoltaic Specialists Conference*, 1997, 295–300. https://ieeexplore.ieee.org/document/653910?denied=
[36] Zhou, C. Z., et al., *Proceedings of the 26th IEEE Photovoltaic Specialists Conference*, 1997, 287–290. https://ieeexplore.ieee.org/document/653910?denied=
[37] 2023 CdTe Manufacturing and Marketing Data: www.pv-tech.org/first-solar-thin-film-modules-sold-through-2026-backlog-passes-70gw/

[38] Scarpulla, Michael A., et al., Solar Energy Materials and Solar Cells, Vol 255, 15 Jun 2023 Elsevier 112289: www.nrel.gov/docs/fy23osti/85320.pdf
[39] Luque, A., et al., *Proceedings of the 29th IEEE Photovoltaic Specialists Conference*, 2002, 1190–1193. https://publica.fraunhofer.de/entities/mainwork/adfd7ccf-5de7-462a-8c3a-e9c1848f0d54/details
[40] Zahler, J. M., et al., *Proceedings of the 29th IEEE Photovoltaic Specialists Conference*, 2002, 1029–1032. https://publica.fraunhofer.de/entities/mainwork/adfd7ccf-5de7-462a-8c3a-e9c1848f0d54/details
[41] University of Houston, Institute for Nanotechnology, Organic Solar Cells: http://ine.uh.edu/research/organic-solar-cells/index.php2016
[42] Jacoby, Mitch, "The Future of Low Cost Solar Cells," *Chemical & Engineering News*, vol. 94, no. 18, pp. 30–35, May 2, 2016.
[43] Philipps, Simon, Bett, Andreas, Horowitz, Kelsey and Kurtz, Sarah, *Current Status of Concentrator Photovoltaic (CPV) Technology*, Version 1.2, February, 2016. Fraunhofer Institute for Solar Energy Systems ISE in Freiburg, Germany; National Renewable Energy Laboratory (NREL) in Golden, Co.
[44] Perovskites and Perovskite Solar Cells: An Introduction: www.ossila.com/pages/perovskites-and-perovskite-solar-cells-an-introduction (accessed 6–2016).
[45] Stranks, Samuel D., et al., "Electron-Hole Diffusion Lengths Exceeding 1 Micrometer in an Organometal Trihalide Perovskite Absorber," *Science*, vol. 342, no. 6156, pp. 341–344, October 18, 2013.
[46] Fu, Fan, Li, Jia, et al., "Monolithic Perovskite-Silicon Tandem Solar Cells: From the Lab to Fab?" *Advanced Materials*, vol. 34, p. 2106540, 2022.

SUGGESTED READING

Berst, Jesse, "For a Smart Grid, Look to Smart States," *Solar Today*, vol. 23, no. 4, May 2009, p. 30 (solartoday.org).

Burns, Cameron M., "The Smart Garage (V2G): Guiding the Next Big Energy Solution," *RMI Solutions Journal*, vol. 1, no. 1, July 2008, p. 22, Wiley-VCH GmbH.

Hendricks, Bracken, "Wired for Progress," *Solar Today*, vol. 23, no. 4, May 2009, p. 26 (solartoday.org).

Index

A

absolute temperature, 17
absorption of light, 317–320
    absorption constant, 320
    absorption equation, 320
acid rain, 285
AC submersible pump, 267
aesthetic considerations, 157
aesthetics, 97
"a" factor, 224
agrivoltaics, 128
air mass, 20, 21
Albany, NY, 220
AlInP/GaInP/GaAs structure, 362
all back contact cell, 360
aluminum, 153–155
    aluminum alloys, 154
    aluminum oxide, 154
    cast aluminum, 154
    wrought aluminum, 154
amorphous CdTe, 1
antenna - loops in module string wiring, 126
antireflective coating, 360
arc fault circuit interrupters (AFCI) - DC, 80
array mounts, 31
    backtracking, 31
    fixed, 31
    tracking, 31
array orientation, seasonal dependence, 30
array seasonal optimization, 31
array selection and installation, 101
array series *vs.* parallel connection, 48
array special orientation considerations, 31
array tilt angle - optimizing, 273
array voltage, 48
atmospheric effects on sunlight, 19
    absorption, 19
    air mass, 19, 20
    reflection, 19
    scattering, 19
automatic switching between charging and discharging, 209
axisymmetrical hill, 164

B

backup, 233
    full, 230
    partial, 233
backup combiner, 243
backup - whole house (full), 230–236

battery AC and DC interfaces, 208–209
battery backup systems, 214–217
    AC coupled, 214
    DC coupled, 216
battery technology, 192–201
    battery management systems, 193–194
        current sensing, 193
        failsafe MCU, 193
        fuse, 193
        monitoring sensors, 193
        overcurrent protection module, 193
        watchdogging, 193
battery terminology, 191
    ampere-hours (Ah), 191
    charging rate, 191
    discharging rate, 191
    kilowatt-hours (kWh), 191
    lithium - LFP & 5 other chemistry variations, 192–195
    maximum energy, 192
    maximum power, 192
    round-trip efficiency, 192
Bell Labs, 1
bi-directional transmission, 2
Bill Nye, 307
bio-absorption, 7
black and white pyranometer, 28
Boise, ID, 231
Boltzmann's constant, 17
Boulder, CO, 265
bridging (grid), 244
buckling, 149
building energy use calculations, 279–281
bulk charge neutrality, 326
busbar - DC, 242

C

cadmium telluride (CdTe), 367
    2023 market share, 368
    area-related costs, 368
    cell structure, 367
    cell structure and fabrication improvements, 368
    escrow account for recycling, 367
    production cost, 367
    recycling, 367
California Hazardous Waste Control Law, 311
    soluble threshold limit concentration test (STLC), 311
    total threshold limit concentration (TTLC), 311
    Waste Extraction Test (WET), 311

cantilever, 173, 182
carbon credits, 293
Cat 5 cables, 240
category II building, 224
cell resistance losses, 342–344
  series losses, 342, 343
  shunt losses, 344
cells to modules, 361
  backplanes, 361
  bifacial, 361
  encapsulants, 361
  interconnects, 361
charge controllers, 67–70
  charging rate, 68
  charging stages (modes), 68, 69
  hysteresis, 69
  lead acid batteries, 67, 68
  lithium batteries, 67, 68
  temperature dependence, 69
  Thevenin equivalent circuit, 67
CIGS cell processes, 364
circuit breakers, 79
circuit breakers - adjustable, 137
Clean Air Act, 307
climate change, 3
codes, structural, 155
coefficient of utilization (CU), 261
color rendition index (CRI), 260
color temperature, 260
Columbia River oil slick, 307
communications, 76
  purpose, 76
  wired, 76
  wireless, 76
compatible, 224
  downward, 224
  upward, 224
complex plane, 303
compressed air storage, 204
  availability, 204
  calculations, 204
  history, 234
  siting, 204
compressive loading, 149
concentration gradient, 326
conduction band, 320
conductivity, 322
  mobility - electrons, 322
  mobility - holes, 322
conservation of energy, 319
conservation of momentum, 319
consumer price index (CPI), 286
COP28, 5
copper indium (gallium) diselenide (CIGS), 364
  carrier diffusion length, 364
  CdS/p-type CIS heterojunction, 364
  transparent conducting oxide (TCO) top contacts, 364

corrected loads, 269, 272
corrosion and other degradation processes, 98
corrosion types, 150–152
  anodic metal, 150
  cathodic metal, 151
  crevice corrosion, 151
  dezincification, 152
  dry corrosion, 152
  erosion corrosion, 152
  galvanic corrosion, 150
  intergranular corrosion, 151
  pitting, 151
  selective leaching, 151
  stress corrosion cracking, 152
  ultraviolet degradation, 152
  uniform attack, 150
critical and noncritical loads, 50–54
  downtimes due to weather, 51
  effect of amount of storage, 52
  effect of array size, 52
  effect of geographical location, 54
  system output - estimated, 51
crystal defects, 336
cumulative present worth factor, 288

D

days of autonomy, 52, *see also* storage days
DC:AC ratio, 108, 130
declination, 23
  Tropic of Cancer, 23
  Tropic of Capricorn, 23
decommissioning, 311
degradation factors, 365
density, examples
  atmosphere, 159
  charge carriers, 325
  current, 325
  energy, battery, 191
  energy, sunlight, 18
  excess minority carrier, 331
  recombination centers, 336
Department of Energy Sunshot, 293
design wind pressure, 165
diffusion, 325
  electron diffusion constant, 326
  hole diffusion constant, 326
diffusion length, 330
diode equation, 333
disconnects - lockable, 224
discount rate, 285
dispatch and transmission, 2
distributed generation, 2
distribution level (7860/13,600V), 244
dopants, 309
Dow-Jones Industrial Average (DJIA), 286

Index

drift velocity, 325
    electron mobility, 322
    hole mobility, 322
duplex outlet, 269

E

edge factor, 165
effective funds rate, 286
effective mass, 334
Einstein's formula - mass to energy, 17
either, or derating method, 125, 242
electric arc, 140
    calculation, 140
    power, 140
    temperature, 140
electromagnetic interference, 97
    conducted, 97
    radiated, 97
electron-hole pairs (EHP), 322
    excess carrier lifetime, 322
    generation, 322
    nonthermal equilibrium, 322
    recombination, 322
electrons, 322
energy band gap, 319, 320
    direct, 319
    indirect, 319
energy crisis, 1
energy density of sunlight, 18
    extraterrestrial, 18
    terrestrial, 18
energy return on investment (EROI), 308
energy star rating, refrigerator, 218, 265
energy storage, 49
energy unit equivalencies, 11
environmental costs, 2
environmental costs of PV module production, 308
    cleaning solvents, 308
    CO_2, CH_4, B_2H_6, ASH_3, H_2Se, Cd, Li, Te, 308
    fabrication materials, 308
    glass, 308
    mining, 308
epitaxial growth, 362
epitaxial lift-off (ELO), 362
escarpment, 164
ESS comparison with simple battery, 213
Euler's formula, 150
evaluating system options, 220
excess electrical production, 276
excess minority carriers in neutral regions, 329
exposure category, 162
exposure coefficients, 162
externalities, 285
externally applied voltage, 331
external pressure coefficients, 165

F

federal incentive programs, 287
feeder conductor, 105
Fermi levels, 323, 347
fill factor, 333
fire, floods and high water zones, 99
fire concerns, 309
fire protection codes, 99
first law of thermodynamics, 8
First Solar Corporation, 1
flow battery specifics, 197–200
    advantages and disadvantages, 197
    electrolyte energy storage, 197
    hybrid redox anode and cathode electrolytes, 197
    hybrid redox storage mechanisms, 197
    ion transfer membranes, 197
    optimizing for load conditions, 197
    recycling, 198
flow battery technologies, 198–200
    all iron redox, 200
    hybrid flow batteries, 200
    organic flow batteries, 197, 200
    sodium sulfur, 198
    vanadium redox, 199
    zinc bromide, 199
flywheel energy storage calculations, 206
foot candles, 259
forces associated with PV systems, 157–159
    dead loads, 158
    live loads, 158
    wind loading, 159–167
        empirical formula base, 159
        friction forces, 159
        pressure forces, 159
fossil fuel generators, 77
    stand-alone systems, 77
        grid-connected systems, 77
fuel cell comparative efficiencies with other technologies, 202
fuel cells - advantages and disadvantages, 202
fuel cells - sizes and applications, 202
fuel cell technologies, 202
    alkaline (AFC), 202
    molten carbonate cell (MCFC), 202
    phosphoric acid (PAFC), 202
    proton exchange membrane (PEMFC), 202
    solid oxide cell (SOFC), 202
fuel transport, 307
fuses, 79

G

GaAs testing, 363
    bending, 363
    contact integrity, 363
    high temperature vacuum, 363

humidity, 363
gallium arsenide, 362
    bandgap, 362
    concentrating collectors, 362
    extraterrestrial, 363
    production cost, 362
gateways, 76
Glasgow Climate Pact - 2021, 5
global climate change, 285
global temperature rise, 8
global warming, 3, 20
green book, 307
greenhouse gases, 3, 7, 8, 20, 285
greening, 301
Greenland ice cap, 3
green pricing, 304
grid connections, 104–107
    line side connections, 104–105
    load side connections, 105–107
grid-forming inverter, 223
grid of the future, 374
    grid restoration via electronic inverters, 374
    micro grids, 374
grid stabilization, 244
ground fault detection and interruption (GFDI) - DC, 79
ground mount arrays, 178–187
    backtracker, 186
    damping coefficient, 187
    defining parameters, 180
    dynamic analysis, 182
    fixed, 180–184
    tracking, 184
    wind loads - pressures and moments, 182
    wind zones, 182
ground - single point, 235
guesstimate, 231
guesstimate process, 218
Gulf Stream, 3

H

Heisenberg uncertainty principle, 349
history of PV development, 10
    bipolar transistor, 10
    Edmond Becquerel, 10
    photogalvanic effect, 10
    solar cell, 10
holes, 322
hybrid inverter, 215
hybrid water heater - operating parameters, 266, 268
hydrogen storage, 201
    advantages and disadvantages, 201
    conversion efficiencies, 201
    energy density - comparison with other fuels, 201
    energy density - liquid and gaseous hydrogen, 201
    hydrogen - production, transportation and storage, 201

I

ideal thermal efficiency, 9
IEA World Energy Outlook scenarios, 5
    APS, 5
    NZE, 5
    STEPS, 5
IEEE Standard 1547–2018, 90–97
    anti-islanding requirements, 93
    development process, 91
    grid-connected inverter performance requirements, 91
    grid stabilization, 92
    power factor correction, 92
    power quality requirements, 93
    purpose, 90
    ride-through requirements and categories, 92
    scope, 90
Illumination Engineering Society (IES), 259
impurities, 323
    acceptor (p-type), 324
    donor (n-type), 323
inflation rate, 285
inflation reduction act (IRA), 287
integrated arrays (BIPV), 178
integrated storage, 2
intermittent, 2
intrinsic carrier concentrations, 322
inverter bypass switches, 81
inverters, 55–66
    AC modules, 66
    comparator, 61
    current source, 66
    efficiency, 56
    grid-forming, 66
    harmonic distortion, 56
    insulated gate bipolar transistor, 58
    islanding, 65
    microinverters, 66
    modified sine wave, 56
    multilevel H-bridge, 57
    multimode, 65
    multivibrator, astable, 56
    performance, 55
    power MOSFET, 58
    pulse width modulated, 60–64
    rapid shutdown, 66
    regenerative comparator, 69, 70
    search mode, 64, 65
    silicon-controlled rectifier, 58
    square wave, 56
    stand-alone, 65
    surge capacity, 55

Index

transformerless, 64, 235
ungrounded, 56
utility interactive, 65
voltage source, 66
inverter selection, 102–103
    inverter selection considerations, 102–103
    night time loads and storage options, 102
investment tax credit, 302
irradiance, 21
irradiation, 21, *see also* peak sun hours
islanding and grid stability analysis, 95–97

J

junction box - rooftop, 224, 228

L

lag screw pull strength, 174
landscape module orientation, 113
large battery storage, 210
    operating current, 210
    operating power, 210
    operating voltage, 210
LED lighting, 218
levelized cost of electricity (LCOE), 114, 292
levelized cost of storage (LCOS), 294
level playing field, 304
life cycle costing, 285–297
lifted weight storage calculations, 206
    system specifics, 206
    up-down efficiency, 206
lightning rods, 90, 126
linear current boosters (LCB), 70, 71
    buck-boost converter, 72
    buck converter, 72
liquid encapsulated Czochralski (LEC) method, 362
load centers, 104, 126, 137
load curtailment, 219, *see also* load management control; load shedding
load leveling (grid), 244
load management control, 2
loads, determining, 250
load shedding, 216
load side connection types, 106
    100% rule, 106
    120% rule, 106
Lubbock, TX - weather data, 257
lumens, 259
luminous efficacy, 260

M

maintenance factor (MF), 261
Mauna Loa CO_2 data, 6, 7
maximum fault current calculation, 140
maximum power hyperbola, 50

maximum power point tracking (MPPT), 49, 70, 71
mechanical rotating generators, comparison with electronic inverters, 94
microinverter input and output limits, 110
MID (microgrid interface device) switch, 223, 231
mineral depletion allowance, 304
minimizing reverse saturation current, 334
mini-split air conditioner, 218
mini-split heat pump, 269
minority carrier lifetime, 330
mitigation strategies - utility scale systems, 310–311
    agrivoltaics, 311
    chemical vegetation control, 310
    fencing, 311
    local wildlife, 310
    soil erosion, 310
MOCVD process, 362
module degradation over time, 100
module selection criteria and process, 108
molecular beam epitaxy (MBE), 362
molten salts
    boiling point, 205
    ideal properties, 205
    melting point, 205
    specific heat capacity, 205
    thermal mass, 205
monitoring systems, 139
monolithically connected minimodules, 365
MPPT, 81
MPPT input - string level, 217
multimode inverter, 215, *see also* hybrid inverter
    5 modes of operation, 215

N

National Electrical Code (NEC), 86–90
    deratings, 90
    overcurrent protection, 89
    voltage drop, 87
    wire and conduit types, 89
    wire sizing, 88
net impurity concentration, 324
net pressure coefficients, 176
neutral forming transformer, 223
nickel - NiCd & 2 other chemistry variations, 195–197
nonattainment levels, 307
    carbon monoxide, 307
    lead, 307
    nitrogen dioxide, 307
    ozone, 307
    particulate matter, 307
    sulfur dioxide, 307
normal incidence pyrheliometer, 28

NREL SAM - estimating system energy output, 100
NREL SAM simulation methods, 107
N-type TOPcon technology, 361
nuclear safety, 307

O

ohmic contacts, 359–360
    back contact, 359
    front contact, 359
Ohns/kft (W/kft), 242
open circuit voltage, 333
optimizers, 74
optimizer selection, 237
optimizer systems, 112–119
    determining device ratings, 115–117
    determining maximum string length, 116
    determining operating voltage and current dependence on module output power, 117
orbit and rotation of the earth, 22–24
organo-metallic vapor phase epitaxy process (OMVPE), 362
other emerging technologies, 369–374
    concentrating PV, 370
        cell efficiencies, 371
        disadvantages, 371
        high concentration PV (HCPV), 371
        low concentration, PV (LCPV), 371
    hot carrier cells, 369
    intermediate band cells, 369
    optical up-down conversion, 369
    organic cells, 370
other junction types, 344–349
    graded, 344–345
    heterojunctions, 345
    multijunctions, 348
    Schottky junctions, 346
    tunnel junctions, 349

P

panelboard, switchgear, 137, *see also* load centers
Paris Agreement - 2015, 5
passivation emitter rear contact (PERC) process, 360
peak sun hours, 21, *see also* irradiation
percent voltage drop - 3-phase systems, 135
performance clause, 284
perovskites, 372–374
    active layers, 374
    basic cell structure, 372
    efficiency history, 372
    organometal trihalide perovskite absorber, 373
    perovskite layer thickness, 373
    Perovskite-Si tandem cell, 372
    stability issues, 373

phase sequence, 243
photoconductor, 322
    sensitivity, 323
    speed, 323
photons, 319
    energy, 319
    frequency, 319
    wavelength, 319
photosynthesis, 20
PIN junction, 339
Planck's blackbody radiation formula, 17
Planck's constant, 17
plan view *vs.* actual dimension, 109
PN junction, 326–329
    built-in junction voltage, 328
    depletion region, 328
    diffusion current, 327
    drift current, 327
    effect on photon-generated EHPs, 338
    Einstein relationship, 328
    Gass' Law, 327
    junction width, 338
    Kirchhoff's current law, 327
    law of detailed balance, 327
    space charge layer, 329
population forecast models, 3
portrait module orientation, 109
power grid, 2
present worth factor, 287
pressure coefficients, 225
pressure tank, 268
public health economic cost, 307
pulse width modulation (PWM), 49
pumped hydro storage, 203
    availability, 203
    calculations, 203
    history, 203
purchasing power, 285
PV cell, 41–43
    anti-reflective coatings, 46
    arrays, 42
    cell power *vs.* cell voltage curves, 44
    cell power curve, 44
    configurations, 41, 42
    diode equation, 43
    electron-hole pairs (EHPs), 43
    fill factor, 45
    ideal performance limits, 43
    I-V characteristics, 43
    maximum power point, 44
    open circuit voltage, 43
    PN junction, 43
    Schottky barrier devices, 43
    short circuit current, 43
    spectral dependence of photocurrent, 46
    temperature dependence of cell power, 45

Index

PV cell generations, 353
   first generation, 353
   second generation, 353
   third generation, 353
PV impact on grid, 54, 55
PV module, 46
   bypass diodes, 47
   charge controllers, 46
   half cells, 47
   maximum open circuit voltage, 46
   module efficiency, 48
   nominal operating cell temperature (NOCT), 47
   temperature dependence, 48
PV-powered fan, 251–253
   available power *vs.* power used, 253
   component selection procedure, 251
   I-V characteristic, 252
   MPPT, 253
   operating point, 252
   performance characteristics, 252
   stalled rotor current, 252
   starting torque, 252
PV-powered water pumping system, 253–258
   array size, 257
   battery storage capacity, 257
   determining operating options, 253
      pumping height, 253
      pumping rate, 253
      water storage, 253
   high head pumps, 255
   horsepower conversion factors, 254
   medium head pumps, 255
   MPPT *vs.* LCB, 235
   piping friction loss, 254
   pressure *vs.* flow, 254
   water weight, 254
PV system loads - efficiency *vs.* costs, 49, 50
pyranometer, 28

Q

quartz to silicon, 355
   energy cost, 305, 308, 356
   energy payback time, 356
   initial refining to metallurgical grade Si, 356
   refining to solar grade, 356
   removal of oxygen, 355

R

radiation, 18
   albedo, 19, 28
   blackbody radiator, 18
   carbon dioxide effect on solar spectrum, 19
   diffuse, 19, 28
   direct beam, 28
   global, 19, 28
   perfect absorber, 18
   perfect radiator, 18
   spectral characteristics of light sources, 19
   spectral testing conditions, standardized, 19
radiation, daily cumulative, 30
radiation resistance, 363
rain, snow, hail, ice and earthquakes, 98, 167
rapid depreciation, 302
rapid shutdown, 76
   definition, 77
   implementation, 77
rapid shutdown (RS) with microinverters, 109
receptacles, 79
recombination centers, 336
regulated attainment levels, 307
remote applications, 1
Resource and Conservation Recovery Act, 311
risk category, 162
roof attachments (mounting feet), 225
roof zone determination - "a" factor, 171
roof zones, 162
room cavity ratio (RCR), 261
RS (rapid shutdown) devices, 234

S

second law of thermodynamics, 8
seismic effects, 168
semiconductor, 319
   extrinsic, 323
   intrinsic, 322
sequestering CO_2, 305
   plankton, 306
   trees, 305
service rated AC disconnect, 105
shading, 33
   computational methods, 33
   effect on array performance, 32
   field measurements, 33
   solar pathfinder, 33, 34
shadow band stand, 28
sharp cutoff, 261
short circuit current, 333
silicon crystal to silicon wafer, 356
   etching, 357
   junction diffusion, 357
      drive-in diffusion, 358
      predeposition diffusion, 357
   net impurity concentration, 324, 358
   sawing, 357
skin depth, 138
smart energy management, 127
snow loads - roof effects, 167
solar access laws and restrictions, 99
solar energy skeptics, 2
solar noon, 23
   analemma, 23
   clock noon, 23

solar panel pressure equalization factor, 165
solar powered wristwatches, 1
solar spectrum, 17, 319
    blackbody radiation, 319
    peak, 319
source circuit combiner boxes, 81
source connection *vs.* load connection, 104
South pole, 3
span, 225
spectral components - discrete, 260
split-phase, 234
stability analysis, grid, 129
stacking cells, 363
    air mass consideration, 363
    establishing equal layer current production, 363
    minimize thermal losses, 363
stand-alone and grid-connected PV systems with storage, 250
stand-alone systems, 50
standard candle, 259
standards, structural, 155
standoff roof mounting considerations, 169
starting current, 219
steel, 152
    alloy steel, 153
    anodic protection, 153
    carbon steel, 152
    cathodic protection, 153
    galfan, 153
    galvalume, 153
    hot dip galvanizing, 153
    spangle, 153
storage, 243–246
    importance of, 55
    long term, 244
    medium term, 244
    short term, 244
storage days, 52–54, 257
string inverter, 119–122
    advantages and disadvantages, 119
    input and output current, voltage and power limitations, 120
    rapid shutdown implementation, 121
    string lengths, determining maximum and minimum, 120–121
    typical input and output configurations, 119
structural design considerations, 149
structural standards - ASCE 7, 98
SunShot Program (USDOE), 364
super capacitors - voltage stabilization, 27
superconducting magnet - current stabilization, 207
surface recombination velocity, 339, 340
surge protection, 80
    DC and AC, 80
    types of devices, 80
switches, 79
system commissioning procedure, 141–143

system controller, 217
system costs, 284
    acquisition, 284
    decommissioning, 284
    maintenance, 284
    operating, 284
    replacement, 284
system monitoring, 245
    BESS state of charge, 245
    fiber optics, 245
    insolation, 245
    production/consumption metering, 245
    weather, 245

T

tandem inverter connection, 239
tap conductor, 105
tariffs, 304
temperature extreme exposure, 363
Tennesee Valley Authority (TVA), 114
terminators - microinverter cable, 226
thermal expansion, 150
thermal velocity, 325
third party ownership, 284
time value of money, 285
Toxicity Characteristic Leaching Procedure (TCLP), 311
toxic waste as an externality, 308
tracking the sun, 24
    azimuth angle, 25
    daylight hours calculation, 26
    hour angle, 25
    solar altitude, 25
    sun position calculation, 26
    sunset angle calculation, 26
    time zones, 24
    zenith angle, 26
trade-offs, 365
transformer current - primary and secondary, 139
transformer taps, 246
transmission level, 245
typical meteorological year (TMY), 28, 108

U

UL 1741 requirements on battery energy storage systems, 213
utility class II system, 105

V

valence band, 320
vanguard satellite, 1
vehicular battery packs, 243–244
velocity pressure, 162

voltage drop - optimizer array, 125
voltages - distribution level, 128

W

wind loading factors, 159
    wind speed, 159
wind loads, 98
    density of air, 159
    elevation above ground, 159
    gust effects, 159
    orientation, 159
    safety factors, 159
    shape and surface area, 159
    topographical effects, 159
    wind speed contours, 159
wind tunnel tests, 159

X

xenon, 260

Z

zoning or permitting issues, 104

For Product Safety Concerns and Information please contact our
EU representative GPSR@taylorandfrancis.com Taylor & Francis
Verlag GmbH, Kaufingerstraße 24, 80331 München, Germany